III. AIR QUALITY CONTROL 123

Monitoring Air Quality in Urban Areas 125
    I. Allegrini and F. Costabile

Sustainable Mobility: Mitigation of Traffic Originated Pollution 141
    M. Mazzon

IV. WASTE MANAGEMENT 157

Integrated Waste Management. Technologies and Environmental Control 159
    L. Morselli, I. Vassura and F. Passarini

Economic Analysis of Waste Management Systems in Europe 171
    A. Massarutto

Hospital Waste Management 187
    I. Pavan, E. Herrero Hernández and E. Pira

V. ENERGY EFFICIENCY AND RENEWABLES 193

V.1. RENEWABLE RESOURCES 193

Solar Energy 195
    R. Barile

Geothermal Energy 207
    R. Bertani

Renewable Energy from Biomass: Solid Biofuels
and Bioenergy Technologies 221
    M. Chiadò Rana and R. Roberto

Wind Energy 237
    L. Pirazzi

Power Electronics in Distributed Power 249
    J. Enslin and F. Profumo

V.2. ECO-BUILDING 263

Energy Optimization of a Building-plant System 265
    L. Schibuola

The Sino-Italian Environment & Energy Building (SIEEB):
A Model for a New Generation of Sustainable Buildings 279
    F. Butera

V.3. ENERGY POLICY 285

New Policy Schemes to Promote End-use Energy Efficiency
in the European Union 287
   M. Pavan

Subsidies and Market Mechanisms in Energy Policy 301
   G. Pireddu

CDM – A Policy to Foster Sustainable Development? 317
   B. K. Buchner

VI. SUSTAINABLE INDUSTRIAL DEVELOPMENT 333

Industrial Ecology in a Developing Context 335
   M. R. Chertow

Strategic Environmental Assessment (SEA): Integration and Synergy
with EMAS Management Systems on the Territory 351
   G. Chiellino

VII. SUSTAINABLE URBAN DEVELOPMENT 363

General Aspects of Sustainable Urban Development (SUD) 365
   A. Costa

A New Perspective for the Future: Environmental Protection
and Sustainability in Urban Planning 381
   M. Savino

Brownfields Remediation and Reuse: An Opportunity
for Urban Sustainable Development 397
   M. Turvani and S. Tonin

VIII. AGRICULTURE AND NATURAL RESOURCE MANAGEMENT 413

Ecological Agriculture: Human and Social Context 415
   F. Caporali

Sustainable Agriculture in the Frame of Sustainable Development:
Cooperation between China and Italy 431
   M. L. Gullino, A. Camponogara and N. Capodagli

Economics and Policy of Biodiversity Loss 451
   S. Dalmazzone

Forestry and Rural Development: Global Trends and Applications
to the Sino-Italian Context 467
   G. Scarascia-Mugnozza and M. E. Malvolti

Index 483

PREFACE

This book stems from a four-year experience of a Training Programme addressing members of several Chinese governmental Institutions which, given the moment of extremely intense and fast development of their country, consider the issues of environment and sustainable growth among the foremost priorities.

In particular, they expressed the need and will to develop policy and management tools that could lead to a strategy of sustainable growth from an economic, social and environmental point of view.

The Programme turned out to be a success (it involved, up to June 2007, more than 2000 trainees from almost all the Provinces of China) precisely because the forces that answered those needs are extremely diverse as to include the academia, national and local governments, public institutions, private companies and international agencies. Following this feature, the book's contributors have been selected among more than 300 professors, researchers, policy makers, and entrepreneurs involved in the Training activities, thus offering different approaches to the key questions of environmental management.

The Programme was made possible thanks to the support and commitment of the Italian Ministry for the Environment, Land and Sea, which had started a broad bilateral cooperation programme in China some years ago; the Venice International University, whose peculiar character of international and interdisciplinary association allowed the involvement of experts from a variety of Institutions and countries, and hosted the Chinese trainees in a beautiful location in the Venice lagoon; and the University of Turin, that through its Center of Competence Agroinnova provided its expertise in the agro-environmental field.

The Programme's aim of building capacity in sustainable development and environmental management for the Chinese trainees is contained in this book as it addresses the key aspects of environmental sustainability from the point of view of its policies, economic instruments and social implications, by providing theory and case studies from the authors' different experiences and backgrounds. Water Management, including resources, monitoring and waste water treatment; Air Quality, in terms of monitoring and pollution control; Waste Management, from the point of view of technology and economics; Energy, conceived as efficiency and promotion of renewables; Sustainable Industrial Development, in terms of tools offered by the Industrial Ecology framework; Sustainable Urban Development, based on new perspectives of planning; Agriculture and Natural Resource Management, including ecological, economic and social aspects, with examples of the successful cooperation in China; these are all the aspects that shape the book.

The variety of issues covered and the wide approach used make this book a useful tool for readers with different backgrounds and levels of knowledge on environmental and sustainability sciences.

ACKNOWLEDGEMENTS

A special thanks goes to: Alessandra Fornetti, Ilda Mannino, Elisa Carlotto, Francesca Zennaro, Francesca Radin, Denise Tonolo, Selina Angelini, Lisa Botter, Lorenza Fasolo, Eleonora Chinellato, Enzo Casulli (the TEN Center – Thematic Environmental Networks – of the Venice International University), their devotion and carefulness made the Training just perfect; to Daniela Adami, Laura Cassanelli, Raffaella Gallio and the other Chinese interpreters, strongly responsible for the success of the Programme; to all institutions that supported the program for their untiring cooperation.

We also would like to thank: the Project Management Offices of the Italian Ministry for the Environment, Land and Sea in Beijing and Shanghai; the Chinese Academy of Social Sciences; the State Environmental Protection Administration of China; the Chinese Ministry of Science and Technology; the Environmental Protection Bureaus of the Municipalities of Beijing and Shanghai; Tsinghua University, Beijing; Tongji University, Shanghai and all Chinese trainees.

C. C.
I. M.
M.L. G.

C. CLINI

THE EUROPEAN WAY TO SUSTAINABLE DEVELOPMENT

The integration of the environmental dimension within development strategies, the positive and necessary role of the business community and enterprises in combining economic growth and environmental protection are by now universally known; so much so to hope that we are closer to ending the theory and practice of conflict between environment and development once and for all.

The European Union is consolidating this vision, primarily through new environmental policies, based on "positive actions" rather than on constraints and prohibitions. Such actions enhance the conservation of resources and the protection of the environment considered as a "development engine" and as an "opportunity".

However, to concretize this perspective it is necessary to modify the "culture" and the instruments of traditional environmental policies. The improvement of the economy's environmental performances, technological innovation and the diffusion of new knowledge and skills cannot be obtained as a result of constraints and prohibitions, but as a shared process between public authorities, enterprises, business community and consumers' associations.

To this end, the identification and development of a regulatory framework and market mechanisms, including fiscal ones is highly important. These, in turn, encourage the expansion of best practices and clean technologies for the benefit of businesses and consumers, who choose the environmental sustainability of processes and products as reference criteria.

The European strategy for sustainable development must be reinforced through a determined cross-integration of the environmental dimension in sector policies. The integration between environmental protection and the preservation of natural resources together with Europe's energy security and the sustainable mobility of passengers and freight are particularly important.

However, the European way to policy integration is still long and controversial since national protectionisms withdraw resources from investments for technological innovation which are necessary both to win environmental challenges and to support European economic competitiveness.

At a national level, an innovative process must be started urgently to modify the objectives and instruments of environmental policies. This type of process should be particularly orientated towards supporting – within the European regulatory framework – stimulating measures for the development of best practices and clean technologies for the benefit of public administrations, businesses and consumers who choose environmental sustainability and as reference criteria for the management of resources and for processes and products.

is the whole set of compliance assistance programs, to help in particular small and medium sized enterprises to better understand European environmental regulations and to better comply with environmental legislation, and to get information about best practice clean technologies in different business sectors.

A recently suggested method to strengthen partnership between environmental regulators and business communities which is especially relevant for large firms is the *voluntary environmental agreements* (Segerson, 2000b). They are considered as a way to respond to the high costs of command and control type of environmental regulation: under a voluntary agreement, polluters voluntarily undertake pollution control measures rather than undertaking them because of the existence of regulatory requirements.

Voluntary agreements can take the form of unilateral initiatives by firms or industries, without the government being actively involved: similar initiatives are strongly related to the application of corporate social responsibility to the environmental field. Or they can take the form of bilateral agreements between a regulatory agency and a group of firms, emerging from the recognition of the importance of reaching such an agreement rather than rely on environmental regulation. Or they can be designed by the regulatory agency to induce participation by individual firm; in this case the designing agency should make explicit the obligations of the participating firms and their rewards with respect to the use of mandatory environmental regulation

These voluntary programs can indeed be effective, but their success is not automatically guaranteed. Success is most likely where firms have a strong inducement to participate in the agreement. Reasons to participate are the perception that consumers are increasingly aware of the importance of environmental effects of products and productive processes, the possibility of getting strategic benefits by avoiding the costs of a more stringent environmental regulation, the possibility of receiving government financial support.

Hence, governments can induce participation of firms through "carrots" such as financial subsidies or through "sticks" i.e. the threat of a stricter regulation. Subsidies must however be used with great care, both because of their implications on the public budget and their potentially distorting effects; they should be oriented to support R&D and innovation activities included in the voluntary agreement.

The mere signing of voluntary environmental agreements does not automatically mean that they will be enforced and that they are going to produce a significant environmental improvement. Monitoring the way the agreement is carried on is important; and the threat of implementing a mandatory regulation in the case the agreement has not produced the expected results remains crucial. For this reason voluntary agreements should not be considered as substitute for environmental regulation: voluntary and regulatory approaches should be considered as complementary tools in an integrated environmental policy.

Very important in promoting participation to environmental improvement is the role of consumers. Consumers can make the market mechanism work directly in improving the quality of the environment. Consumers' preferences may change so as to determine a higher willingness to pay for environment friendly products and processes, and to reduce the willingness to pay for polluting products and processes.

Firms will be ready to take the market messages deriving by these more environment friendly consumers' preferences.

Consumers' appropriate information is essential to this task. An important instrument in this direction is the eco-label scheme that has been developed at the European and State level with the aim of affecting consumers' choices. These eco-label schemes are also part of the IPP approach. The financial sector could play an important role offering Green Investment Funds to those investors who want to be reassured about production in an environmentally and socially responsible manner.

A fundamental role is played by environmental education, which is probably best provided at local, regional and national level and by a range of organizations, such as NGOs, commanding the required respect and trust.

Ignazio Musu
Department of Economics,
Ca' Foscari University of Venice, Italy
Thematic Environmental Networks Center,
Venice International University, Italy

REFERENCES

Brock W. and Scott Taylor M., 2004. *Economic Growth and the Environment: a Review of Theory and Empirics,* NBER Working Papers Series, No.10854.
Callan S. and Thomas J., 1996. *Environmental Economics and Management,* Irwin, Boston.
Common M., Ma Y., McGilvray J. and Perman R., 2003. *Natural Resource and Environmental Economics,* Pearson, Addison Wesley, Harlow.
Common M. and Stagl S., 2005. *Ecological Economics: an Introduction,* Cambridge University Press, Cambridge.
Goulder L. *Environmental taxation in a second-best world,* in H. Folmer and T. Tietenberg (eds.), 1998. *The International Yearbook of Environmental and Resource Economics 1997-98,* Edward Elgar, Cheltenham, pp. 28-54.
Kolstad C., 2000. *Environmental Economics,* Oxford University Press, Oxford.
Kraemer L., 2003. *EC Environmental Law,* Sweet & Maxwell, London.
Krugman P. and Wells R., 2005. *Microeconomics,* Worth Publishers, New York.
McCormick J., 2001. *Environmental Policy in the European Union,* Palgrave, MacMillna, London.
O'Riordan T. (ed.), 1997. *Ecotaxation,* Earthscan, London.
Segerson K., 2000a. *Liability for environmental damages,* in H. Folmer, H. Landis Gabel, *Principles of Environmental and Resource Economics,* Edward Elgar, Cheltenham, 2000, pp. 420-446.
Segerson K., 2000b. *Voluntary approaches to environmental protection,* in H. Folmer, T. Tietenberg (eds.), *The International Yearbook of Environmental and Resource Economics 1999-2000,* Edward Elgar, Cheltenham, 2000, pp. 273-306.
Siebert H., 1998. *Economics of the Environment,* Springer, Heidelberg.
Tietenberg T., 2006. *Environmental and Natural Resource Economics,* Pearson, Addison Wesley, New York.
World Commission on Environment and Development, 1987. *Our Common Future,* Oxford University Press.

The issue of the distinction between legal principles and the other two main sources of international law, and in particular the one between customary rules and legal principles, has been addressed by several legal writers, but it is in particular the work of Cheng which may be of a major help for the sake of our analysis. According to him, in particular,

> while conventions can be easily distinguished from the two other sources of international law, the line of demarcation between custom and general principles of law recognized by civilized nations is often not very clear, since international custom or customary law, understood in a broad sense, may include all that is unwritten in international law. In Article 38, however, custom is used in a strict sense, being confined to what is a general practice among States accepted by them as law. General practice among nations, as well as the recognition of its legal character, is therefore required. It should be observed that the emphasis in the definition of what constitutes a custom lies not in the rule involved in the general practice, but rather in its being part of objective law as a whole. In the definition of the third source of international law, there is also the element of recognition on the part of civilized peoples but the requirement of a general practice is absent. The object of recognition is, therefore, no longer the legal character of the rule implied in an international usage, but the existence of certain principles intrinsically legal in nature. This part of international law consists in the general principles of that social phenomenon common to all civilized societies which is called law. (Cheng, 1953)

In more general terms, the difference among legal principles and legal rules is a "logical distinction", according to the legal philosopher Dworkin. In his opinion, in fact,

> the difference between legal principles and legal rules is a logical distinction. Both sets of standards point to particular decisions about legal obligation in particular circumstances, but they differ in the character of the direction they give. Rules are applicable in all-or-nothing fashion. If the facts a rule stipulates are given, then either the rule is valid, in which case the answer it supplies must be accepted, or it is not, in which case it contributes nothing to the decision. (Dworkin, 1977)

From the reasoning of Dworkin, it emerges that the flexibility which characterizes legal principles may be an advantage in all those circumstances in which the application of the very rigid instrument of legal rules may run the risk of not rendering effective justice in the concrete case at stake. This may be further clarified by the following excerpt from Cheng:

> Since principles express general truth, general principles of law express general juridical truth. They form the theoretical bases of positive rules of law. The latter are the practical formulation of the principles and, for reasons of expediency, may vary and depart, to a greater or lesser extent, from the principle from which they spring. The application of the principle to the infinitely varying circumstances of practical life aims at bringing about substantive justice in every case; the application of the rules, however, results only in justice according to law, with the inescapable risk that in individual cases there may be a departure from subjective justice. (Cheng, 1953)

A similar line of reasoning, to the ones expressed by Cheng and Dworkin, was already made in very clear terms, a long time ago, in the decision rendered by the Arbitral Tribunal in the *Gentini Case*, back in 1903. In such a circumstance, in fact, the Umpire had held that:

A rule... "is essentially practical and, moreover, binding..."; there are rules of art as there are rules of government", while "a principle expresses a general truth, which guides our action, serves as a theoretical basis for the various acts of our life, and the application of which to reality produces a given consequence". (Gentini Case 1903)

What said above makes clear that the fundamental and essential difference among legal principles and legal rules is an ontological distinction, insofar legal principles constitute general and abstract provisions, which express general and always valid truths and represent the theoretical basis for the definition of legal rules, whereas legal rules, which express the practical formulation of the legal principles from which they originate, have an essentially practical nature and normally entail binding consequences for their addressees.

In conclusion, therefore, we can affirm that despite the fact that legal principles constitute general and abstract provisions, which often do not possess the sufficiently precise character which characterizes legal rules, this does not by any means prevent such principles from having an important role as sources of international law, beside treaty and customary rules, as the third main source of law to which the International Court of Justice (ICJ) may refer when rendering their decision based on article 38 of the ICJ Statute (Cheng, 1953).

This is particularly true in the field of international environmental law, characterized by a great flexibility and variability of situations which is often not very well served by the rigidity of legal rules, but rather tends to receive a greater benefit in terms of effectiveness from the flexibility of legal principles.

2. THE ENVIRONMENTAL LEGAL PRINCIPLES

It is only in the last few decades that humanity has begun to become more and more aware of the consequences for the environment that invasive and unsustainable patterns of human development, which for a long time were subject to very little control, are posing for the global environment and ultimately for the very survival of our planet. An institutional and legal response to tackle the issue of the "environmental challenge" has been devised by the international community of States around the world, starting from the 1970s, based on the belief, aptly explained by the International Court of Justice, that

the environment is not an abstraction, but represents the living space, the quality of life and the very health of human beings, including generations unborn. (International Court of Justice, 1996)

Actually, the record of the international practice also shows some early cases with "environmental" features which were decided well before humanity has become aware of the "environmental challenge" which we are presently facing. For instance, the 1941 *Trail Smelter Case*, which dealt with an issue of trans-boundary air pollution, is often cited as an example of an early "environmental" case, which involved two States at international level, although it concretely resulted in the application to the relationship between two sovereign States of the traditional rules of "good neighbourliness" and "nuisance" between private persons, which exist in most national legal orders (Trail Smelter Case, 1941).

Apart from this one, and a very few other cases, however, it is not until the beginning of the 1970s that the international community started to approach the "environmental challenge", by devising a proper legal framework, including a specific set of environmental legal principles[2]. The starting date of what has been then called international environmental law is commonly reported back to 1972, when the first world conference for the protection of the environment was held in Stockholm, under the auspices of the United Nations. The 1972 Stockholm Conference, officially named *UN Conference on the Human Environment*, had among its merits to start discussing the main environmental issues at stake in a well framed way and promoting the adoption of instruments and the establishment of institutions to effectively tackle those issues. To this effect, the States gathered in Stockholm adopted a well-known Declaration which crystallizes some of the most important

> common principles [which ought] to inspire the peoples of the world in the preservation and enhancement of the human conditions (UN, 1972).

The Principles listed in the Stockholm Declaration represent the first example of the concrete effort of the international community to establish some guidelines which can be applied in order to try and find an adequate and successful balance between the need to foster economic and social development and the opposite necessity to promote and improve the environment, to maintain or restore a quality of the environment at a level which permits a life of dignity and well-being for the human beings, with regard to both present and future generations.

In particular, among the twenty-six Principles listed in the Stockholm Declaration, it is Principle 21 which represents a landmark. In fact, such a Principle, which has been commonly considered since its first formulation one of the cornerstones upon which international environmental law is based, provides that States have the sovereign right to exploit their own natural resources according to their national policies and the correspondent responsibility to ensure that activities carried out within their jurisdiction or control do not cause damage to the environment of other States or of areas beyond the limits of national jurisdiction.

Principle 21 of the Stockholm Declaration probably represents the first attempt of the international community to try and find instruments and patterns to achieve a balance between the right to economic development and the protection of the environment. In this case, the right to development is enshrined in the right for each State to exploit its own natural resources according to its own needs and policies, which States have a sovereign right to pursue, while at the same time they must ensure that economic development does not damage the environment, particularly the environment beyond the borders of the State within whose jurisdiction the potentially damaging activities are performed. In fact, in Principle 21, the attempt to strike a balance between the right to development and the duty to protect the environment is still framed in the very traditional terms of limiting trans-boundary pollution which may give rise to international controversies among sovereign States. However, it cannot be denied that in concrete terms the establishment of Principle 21 probably paved the way for the appearance on the international scene of the concept of sustainable development, which was firstly launched in the 1980s and now represents the basis upon which most of international environmental laws stand.

The genesis of the concept of sustainable development is commonly reported to the 1987 *Brundtland Report*, which contains the well-known definition of "sustainable development" as

> development that meets the needs of the present without compromising the ability of future generations to meet their own needs (UN, 1987).

The characteristics of the principle of sustainable development will be analyzed in greater detail in the next paragraph of the present contribution. For now it is sufficient to recall that the principle of sustainable development, since its first formulation in the Brundtland Report, has immediately become the common keyword for the promotion of all national and international efforts aimed at promoting patterns of development wishing to pay a closer attention to the needs related to the protection of the environment.

It is certainly not a coincidence that the 1992 Rio Conference promoted by the UN General Assembly to assess the state of the art regarding such a world-wide effort towards a sustainable and environmentally sound development and to start developing common strategies and adopting specific measures to halt and reverse the adverse effects of humanly-induced environmental degradation on the planet, was named *UN Conference on Environment and Development*. The name chosen for the 1992 Rio Conference stressed in fact the importance of the overarching objective of the balance between the right to pursue economic development for all peoples on earth with the competing common interest of mankind to promote sustainable patterns of development and abandon practices and behaviours which are deemed to be too risky for the environment.

The *Rio Declaration* adopted at the 1992 Rio Conference contains twenty-seven Principles, which although are *per se* generally non binding, certainly represent the main framework upon which most of the global efforts to affirm, both at international and national levels, an increased recourse to the principle of sustainable development are based since then. In the Rio Declaration, a particular stress is placed on the need that States co-operate in good faith in order to fulfil their right to development without endangering the environment and to further promote the establishment of a full body of international law in the field of sustainable development.

Several Principles listed in the Rio Declaration deserve a special attention: first of all, one may recall Principle 2 of the Rio Declaration, which substantially reproduces Principle 21 of the Stockholm Declaration, with a small, but quite significant modification. In fact, while the text of Principle 21 states that "States have the sovereign right to exploit their own resources pursuant to their own environmental policies" (UN, 1972), twenty years later, probably under the influence of the principle of sustainable development, the text of Principle 2 affirms that States have the right to exploit their own resources pursuant not only to their environmental policies, but also to their "developmental policies" (UN, 1992).

As one can see, this slight, but not irrelevant modification probably reflects a significant change in the perspective towards the protection of the environment which took place during the 1990s. The new perspective is dominated by the concept of sustainable development, which aims at reconciling the interest of economic development with the one of environmental protection, on a case by case basis, without trying to give an *a priori* primacy to one of those interests over the other.

Beside Principle 2, there are at least two more principles in the Rio Declaration which deserve a special mention. The first one is Principle 15, embodying the precautionary principle, which calls for an anticipatory approach to environmental problems, also in those circumstances where there is no scientific certainty on the possible negative consequences which may derive from an incumbent situation of risk for the environment. According to the definition contained in the Rio Declaration,

> where there are risks of serious and irreversible damage, lack of full scientific certainty shall not be used as a reason for postponing cost-effective measures to prevent environmental degradation. (UN, 1992)

In brief, since its incorporation in the Rio Declaration the precautionary principle has received a broad support by most States of the international community and nowadays represents one of the most striking features of international environmental law, despite the remaining uncertainties over the precise consequences which may derive from its widespread application that could be assessed only on a case by case basis.

The other remaining principle contained in the Rio Declaration that deserves a special consideration in the present context is Principle 27, which refers to the co-operation principle. The definition of such a principle, which is largely based on the former Principle 24 of the Stockholm Declaration, reads as follows:

> States and peoples shall co-operate in good faith and in a spirit of partnership in the fulfilment of the principles embodied in this Declaration and in the further development of international law in the field of sustainable development. (UN, 1992)

The relevance of the co-operation principle is two-fold. On the one side, it may provide the relevant legal underpinning to the activities of States directed to the adoption and implementation of international legal obligations for environmental protection. In addition to that, on the other side, it may further provide the legal basis for drafting more precise legal rules, instruments and techniques aiming at rendering more effective the general duties of information and participation in the environmental field.

After the 1992 Rio Conference, international developments regarding the evolution and further classification and advancement of the environmental legal principles have become more rare and less relevant and incisive. A reference, however, ought to be made to the third global environmental legal conference, which followed those of Stockholm and Rio, and which greatly influenced the further elaboration, interpretation and application of the principle of sustainable development. It is the Johannesburg Conference, which was convened by the United Nations in 2002 in order to assess the progresses made by the international community in the process for the development of an international law in the field of sustainable development and which not surprisingly was named *UN Conference on Sustainable Development*.

In concrete terms, however, the 2002 Johannesburg Conference did not produce the same interesting results as the previous two 1972 and 1992 Conferences, as far as the adoption of declarations of principles and international treaties are concerned. At the Johannesburg Conference, only a *Political Declaration* and a corresponding *Plan of Action* were adopted. These documents do not represent any advancement in the progressive development of international law in the field of sustainable development, but merely recall the concepts already elaborated in the previous Stockholm and Rio Conferences. In the *Political Declaration*, in fact, all States and peoples of the world gathered in Johannesburg merely reaffirmed their general commitment to the principle of sustainable development, as the key-word to pursue the objective of a better future for all humankind, and in the *Plan of Action* launched a very broad working agenda for the next few years.

3. THE PRINCIPLE OF SUSTAINABLE DEVELOPMENT: ORIGINS AND MAIN CONSTITUTIVE ELEMENTS

We have mentioned in the previous paragraph that the first and commonly accepted basic definition of sustainable development is the one contained in the 1987 *Brundtland Report*, where "sustainable development" is defined as

> development that meets the needs of the present without compromising the ability of future generations to meet their own needs. (UN, 1987)

The definition contained in the *Brundtland Report* is based on the two opposed concepts of "needs" and of "limits" (Sands, 2003). On the one side, it says that the objective of sustainable development can be reached only if the "needs" of the present and future generations are satisfied in a fair and equitable way. On the other side, however, it implies that no development can be considered to be "sustainable" if no "limits" are placed on the exploitation of the environmental resources in the name of economic development. In other words, development can be sustainable only if some "limits" are posed upon it and if it does not pose an excessive burden on the capacity of the environment to sustain the human pressure, so as to permit a fair and equitable satisfaction of the needs of the present and future generations.

In spite of the fact that the concept of sustainable development has emerged as such just in the last fifteen years, the idea of the "sustainability" of certain human activities potentially dangerous for the natural environment can be traced back to the XIX century. An early example of a claim based on the idea of sustainability can be found, for instance, in the *Fur Seals Case* (1893), which opposed the United States and Great Britain. In such a case, the attempt of the United States to act unilaterally so as to prevent an excessive and unregulated fishing of fur seals located in the Behring Sea, operated mainly by British fishermen, was considered to be not acceptable by the Arbitral Tribunal called upon to judge on the question. The Tribunal, in fact, held that the United States in principle had no right to regulate fishing outside their territorial sea. However, responding to the quest of sustainability of the US, the Tribunal partially endorsed the US position and promoted the conclusion of a fishing agreement between the two States, with the aim of preventing an excessive fishing of fur seals and therefore avoiding the risk of an extinction of the said species (Fur Seals Case, 1893).

The main elements which constitute the principle of sustainable development, in the framework of its basic and traditional definition contained in the *Brundtland Report*, are essentially four. The first constitutive element is represented by the concept of the prudent and rational use of natural resources. Such a concept is the basic guideline which must inform the conduct of all States of the international community, when they are determining and implementing their national policies on economic development and environmental protection. Moreover, the aim of the prudent and rational use of natural resources must be taken into account by States also when they co-operate for the conclusion of international agreements in the field of economic development and environmental protection, so as to promote sustainable patterns of development (Sands, 2003).

The second constitutive element of the concept of sustainable development is represented by the principle of inter-generational equity. Such a principle imposes upon States the duty to take into account not only the needs of the present generation, but also the plausible needs and the overall benefit of future generations. The wide reach of the concept of inter-generational equity may be well explained by making reference to the legal concept of "trust". In fact, if we keep in mind that as "members of the present generation, we hold the earth in trust for future generations" (Brown Weiss, 1990), it follows that we are compelled to make a prudent and rational use of the natural resources available, so as not to deplete excessively the stock of natural resources left to us by the previous generation and leave to the next generations the same possibilities we have to meet their own needs (Sands, 2003)[3].

The third constitutive element of the concept of sustainable development is represented by the principle of intra-generational equity. According to such a principle, each State in the definition of its developmental and environmental policies must take into account the benefit of the other peoples and States around the world. The principle of intra-generational equity is closely related to the principle of inter-generational equity. Both principles have in common a fundamental "social dimension". However, while the inter-generational perspective adopts a dynamic and inter-temporal approach, the intra-generational perspective tries to promote a fair and equitable distribution of the world resources with reference to a static approach, which considers the present historical situation. Moreover, the principle of intra-generational equity is closely linked also to the principle of common but differentiated responsibilities, as enshrined in Principle 7 of the Rio Declaration, which affirms that all States have a duty to co-operate in a spirit of global partnership to tackle the most serious global environmental problems. However, not all States have the same duties and responsibilities to act for the benefit of the Earth's ecosystem. On the contrary, in many circumstances they ought to be subject to differentiated responsibilities, in the framework of a common general objective. In such a context, obviously, in order to define precisely such respective responsibilities, due account should be taken of their different contributions to the occurrence of the specific cases of environmental degradation.

The fourth constitutive element of the concept of sustainable development is represented by the principle of integration. Pursuant to such a principle, environmental considerations must be integrated into economic development projects, plans and programmes, so as to promote an environmentally friendly approach to economic development. The principle of integration is defined in the following terms in Principle 4 of the Rio Declaration:

> In order to achieve sustainable development, environmental protection shall constitute an integral part of the development process and cannot be considered in isolation from it. (UN, 1992)

It should be underlined that, in absence of any specific indication to the contrary, the principle of integration is meant to drive both the national and international policies of States towards a more sustainable approach; in fact in recent years it has been incorporated in several national and international legal instruments to this effect. Moreover, the principle of integration ought to play a decisive role if the concept of "sustainable development" is really destined to be a privileged instrument to be used in order to find a balance, on a case by case basis, between the right to development and the protection of the environment (Sands, 2003).

The basic and traditional definition of the principle of sustainable development just described above, based on four constitutive elements, was partially reviewed and updated at the Johannesburg Conference. In fact, the *Johannesburg Political Declaration* underlines that the principle of sustainable development is based on three

> "interdependent and mutually reinforcing pillars", namely "economic development, social development and environmental protection", which must be collectively promoted and advanced "at the local, national, regional and global levels" (UN, 2002).

In more general terms, it can be noted that while the Stockholm and Rio Conferences had mainly focused on environmental issues, at the Johannesburg Summit the attention gradually shifted from environmental to economic and social issues.

Therefore, in the Johannesburg Declaration the Parties recognize that

> poverty eradication, changing consumption and production patterns, and protecting and managing the natural resource base for economic and social development are overarching objectives of, and essential requirements for sustainable development. (UN, 2002)

In such a context, the centrality of the more traditional objective of the protection of the environment as such seems to be destined to be gradually replaced by the more dynamic objective of the protection and sound management of natural resources intended as pre-requisites for economic and social development.

In an era of increasing economic globalization, the traditional environmental and social challenges seem to have gained a new dimension which must be taken into account. Pursuant to the Johannesburg Declaration,

> the rapid integration of markets, mobility of capital and significant increases in investment flows around the world have opened new challenges and opportunities for the pursuit of sustainable development (UN, 2002).

Therefore the priority for the international community must be "the fight against the world-wide conditions that pose severe threats to the sustainable development" of all the people[4].

In sum, the wording of the Johannesburg Declaration seems to build up on the previous Stockholm and Rio Declarations, but with two important differences. The first one is that the present Declaration does not contain legal principles, but merely political statements. The second one is that the present Declaration, departing from the previous ones, shifts its main focus on economic and social development themes, whereas the protection of environment seems to be left aside, almost as the less important pillar among the three ones on which the concept of sustainable development is based.

4. THE PRINCIPLE OF SUSTAINABLE DEVELOPMENT: AN ANALYSIS OF THE MOST RELEVANT LEGAL PRACTICE

Once described the four constitutive elements which compose the principle of sustainable development, as deriving from the development of international environmental law marked by the Rio and the Johannesburg Conferences, it is now time to briefly analyze how the principle of sustainable development has been incorporated in international treaties and has been relied upon in some relevant decisions of international courts and tribunals.

If one exempts the broad reference to the concept of sustainability contained in the 1982 Convention on the Law of the Sea, with specific reference to the issue of the sound management of international fisheries (UN Convention on the Law of the Sea, 1982), the first two international treaties which made an explicit reference to the concept of sustainable development as one of the guiding principles to which such treaties are based, are the 1992 *Framework Convention on Climate Change* and the 1992 *Convention on Biological Diversity*, both signed during the Rio Conference on Environment and Development.

In particular, as regards to the Climate Change Convention, the principle is enshrined in article 3, which reads as follows:

> The Parties have a right to, and should, promote sustainable development. Policies and measures to protect the climate system against human-induced change should be appropriate for the specific conditions of each Party and should be integrated with national development programmes, taking into account that economic development is essential for adopting measures to address climate change. (UN Framework Convention on Climate Change, 1992)

Correspondingly, in the Biodiversity Convention, "sustainability" is one of the key-words which ought to regulate the sound management of biological resources and although the term "sustainable development" is not explicitly used in the text of the Convention, a certain relevance is given to the concept of "sustainable use" of biological diversity. For instance, article 1 of the Biodiversity Convention states that:

> The objectives of this Convention, to be pursued in accordance with its relevant provisions, are the conservation of biological diversity, the sustainable use of its

components and the fair and equitable sharing of the benefits arising out of the utilization of genetic resources. (UN Convention on Biological Diversity, 1992)

As to the concept of sustainable use, in particular, this is defined as such by article 2 of the Convention:

> Sustainable use means the use of components of biological diversity in a way and at a rate that does not lead to the long-term decline of biological diversity, thereby maintaining its potential to meet the needs and aspirations of present and future generations. (UN Convention on Biological Diversity, 1992)

Even more interestingly, during the 1990s the concept of sustainable development was incorporated in a series of treaties not specifically dealing with the protection of the environment, such as the 1994 Treaty Establishing the World Trade Organization (WTO), whose preamble contains an explicit reference to the principle of sustainable development, when it affirms that the Parties recognize that:

> Their relations in the field of trade and economic endeavour should be conducted with a view to raising standards of living, ensuring full employment and a large and steadily growing volume of real income and effective demand, and expanding the production of and trade in goods and services, while allowing for the optimal use of the world's resources in accordance with the objective of sustainable development, seeking both to protect and preserve the environment and to enhance the means for doing so in a manner consistent with their respective needs and concerns at different levels of economic development. (UN Treaty Establishing the World Trade Organization, 1994)

As for the reference to the principle of sustainable development contained in the decisions of international courts and tribunals, two cases in particular ought to be mentioned. Such cases are the 1997 *Gabcikovo-Nagymaros Case*, decided by the International Court of Justice, and the 1998 *Shrimps-Turtles Case*, decided by the WTO dispute settlement authorities.

For instance, in the *Gabcikovo-Nagymaros* case, which regarded a project of dams to be built on the Danube for hydro-electric purposes and whose construction opposed Hungary to Slovakia mainly on environmental protection grounds, the concept of sustainable development was referred to by the ICJ as the key concept which aptly expresses in a single framework the need to balance and reconcile the two conflicting interests of economic development and environmental protection. The Court, in particular held that:

> Throughout the ages, mankind has, for economic and other reasons, constantly interfered with nature. In the past, this was often done without consideration of the effects upon the environment. Owing to new scientific insights and to a growing awareness of the risks for mankind — for present and future generations — of pursuit of such interventions at an unconsidered and unabated pace, new norms and standards have been developed, set forth in a great number of instruments during the last two decades. Such new norms have to be taken into consideration, and such new standards given proper weight, not only when States contemplate new activities but also when continuing with activities begun in the past. This need to reconcile economic development with protection of the environment is aptly expressed in the concept of sustainable development. (ICJ, 1997)

In the *Gabcikovo-Nagymaros Case*, then, the importance of the concept of sustainable development was further stressed by Judge Weeramanrty, who provided for a thorough examination of the principle in its Dissenting Opinion (ICJ, 1997). In particular, as to the role of the principle of sustainable development in general, he held that:

> The Court must hold the balance even between the environmental considerations and the developmental considerations raised by the respective Parties. The principle that enables the Court to do so is the principle of sustainable development. (ICJ, 1997)

Moreover, as to the applicability of the principle in the case at stake as a balancing tool between the conflicting interests related to the right to development on the one side and the right to environmental protection on the other side, the Court stated that:

> When a major scheme, such as that under consideration in the present case, is planned and implemented, there is always the need to weigh considerations of development against environmental considerations, as their underlying juristic bases — the right to development and the right to environmental protection — are important principles of current international law. (ICJ, 1997)

Judge Weeramantry then observed that in the case under scrutiny both Parties, namely Hungary and Slovakia, agreed on the existence and applicability of the principle of sustainable development to the case and their disagreement was merely as the way in which the principle was to be concretely applied (ICJ, 1998).

As mentioned above, another recent case which bears a particular relevance for the issue of the progressive incorporation of the principle of sustainable development in the recent international practice is the *Shrimps-Turtles Case*, decided in the framework of the WTO dispute settlement mechanism (Shrimps-Turtles Case, 1998). The *Shrimps-Turtles Case* concerned a US regulation which unilaterally imposed to the fishing activities occurring outside the jurisdiction of the United States certain fishing technologies to prevent accidental take of sea turtles. The US regulation was opposed by several Parties and was ultimately judged on the basis of the exception contained in article XX(g) of the General Agreement on Tariffs and Trade (GATT). For what concerns our analysis, in particular, it should be recalled that the WTO Appellate Body, when analyzing the term "exhaustible natural resources", contained in article XX(g) GATT, affirmed that since:

> The words of Article XX(g), "exhaustible natural resources", were actually crafted more than 50 years ago. They must be read by a treaty interpreter in the light of contemporary concerns of the community of nations about the protection and conservation of the environment. (Shrimps-Turtles Case, 1998)

The WTO Appellate Body then held that, in order to achieve such an "evolutive interpretation" of the existing norms of the GATT Treaty[5], a particular relevance had to be recognised to the principle of sustainable development, whose importance is explicitly acknowledged by the WTO Agreement in its preamble (UN Treaty Establishing the World Trade Organization, 1994).

5. CONCLUSION

The present analysis, specifically focused on the principle of sustainable development, has shown that environmental principles may be a useful tool for environmental management, since they are more flexible and can adapt to different situations and changing circumstances better than legal rules.

Before concluding the analysis, however, it seems necessary to throw a brief look at the classification of the different functions environmental principles may play in international environmental law, in order to better assess their effective actual and potential contribution to environmental management[6].

On the basis of the relevant legal theory, principles falling within the concept of "general principles of law", to which also environmental legal principles pertain, may be classified under three different categories, on the basis of their function. Firstly, there are those principles which have a precise and well defined legal content and may have in some specific circumstances an integrative function with respect to the other sources of international law, namely treaties and customary law provisions. Secondly, there are those principles which, although generally recognized by the majority of States, do not possess a well defined normative content. Such principles may only have an interpretative function with regard to the existing legal rules, so as to fill existing gaps or help adapting them to the progress of contemporary international law. In the third place, there are those principles which have a very general and broad scope and a merely programmatic nature. Such principles are too loose in nature to have an integrative or an interpretative function, but may rather contribute to the inspiration and development of new legal rules, instruments and techniques, with a legally binding content.

In more general terms, it should be underlined that the three categories of principles just mentioned above are to be considered dynamic and not static ones, which means that each of the principles which belong to one of those three categories, in the course of its progressive consolidation and development, may depart from the (lower) category of the programmatic principles, in order to ascend partially or totally to the (higher) categories of the integrative and interpretative principles.

We can therefore conclude the analysis by saying that although all the principles we have cited and examined in this contribution may have a relevant role for environmental management, they may be differently classified according to their function. For instance, it seems that the principle of sustainable development may be recognized merely as having a programmatic, and sometimes a limited interpretative function, whereas the other major environmental legal principles, such as principle 21, the precautionary principle and the co-operation principle, in some specific circumstances may also have an integrative function, which may ultimately help to achieve in more effective and concrete terms the general objective of sustainable development.

Massimiliano Montini
Department of Economic Law,
University of Siena, Italy

NOTES

[1] On the issue of the qualification of the general principles of law, or more precisely of the "general principles of law recognized by civilized nations" among the main sources of international law see for instance: Cheng, General Principles of Law as applied by International Courts and Tribunals, London, 1953; McNair, The General Principles of Law Recognized by Civilized Nations, in BYIL; 1957, 1; Fitzmaurice, The General Principles of International Law Considered from the Standpoint of the Rule of Law, in Recueil des Cours, Vol. 92, 1957, II, p. 5; Virally, The Sources of International Law, in Sorensen (ed.), Manual of Public International Law, New York, 1968, p. 126; Bogdan, General Principles of Law and the Problem of Lacunae in the Law of Nations, in NTIR, 1977, p. 37; Lammers, General Principles of Law recognized by Civilized Nations, in Essays in the development of the International Legal Order in memory of H.F. Van Panhuys, The Hague, 1980, p. 53; Mosler, General Principles of Law (1984), in EPIL, Vol. II, 1995, p. 511; Gaja, Principi del diritto (Diritto Internazionale), in ED, Vol. XXXV, Milano, 1986, p. 533; Strozzi, I principi dell'ordinamento internazionale, in CI, 1992, p. 162; Salerno, Principi generali di Diritto (Diritto Internazionale), in Digesto, IV ed., 1996, Vol. IX (Discipline pubblicistiche), p. 524; Degan, Sources of International Law, The Hague, 1997, p. 14.

[2] On international environmental law see in general Sands, Principles of International Environmental Law, Cambridge, 2003; Birnie & Boyle, International Law and the Environment, Oxford, 2002.

[3] On this topic see also Francioni, Per un governo mondiale dell'ambiente: quali norme? quali istituzioni?, in Scamuzzi, Costituzioni, Razionalità, Ambiente, Torino, 1993, p. 443.

[4] According to the Johannesburg Declaration, among the conditions that pose severe threats to the sustainable development of the people are: "chronic hunger; malnutrition; foreign occupation; armed conflicts; illicit drug problems; organized crime; corruption; natural disasters; illicit arms trafficking; trafficking in persons; terrorism; intolerance and incitement to racial, ethnic, religious and other hatreds; xenophobia; and endemic, communicable and chronic diseases, in particular HIV/AIDS, malaria and tuberculosis" (Johannesburg Political Declaration, 2002).

[5] On the principle of the "evolutive interpretation" of existing treaty provisions see Francioni, Environment, Human Rights and the Limits of Free Trade, in Francioni (ed., 2001), Environment, Human Rights and International Trade, Hart, Oxford, p. 23.

[6] On the classification of legal principles according to their function see in general Strozzi, I principi dell'ordinamento internazionale, cited supra, p. 162; Lammers, General Principles of Law recognized by Civilized Nations, cited supra, p. 53; Mosler, General Principles of Law (1984), cited supra, p. 89; Gaja, Principi del diritto (Diritto Internazionale), cited supra, p. 536; Salerno, Principi generali di Diritto (Diritto Internazionale), cited supra, p. 524; for the classification of environmental legal principles in particular see De Sadeleer, Environmental Principles: from political slogans to legal rules, Oxford, 2002, p. 233.

REFERENCES

Brown Weiss E., 1990. *Our Rights and Obligations to Future Generations for the Environment*, 84 AJIL, p. 1999.
Cheng B., 1953. *General Principles of Law as applied by International Courts and Tribunals*, London, p. 23, p. 376.
Dworkin R., 1977. *Taking Rights Seriously*, London, p. 24.
Gentini Case, 1903. Italy-Venezuela Mixed Claims Commission, in Ralston & Doyle (1904), *Venezuelan Arbitrations of 1903*, Washington.
International Court of Justice, ICJ, 1996. Nuclear Weapons Advisory Opinion, ICJ Reports, 1996, p. 66 and p. 226, in ILM, Vol. 35, p. 809 ff., § 29.
International Court of Justice, ICJ, 1997. Gabcikovo-Nagymaros Case, ICJ Reports, 1997, p. 7, in ILM, Vol. 37, p. 162 ff., § 53, § 140, and Dissenting Opinion Judge Weeramantry, section A.
Sands P., 2003. *Principles of International Environmental Law*, Cambridge, p. 198, p. 201, p. 199, p. 205
Shrimps-Turtles Case, 1998. *Report of the Appellate Body*, WT/DS58/AB/R, § 129.
Trail Smelter Case, 1941. US-Canada Mixed Arbitration Commission, in *United Nations Reports of International Arbitral Awards* (UNRIAA), 1947, Vol. III, p. 1905.

UN, 1972. *Stockholm Declaration* (Declaration of the United Nations Conference on the Human Environment).
UN, 1987. *Our Common Future* (Brundtland Report), Oxford University Press, Oxford.
UN, 1992. *Rio Declaration* (Rio Declaration on Environment and Development).
UN, 2002. *Johannesburg Political Declaration,* § 5, §11, §14, § 19.
Fur Seals Case, 1893, in Moore, *International Arbitral Awards,* Vol. I, p. 755.
UN Convention on Biological Diversity, 1992. ILM, reprinted in Vol. 31, 1992, articles 1 and 2, 82 ff.
UN Convention on the Law of the Sea, UNCLOS, 1982. Preamble, reprinted in ILM (1982), Vol. 21, p. 1261 ff.
UN Framework Convention on Climate Change (1992), reprinted in ILM, Vol. 31, 1992, article 3 (4) p. 849 ff.
UN Treaty Establishing the World Trade Organization, 1994. Reprinted in 33 ILM, 1994, preamble, p. 13 ff.

G. MUNDA

VALUATION OF SUSTAINABLE DEVELOPMENT POLICIES: SOCIAL MULTI-CRITERIA EVALUATION

Abstract. Sustainable development is a multidimensional concept, including socio-economic, ecological, technical and ethical perspectives. In making sustainability policies operational, basic questions to be answered are sustainability of *what and whom*? As a consequence, sustainability issues are characterized by a high degree of conflict. The main objective of this Chapter is to show that multiple-criteria decision analysis is an adequate approach for dealing with sustainability conflicts. To achieve this objective, lessons learned from both theoretical argumentations and empirical experience are examined. Guidelines of *"good practice"* are suggested too.

1. ECONOMIC VALUATION AND SUSTAINABLE DEVELOPMENT

In the eighties, the awareness of actual and potential conflicts between economic growth and the environment led to the concept of "sustainable development". Since then, all governments have declared, and still claim, their willingness to pursue economic growth under the flag of sustainable development. The concept of sustainable development has a wide appeal, partly because, in contrast with the "zero growth" idea by Daly (1977), it does not set economic growth and environmental preservation in sharp opposition. Rather, sustainable development carries the ideal of a harmonization or simultaneous realization of economic growth and environmental concerns. For example, Barbier (1987, p. 103) writes that sustainable development implies:

> To maximize simultaneously[1] the biological system goals (genetic diversity, resilience, biological productivity), economic system goals (satisfaction of basic needs, enhancement of equity, increasing useful goods and services), and social system goals (cultural diversity, institutional sustainability, social justice, participation).

This definition correctly points out that sustainable development is a multi-dimensional concept. However, as our everyday life teaches us, it is generally impossible to maximize different objectives at the same time, and as formalized by multi-criteria decision theory, compromise solutions must be found. A legitimate question could be raised: sustainable development of what and whom? Norgaard (1994, p. 11) writes:

> Consumers want consumption sustained, workers want jobs sustained. Capitalists and socialists have their "isms", while aristocrats and technocrats have their "cracies".

Social policies based on principles of compensation and substitution sometimes might be operative, but one should be very cautious in applying such principles as a general guideline. There are allocations without any possibility of transactions in actual or fictitious markets. Who would be willing to accept compensation for the destruction of the Sagrada Familia, the Statue of Liberty or the Coliseum? In this context, from an economic point of view, the only instrument left is cost-effectiveness; given a certain physical target (e.g. the amount of cultural heritage to be preserved or the amount of contamination to be accepted), it is rational to try to get it by means of the lowest possible use of resources (i.e. at the minimum social cost). Obviously there are several targets possible. In general two rankings are possible:

1. According to the lowest cost.
2. According to the physical target (e.g. the more monuments preserved, the better).

Perhaps a discussion would lead to the judgment that the improvement of a physical target to a better one is worth the extra economic cost, or perhaps the opposite judgment will be reached. In both cases, we would have an ordinal ranking of alternatives and "cost-effectiveness" would "fall down" into multi-criteria evaluation, i.e. two criteria and two different rankings must be explicitly dealt with.

From the above discussion the following conclusion can be drawn: to attach prices to non-market assets (such as most of environmental and cultural ones), gives a positive signal to society and may contribute to a more rational use increasing the chances for a better conservation. When one wishes to preserve a monument or a natural area, a fundamental question is: is there any resource which society is willing to assign to this objective? To answer this question the concept of "total economic value" becomes immediately relevant. To attribute monetary values to e.g. historical heritage implies to capture user (actual, option and bequest) and *non*-user (existential, symbolic, etc.) values. Of course, to compute total economic values has nothing to do with the "true" or "correct" value. All monetary valuation attempts will suffer deep uncertainties such as:

- Which monetary valuation technique has to be used?
- Which time horizon has to be considered?
- Which social discount rate?

Moreover, one should remember that the market alone may be successful in efficient allocation of resources, but does not give any guarantee for preservation of the cultural or natural heritage at all. Once something is on the market, it can be bought or sold and so the willingness to accept and the compensation principle may easily cause the destruction of any asset.

As a first conclusion, we could state that monetary compensation is with no doubt the only possible tool when an irreparable and irreversible damage has already occurred. In this way, if an accident with serious contamination occurs – as in the case of Seveso in Italy (1976), Bhopal in India (1984), the Exxon Valdez in Alaska (1989), or more recently the oil-tanker Prestige offshore the coasts of Galicia (2002) – it seems correct and opportune to indemnify the victims of such contamination. However, it stays to verify if, in the long run, compensation is an effective tool to prevent the appearance of enormous social costs, given that it doesn't guarantee the preservation of natural or cultural goods and services. The economic value is different from the environmental or artistic-cultural value. If we had to decide whether to save the Galapagos Islands or the Inside Sea in Holland, which value one should use? The economic one would favour the inside sea, which, since totally eutrophised, offers an important economic service receiving all the nutrients coming from human activity. The ecological one would obviously point out instead the Galapagos Islands. The choice of the values to be considered as socially predominant is a scientific or socio-political issue? The world is characterized by deep complexity. This obvious observation has important implications on the manner in which policy problems are represented and decision making is framed. Each representation of a complex system is reflecting only a sub-set of the possible representations of it. A consequence of these deep indeterminacies is that in any policy problem, one has to choose an operational definition of "value" in spite of the fact that social actors with different interests, cultural identities and goals have different definitions of "value". That is, to reach a ranking of policy options, there is a previous need for deciding about what is important for different social actors as well as what is relevant for the representation of the real-world entity described in the model. One should note that the representation of a real-world system depends on very strong assumptions about (1) the *purpose* of this construction, e.g. to evaluate the sustainability of a given city, (2) the *scale* of analysis, e.g. a block inside a city, the administrative unit constituting a municipality or the whole metropolitan area and (3) the set of dimensions, objectives and criteria used for the evaluation process. A reductionist approach for building a descriptive model can be defined as the use of just one measurable indicator (e.g. the monetary city product per person), one dimension (e.g. economic), one scale of analysis (e.g. the Commune), one objective (e.g. the maximization of economic efficiency) and one time horizon. Thus, instead of focusing on "missing markets" as causes of allocative disgraces, or trying to explain economic values by means of energy or other common measures (clearly a non-sense from an economic point of view) we should focus on the creative power that missing markets have, because they push us away from commensurability (i.e. a reductionist approach), towards a social multi-criteria evaluation of evolving realities[2].

From this brief discussion the following conclusions can be drawn:

1. A proper evaluation of sustainability options needs to deal with a plurality of legitimate values and interests existing in society. From a societal point of view, economic optimization cannot be the only evaluation criterion. As it is well known, not all goods have a market price or often such a price is too low (*market failures*). Environmental and distributional consequences (intra/inter generational and on non-humans) have necessarily to be taken into account too. In this framework multi-criteria evaluation is a very consistent approach.

2. If from a sustainability point of view, it is accepted that society as a whole is an immortal body, a much longer time horizon than the one normally used on the market should be used. Suddenly a contradiction then exists: normal politicians have a very short time horizon (often 4-5 years according to the electoral system) and this causes that sustainability is not considered seriously by them – thus a government failure exists (for an overview of different perspectives on the role of governments in the economic sphere see e.g. Buchanan and Musgrave, 1999). For this reason, I think that the evaluation of public projects should be done by considering the whole "civil society" (and ethical concerns on future generations) and not only mythical benevolent policy-makers. This is the reason why I am developing here the concept of "social" multi-criteria evaluation, whose main objective is to integrate scientific knowledge with social participation in the framework of sustainability in public choice.

2. SOCIAL MULTI-CRITERIA EVALUATION

In empirical evaluations of public projects and public provided goods, the multi-criteria decision theory seems to be an adequate policy tool since it allows taking into account a wide range of assessment criteria (e.g. environmental impact, distributional equity, and so on) and not simply profit maximization, as a private economic agent would do. From an operational point of view, the major strength of multi-criteria methods is their ability to revolve questions characterized by various conflicting evaluations thus allowing an integrated assessment of the problem at hand. The multi-criteria decision theory builds on the following basic concepts[3].

Dimension: is the highest hierarchical level of analysis and indicates the scope of objectives, criteria and criterion scores. For example, sustainability policy problems generally include economic, social and environmental dimensions.

Objective: an objective indicates the direction of change desired. For example, within the economic dimension GDP has to be maximized; within the social dimension social exclusion has to be minimized; within the environmental dimension CO_2 emissions have to be minimized.

Evaluation Criterion: it is the basis for evaluation in relation to a given objective (any objective may imply a number of different criteria). It is a *function* that associates each single alternative with a variable indicating its desirability according to expected consequences related to the same objective. For example, GDP, saving rate and inflation rate inside the objective "growth maximization".

Criterion Score: it is a constructed measure stemming from a process that represents, at a given point in space and time, a shared perception of a real-world state of affairs consistent with a given criterion. To give an example, in comparing two countries, inside the economic dimension, one objective can be "maximization of economic growth"; the criterion might be R&D performance, the criterion score can be "number of patents per million inhabitants". Another example: an objective connected with the social dimension can be "maximization of the residential attractiveness". A possible criterion is then "residential density". The criterion score might be the ratio persons per hectare.

Constraint: it is a limit on the values that criterion scores may assume and can or cannot be stated mathematically.

Goal: a goal (synonymous with target) is something that can be either achieved or not (e.g. reducing nitrogen pollution in a lake, by at least 10%). If a goal cannot be or is unlikely to be achieved, it may be converted to an objective.

Attribute: it is a measure that indicates whether goals have been met or not, given a particular decision that provides a means of evaluating the levels of various objectives.

Multi-Criteria Method is an aggregate of all dimensions, objectives (or goals), criteria (or attributes) and criterion scores used. This implies that what formally defines a multi-criteria method is the set of properties underlying its aggregation convention.

The discrete multi-criterion problem can be described in the following way: A is a finite set of N feasible actions (or alternatives); M is the number of different points of view or evaluation criteria g_m $i=1, 2, ..., M$ considered relevant in a policy problem, where the action a is evaluated to be better than action b (both belonging to the set A) according to the m-th point of view if $g_m(a) > g_m(b)$. In this way a decision problem may be represented in a tabular or matrix form. Given the sets A (of alternatives) and G (of evaluation criteria) and assuming the existence of N alternatives and M criteria, it is possible to build an $N \times M$ matrix P called evaluation or impact matrix whose typical element p_{ij} $(i=1, 2, ..., M; j=1, 2, ..., N)$ represents the evaluation of the j-th alternative by means of the i-th criterion. The impact matrix may include quantitative, qualitative or both types of information.

In general, in a multi-criteria problem, there is no solution optimizing all the criteria at the same time (the so-called ideal or utopia solution) and therefore compromise solutions have to be found. Indeed this sad result is very consistent with the basic principle of scarcity in economics (called the sad science exactly for this reason).

As a tool for conflict management, multi-criteria evaluation has demonstrated its usefulness in many sustainability policy and management problems (Munda, 2005). Let's review some real-world examples briefly. A first case study is on water management. This study was part of a project which was commissioned by the Sicily region (an island in the south of Italy) and executed in the frame of the European Commission DGXVI structural funds. This case study was developed in two years of interaction mainly between a multidisciplinary team and the management body of the water supply system of the city of Palermo (as well as some social actors involved in the final step of the study)[4]. Water resource management is characterized by the presence of a strong competition among different categories of consumptive water uses and, as a consequence, among various interest groups. Such a competition also exists between consumptive uses as a whole and "ecological uses" which aim at limiting water diversion for off-stream uses in order to preserve the ecological equilibrium of ecosystems. This permanent condition of competition may become a real conflict under drought conditions, i.e.

when there is a temporary reduction of available water resources due to a long and severe decrease of rainfall (compared to mean or median natural values). The problem of water shortages due to drought is particularly relevant in Southern Europe. In Sicily, the water distribution issue has deep historical roots. The mafia has its origins in the fight over water control.

Water shortages not only depend on hydrological drought which in turn follows from meteorological drought, but also depend on water supply system characteristics and demand levels, which are both affected by different drought mitigation measures. As a consequence, the pure technical hydrological solutions cannot be separated from their consequences on the socio-economic system. Although this was not evident at the beginning of the project, after a few meetings, hydrologists accepted that an economist could be of some help for this kind of problems. However, it was still very difficult to find a common language and to understand which contribution each could give to progress towards a possible solution (or at least a better understanding) of such a complex problem.

The water system of Palermo provides water to municipal and industrial users as well as agriculture by using surface water and groundwater; a reservoir is also used for energy production. It was agreed that alternative management options under drought conditions can be divided into two main groups:

- alternatives that try to satisfy 100% of the water demands.
- alternatives that do not satisfy completely the water demands.

To specify the alternatives, it was necessary to understand the structure of the Palermo water supply system and, given the technicalities involved, it was immediately clear that this was the job of hydrologists. However, these alternatives had to be evaluated for the longest historic drought experienced in the water supply system (4 years) according to a set of criteria including the economic dimension (e.g. connected financial costs and benefits of the company managing the water supply system, the energy production company, and so on), the social dimension (e.g. hygienic risk and social discomfort) and the environmental dimension (e.g. the in-stream flow requirement defined as the discharge which maintains a stream ecosystem or aquatic habitat). At this point the advantage of the multi-criteria structuring of the problem was evident. Each expert suddenly knew her/his comparative advantage. From the experience of this case study, a first lesson can be learned. The use of a multi-criteria framework is a very efficient tool to implement a multi/inter-disciplinary approach. The experts involved had various backgrounds (mainly in engineering, economics and mathematics). While in the beginning the communication process was very difficult, when it was decided to structure the problem in a multi-criterion fashion, it was astonishing to realize that immediately a common language was created. In terms of inter-disciplinarity, the issue is to find agreement on the set of criteria to be used; in terms of multi-disciplinarity, the issue is to propose and compute an appropriate criterion score. The efficiency of the interaction process may greatly increase its effectiveness too[5]. In the Palermo case study, it was also experienced that taking distribution issues explicitly into account increases the transparency of the study and makes a process of interaction with various social actors possible in an effective way.

For the formation of contemporary public policies, it is hard to imagine any viable alternative to extended peer communities (Funtowicz *et al.*, 1999; Guimarães-Pereira *et al.*, 2003, 2005; Kasemir *et al.*, 2003). They are already being created, in increasing numbers, either when the authorities cannot see a way forward, or know that without a broad base of consensus, no policies can succeed. They are called "citizens' juries", "focus groups", or "consensus conferences", or any one of a great variety of names and their forms and powers are correspondingly varied. But they all have one important element in common: they assess the quality of policy proposals, including the scientific and technical component and their verdicts all have some degree of moral force and hence political influence. Here quality is not merely in the verification, but also in the creation, as local people can imagine solutions and reformulate problems in ways that accredited experts, with the best will in the world, do not find natural. However, even a participatory policy process can always be conditioned by heavy value judgements such as: have all the social actors the same importance (i.e. weight)? Should a socially desirable ranking be obtained on the grounds of the majority principle? Should some veto power be conceded to the minorities? Are income distribution effects important? And so on. The management of a policy process involves many layers and kinds of decisions, and requires the construction of a dialogue process among many stakeholders, individual and collective, formal and informal, local and not. This need has been more and more recognized in a multi-criteria decision-aid (MCDA) framework. Banville *et al.* (1998) offers a very well structured and convincing argumentation on the need to extend MCDA by incorporating the notion of stakeholder. This is the reason why a social multi-criteria process must be as participative and as transparent as possible; although in my opinion, participation is a necessary condition but not a sufficient one. This is the main reason why I propose the concept of "Social Multi-criteria Evaluation" (SMCE) (Munda, 2004, 2007) in substitution of "Participative Multi-criteria Evaluation" (PMCE) or "Stakeholder Multi-criteria Decision Aid" (SMCDA) (Banville *et al.*, 1998).

In real-world applications, a very sensitive point is the synthesis of technical and social information to generate alternatives and evaluation criteria. Let's briefly review an application in the field of renewable energy (for more information, see Gamboa and Munda, 2007[6]). In the last decade, renewable energies, and specially wind energy, have had a big impulse by the authorities. It is presented as one of the strategies to confront global warming and to accomplish the Kyoto Protocol. Although wind energy has a green image, it is not difficult to find unfavourable positions regarding the installation of wind-farms. This opposition can depend on the extensive land use of wind-parks, their possible impacts on birds or their visual impact, as well as NIMBY (Never In My Back-Yard) behaviour. The policy process itself for deciding the location of wind turbines can also be a source of conflict. For these reasons in designing, locating and evaluating wind-park alternative location sites in Catalonia (a region in the north-east of Spain), a real-world social process was implemented including several social actors' visions. There were two project proposals: the first one concerned 16 windmills of 850 KW, while the second project planned the installation of 66 windmills of 660 KW. In addition, there were two other projects planning to construct wind-farms of 75 and 15 windmills respectively, reaching 172 windmills in the area. Early in this process, there were several positions regarding the construction of those wind-parks. On the one side, some people started to raise their voices against the wind-farms. Firstly, they expressed the will of participating in the design of the future of their *comarcas* and, secondly, they saw as territorial inequalities the way Catalonia was planning the energy production scheme. On the other side, some municipalities and some citizens agreed with the construction of these infrastructures. They saw the wind-parks as a good opportunity to increase their economic incomes and to improve social services and to change the declining path that characterized these towns. By developing an institutional analysis study and applying various participatory techniques, the social "atmosphere" that emerged from the debate can be synthesized in Tab. 1.

Table 1. Socio-economic actors, scale of action and their position towards the wind-parks

Social actor	Scale of action	Position regarding the wind-parks
Catalonian government	National	The Catalonian government has launched the Renewable energy plan for the year 2010. It foresees RES to grow from 72.2 MW to 1.073 MW of installed capacity. Actually, the regional government has declared that it intends updating the installed capacity target to 3.000 MW.
Town council of Vallbona de les Monges	Local – Province	The municipality wants the wind-parks to be installed. The economic income is seen as a good opportunity to improve some social services, and/or to create additional ones (like elder nursing).
Town council of Els Omells de Na Gaia	Local – Province	The three municipalities are negotiating together with the companies and trying to obtain equal and better retribution conditions from the promoters. However, some of them say that if the economic income is not enough to overcome the actual social trend, then they do not want the wind-parks to be constructed.
Town council of Rocallaura	Local – Province	
Town council of Senan	Local – Province	The town council is fighting together with the inhabitants of Senan to oppose to the wind-parks. They do not want to be surrounded by windmills, and they see the forest as a very good opportunity to develop tourism in the future.

Social actor	Scale of action	Position regarding the wind-parks
Consell comarcal de l'Urgell	Province – National	The president of the council has offered her mediation to reach a compromise solution. However, she shares the opinion of the mayors in the sense that more economic income is needed to revitalize the towns and to offer more and better services.
Political representatives	Province	Representatives from different political parties have signed a motion asking for a moratorium to the wind-parks *Coma de Bertran* and *Serra del Tallat* to defend the development of economic activities without interference with local initiatives.
Coordinadora por la defensa de a terra (Urgell, Conca de Barberà, Segarra, Garrigues)	Province	They think that it is not necessary to jeopardize the future of the towns to revitalize them. They are not against wind energy, but they do not approve the way the process has been carried out. They think that the solution has to be discussed by all the towns involved, to avoid any town to be harmed.
Plataforma per Senan	Province	They see the projects as an undesirable gift from their neighbours. They do not share the way the process has been carried out, and they say that to reach more equitable decisions the discussion must involve all the towns. (See Town council of Senan above)
GEPEC	National	This is an environmental non-governmental organization, acting at the Catalonian level to redefine the Catalonian Energy Plan, with the participation of some social actors. The organization asks for a decentralized electricity production system next to the consumption places. As regards to the location of the wind-farms, the the organization asks for special attention towards the habitats of rare and threaten species, and to the biologic corridors. GEPEC also asks for the application of the Landscape European Convention and for territorial equity.

Social actor	Scale of action	Position regarding the wind-parks
Energía Hidroeléctrica de Navarra	National	The company is the promoter of one of the wind-parks. It is one of the main energy producers from RES in the Spanish territory, and one of its aims is to construct wind-parks as big as possible to impulse a great change *"in the energy production culture"*.
Gerrsa	National	This is the promoter of the *Coma Bertran* project. It has been impossible to set up a meeting with its representatives due to their reluctance to talk to people external to the government.

One of the main features of the SMCE framework is that criteria and alternatives can be constructed by means of continuous feedbacks with the social and technical actors involved in the policy process. The evaluation criteria in particular are a technical translation of social actors' preferences and desires. Their construction is a very delicate step. In this case, a synthesis of the whole process is presented in Tab. 2.

Table 2. Evaluation criteria as a translation of social actors' desires and preferences

Criteria	Social Desires and Preferences
Possible impact on other economic activities	Some people are worried about the consequences on the tourism sector on the long run and on residential value.
Land owners' income	There is a necessity of additional incomes for the farmers To stabilize economic income To improve the quality of life There is a worry about who is going to get the benefit, will it be local or external owners? To improve the quality of life To avoid wealth concentration
Distribution of income	To avoid the concentration of revenues To propel local development
Municipalities' income	To increase the municipalities' budget To offer more social services as city council To keep rural population
Number of jobs	To attract and to keep people in the region To reactivate the economic dynamics of the region
Visual impact	To avoid the industrialization of mountains To protect tourism in the long run To keep rural identity To avoid land/houses' value to decrease

Deforestation	To minimize the ecosystem's disturbance To avoid soil erosion
Noise annoyance	To protect human health To minimize effects on the fauna's habitat
Avoided CO_2 emissions	To achieve reduction of emissions Commitments
Installed capacity	To promote a larger share of renewable energies in electricity production To warrant economic viability

Policy exercises are normally dynamic learning processes which necessarily must be adaptive in the real-world. This implies continuous feedback loops among the various steps and consultations among all the actors involved. Flexibility to fit real-world situations is one of the main advantages of social multi-criteria evaluation (see for example Vargas-Isaza, 2004 for an application of SMCE in Colombia, where there was the extreme situation of some social actors – the so-called *"actor armado"* – belonging to various informal armies, or Martì, 2005, who carried out a study on indigenous communities in Peru). In the framework of a research project on barter markets and biodiversity conservation in the Peruvian Andes, social multi-criteria evaluation has been applied in the context of indigenous community development, in the region of *"Parque de la Papa"*[7]. The issue tackled was an evaluation of the incommensurability of values associated to the biodiversity of medicinal plants (see Martì, 2005[8]). A multi-criterion evaluation matrix was developed by Quechua women without the use of written language. The matrix building process started by taking into account the socio-cultural, political and ecological specificities of the region. The process included the following steps: (a) identification and inclusion in the process of the women in charge of keeping the knowledge on medicinal plants, (b) generation of evaluation criteria by means of collective deliberation, (c) experimentation and further improvements of the evaluation matrix, and (d) development of the evaluation by quantifying the criterion scores. The identified and involved women constituted the Group of Evaluation of Medicinal Plants (GEPM). For the generation of evaluation criteria and selection of medicinal plants, on the one hand, interviews with women in charge of the use of medicinal plants *"curanderas"* were carried out. On the other hand, consultancy was asked to experts in phytotherapy. Finally, the GEPM selected a total of 37 criteria and 43 medicinal plants to be included in the multi-criterion evaluation matrix. Later on, symbols were chosen to represent each one of the criteria. For this aim, materials and local objects were used such as kitchen utensils, dirt and wools among others. The objects were chosen considering their association with the semantic meaning of the criteria. Objects and materials of vivid colours were chosen in order to maintain the attention of the group, along the evaluation process, as high as possible. For each criterion, an ordinal scale of measurement was assigned between 0 and 5. The criterion score was represented by a number of beans. Once the evaluation system was designed, the multi-criterion matrix was first experimented by the GEPM in two community meetings. In these meetings the construction of the matrix was done on the floor with the help of wools.

Once the experimentation was considered successful, other meetings were organized. They lasted about 5 hours and the medicinal plants were evaluated 20 by 20. When the sample of the plants was definitely chosen, the women assumed different roles and also different positions in the ground space. The mediator located herself inside the matrix, the person recording the results in a lateral position and the group of consultants around the matrix. The mediator showed the plants to the consultants one by one, and asked their evaluation in relation to each criterion. An open discussion continued until the women arrived to a consensus on the number of beans to be granted as a criterion score. In the case of persistent disagreements, the discussion included a comparison with the beans granted to the rest of the species. Once the evaluation was concluded, the final matrix was copied on a paper document and the participants all signed the document to emphasize their role of knowledge keeper in relation to the incommensurability of values associated to the medicinal plants. The described procedure guaranteed that the evaluation was adaptive, locally controlled and able to integrate different types of knowledge at different scales. The final result was the selection of 21 among the species evaluated to produce medicines for the indigenous community.

3. CONCLUSION

When science is used for policy making, an appropriate management of decisions implies including the multiplicity of participants and perspectives. This also implies the impossibility of reducing all dimensions to a single unity of measure.

> The issue is not whether it is only the marketplace that can determine value, for economists have long debated other means of valuation; our concern is with the assumption that in any dialogue, all valuations or "numeraires" should be reducible to a single one-dimension standard (Funtowicz and Ravetz, 1994, p. 198).

When science is used for policy making, the call for citizen participation and transparency is more and more supported at institutional level within the European Union, where perhaps the most significant examples are the White Paper on Governance and the Directive on Strategic Environmental Impact Assessment. The social-multi-criteria evaluation accomplishes the goals of being inter/multi-disciplinary (with respect to the research team), participatory (with respect to the local community) and transparent (since all criteria are presented in their original form without any transformations in money, energy or whatever common measurement unit). As a consequence, multi-criteria evaluation seems to be an adequate assessment framework for sustainability policies.

Giuseppe Munda
Department of Economics and Economic History,
Universitat Autonoma de Barcelona, Spain

European Commission, Joint Research Centre,
Institute for the Protection and Security of the Citizen, Ispra, Italy

NOTES

[1] Emphasis added.

[2] *"There is great pressure for research into techniques to make larger ranges of social value commensurable. Some of the effort should rather be devoted to learning – or learning again, perhaps – how to think intelligently about conflicts of value which are incommensurable"* (Williams (1972) – Morality, Cambridge University Press, Cambridge, p. 103). A call for dealing explicitly with incommensurability can also be found in Arrow (1997) – Invaluable Goods, *Journal of Economic Literature*, Vol. 35, No. 2, pp. 757-763. Social multi-criteria evaluation is defined in Munda (2004) – "Social multi-criteria evaluation (SMCE)": methodological foundations and operational consequences, *European Journal of Operational Research*, Vol. 158/3, pp. 662-677.

[3] These definitions have been developed by elaborating standard definitions in multi-criteria decision literature by means of concepts coming mainly from complex system theory. Discussions with M. Giampietro and M. Nardo have been essential.

[4] For more information on this case study see POP Sicily, full final report European Commission contract No. 10122-94-03 TIPC ISP I or for a shorter version Munda *et al.*, 1998.

[5] Here I refer to the idea of orchestration of sciences as a combination of multi/inter-disciplinarity. Multi-disciplinarity: each expert takes her/his part. Inter-disciplinarity: methodological choices are discussed across the disciplines (this definition has been discussed with R. Strand).

[6] This research was carried out in the framework of the European Union project *"Development and Application of a Multicriteria Decision Analysis softwareTool for Renewable Energy sources (MCDA-RES)"*, Contract NNE5-2001-273.

[7] *Papa* in this context means potato.

[8] Research supported by the programme Sustaining Local Food Systems, Agricultural Biodiversity and Livelihoods of the International Institute of Environment and Development, and the "Asociación Kechua-Aymara para Comunidades Sostenible, ANDES".

REFERENCES

Barbier E.B., 1987. The concept of sustainable economic development, *Environmental Conservation*, 14(2), pp. 101-110.

Banville C., Landry M., Martel J.M. and Boulaire C., 1998. A stakeholder approach to MCDA, *Systems Research and Behavioral Science*, Vol. 15, pp. 15-32.

Buchanan J. M. and Musgrave R.A., 1999. *Public finance and public choice*, The MIT Press, Cambridge.

Daly H.E., 1977. *Steady-State economics*, Freeman, San Francisco, CA.

Funtowicz S.O. and Ravetz J.R., 1994. The worth of a songbird: ecological economics as a post-normal science, *Ecological Economics*, 10, pp. 197-207.

Funtowicz S., Martinez-Alier J., Munda G. and Ravetz J., 1999. Information tools for environmental policy under conditions of complexity, European Environmental Agency, Experts' Corner, *Environmental Issues Series*, No. 9.

Gamboa Jiménez G. and Munda G., 2007. The problem of wind-park location: a social multi-criteria evaluation framework, *Energy Policy*, Vol. 35, Issue 3, March, pp. 1564-1588, Elsevier Ltd.

Guimarães-Pereira A., Rinaudo J.D., Jeffrey P., Blasuqes J., Corral-Quintana S.A., Courtois N., Funtowicz S. and Petit V., 2003. ICT tools to support public participation in water resources governance and planning experiences from the design and testing of a multi-media platform, *Journal of Environmental Assessment Policy and Management*, Vol. 5, No. 3, pp. 395-419.

Guimarães-Pereira A., Corral-Quintana S.A. and Funtowicz S., 2005. GOUVERNe: New trends in decision support for groundwater governance issues, *Environmental Modelling & Software*, 20, pp. 111-118.

Kasemir B., Gardner M., Jäger J. and Jaeger C., (eds.), 2003. *Public participation in sustainability science*, Cambridge University Press, Cambridge.

Martí N., 2005. *La multidimensionalidad de los sistemas locales de alimentación en los Andes peruanos: los chalayplasa del Valle de Lares (Cusco)*, Ph.D. Dissertation, Doctoral Programme in Environmental Science, Univ. Autónoma de Barcelona, Spain.

Munda G., Paruccini M. and Rossi G., 1998. Multicriteria evaluation methods in renewable resource management: the case of integrated water management under drought conditions, in Beinat E. Nijkamp P. (eds.), *Multicriteria evaluation in land-use management: methodologies and case studies*, Kluwer, Dordrecht, pp. 79-94.

Munda G., 2004. "Social multi-criteria evaluation (SMCE)": methodological foundations and operational consequences, *European Journal of Operational Research*, Vol. 158/3, pp. 662-677.

Munda G., 2005. Multi-Criteria Decision Analysis and Sustainable Development, in J. Figueira, S. Greco and M. Ehrgott (eds.), *Multiple-criteria decision analysis. State of the art surveys*, Springer International Series in Operations Research and Management Science, New York, pp. 953-986.

Munda G., 2007. Social multi-criteria evaluation for a sustainable economy, Economics Series, Springer, Heidelberg.

Norgaard R. B., 1994. *Development Betrayed*, Routledge, London.

Vargas-Isaza O.L., 2004. *La evaluación multicriterio social y su potencial en la gestión forestal de Colombia*, Ph.D. Thesis, Doctoral programme in Environmental Sciences, Universitat Autonoma de Barcelona.

II. WATER MANAGEMENT

1. WATER RESOURCE MANAGEMENT AND POLICY

G. M. ZUPPI

THE GROUNDWATER CHALLENGE

Abstract. Groundwater is the most extracted natural resource in the world. It provides more than half of humanity's freshwater for everyday uses such as drinking, cooking, and hygiene, as well as thirty percent of irrigated agriculture and industrial development. Given the considerable dependence of the world on this finite and, thus, precious resource, it is reasonable to assume that international attention to groundwater would be substantial. Despite the growing dependence, scientific and subsequently regulatory attention to groundwater, resources have long been secondary to surface water, especially among legislatures and policymakers. Given the uncertainty in defining groundwater flow, coupled with the uncertainty of the hydraulic connection between groundwater and surface water resources, conflicts over water quantity and quality are certain to escalate with increased reliance on groundwater to meet demands for drinking water, agricultural and industrial uses, and maintenance of "green" reservoirs – the highland forests and wetlands.

1. INTRODUCTION

Water is a basic need for human existence and even under the present climatic conditions, freshwater is a very limited resource on earth. Less than 0.01% of all water is freshwater that is easily accessible for human consumption. The demand for water does not stop and continues to grow with the increase of population and the improvement of the standards of living. To overcome this situation, all countries resorted, during the last decades, to a massive mobilization of their water resources, which required big investment efforts. At present, not only almost all renewable water resources are already put in use, but also many countries have resorted to their non-renewable resources and to the use of non-conventional resources such as treated waste and low-quality water.

Groundwater provides a vast strategic reserve of freshwater storage on planet, around 30% of the global total (including ice caps) and as much as 98% of all water in liquid form. The biggest aquifers underlie vast land areas (Fig. 1) and contain more water than all the reservoirs and lakes in the world. Unlike lakes and reservoirs, they loose only small amounts by direct evaporation. Groundwater is emerging as a formidable poverty reduction tool. However, developing and managing this resource in a sustainable way poses many challenges (Custodio and Gurgui, 1989; IWMI, 2000; UNESCO-ISARM, 2004).

The paradox distinguishing several regions of the world characterized by large precipitations and by crystalline basement rocks is the unavailability of underground resources. These regions include the tropical and equatorial basins. Low-productivity aquifers are widely, but rather unpredictably, present in this formation. They yield small water supplies for domestic purposes and for livestock watering of the rural population (Custodio and Gurgui, 1989; Chilton and Foster, 1995; Llamas,

2004). The challenge in these areas is to promote sustainable and equitable development of the resource.

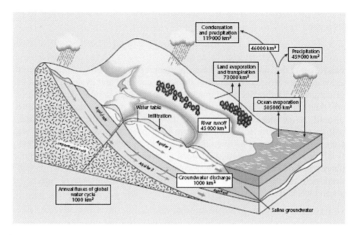

Figure 1. The hydrologic cycle (source: Morris et al., 2003). Precipitation falls to the earth's surface, runs off or infiltrates into the ground, then moves back into the atmosphere through transpiration or evaporation.

Vice-versa in tropical karst aquifers recharge is dominated by discrete infiltration through karst features, such as sinkholes and dry valleys. Infiltration requires enough rainfall to generate runoff to transport water to the karst features (Jones and Banner, 2003). Without such transport, it is likely that rainwater will be taken up by evapotranspiration. These aquifers are characterized by groundwater residence times of years to tens of years. Consequently, these aquifers are fragile systems that respond rapidly to natural and anthropogenic processes. Groundwater quantities in these aquifers usually respond to short- and long-term climatic fluctuations that influence the amount of recharge and therefore the amount of groundwater available for use (Jones and Banner, 2003).

In most arid zones (Sahara, Karroo, Gobi, Australia) there are limited but even less-developed groundwater systems. Here the first challenge is to identify areas with groundwater potential and to develop strategies for a sustainable exploitation of these resources to benefit local population.

While developing groundwater resources promises to help alleviate poverty in many areas, the most formidable groundwater challenge is to attain the sustainable use and management of groundwater in vast and growing regions where the resource is under threat. The over-depletion of groundwater is becoming a major problem everywhere in the world (IWMI, 2000).

Sustainable human development is dependent on the availability of water. It is estimated that more than one third of the global food production is based on irrigation, a significant portion of which may rely of unsustainable groundwater sources (Gibson and Aggarwal, 2001). Despite progress in the last two decades to improve access to safe drinking water, some 1.1 billion people today go without it.

Areas of water scarcity and stress are increasing, particularly in North Africa and West Asia. In the next two decades, total water demand is expected to increase by 40 per cent. By 2025, two-thirds of the world's population may live in countries with moderate or severe water shortages (Foster and Chilton, 2003; UNESCO-WATER PORTAL www.unesco.org/water/). The challenge is how to manage this finite resource, today and in the future. Given that freshwater resources are very often shared by more than one country (Aureli and Ganoulis, 2005; Puri and Aureli, 2005) within a region (Chaco and Guaranì Aquifers in South America, Nubian Sandstone, and Continental Intercalaire aquifers in Northern Africa, Iullemeden Aquifer System in Sub-Sahelian Africa, Karoo Aquifer in Southern Africa, Vechte Aquifer in Western Europe, Slovak Karst-Aggtelek Aquifer and Praded Aquifer in Central Europe, Gaza Strip and Disi aquifer in Middle East).

International conventions on transboundary waters (Fig. 2) should include provisions for the monitoring and assessment of transboundary waters, including measurement systems and devices and analytical techniques for data processing and evaluation. The complexities of groundwater law have been described by many authors in the technical literature (Hayton, 1982; Hayton and Utton, 1989). Overpumping can cause groundwater quality to deteriorate through salinity problems, either by seawater intrusion or evaporation-deposition. Overpumping of groundwater in one country can endanger the future freshwater supplies of another country. The importance of transboundary groundwater resources becomes most apparent when there is an increased pressure for economic development and water related activities on either side of the border (Campana, 2005, Jarvis et al., 2005). Joint management of internationally shared aquifer resources is not only a scientific or technical problem (Puri and Aureli, 2005). It should also involve joint institutions, common monitoring networks, information and data sharing and a common vision for sustainable development of the entire river catchment. The political linkages in transboundary aquifer management are important and involve wider regional aspects as "water for cooperation" (Aureli and Ganoulis, 2005; Puri and Aureli, 2005).

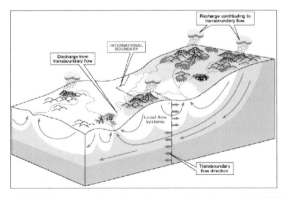

Figure 2. Schematic block diagram of transboundary aquifer (source: UNESCO-ISARM, 2004).

2. HYDROGEOLOGICAL UNCERTAINTY

Uncertainty affects the evaluation of both surface and groundwater resources, since many hydrogeological variables and parameters are uncertain. However, this is not always recognised and uncertain figures, water balances, or other calculations based on them, are often illusorily presented as accurate. It happens that the specialist is often pushed to present very uncertain figures as accurate because non-specialists and politicians tend to think that doing otherwise means poor knowledge or malpractice (Custodio, 2002).

Uncertainty means not only the introduction of errors, due to shifts in average values, but also a wide dispersion. It is the combination of poor knowledge of the phenomenon (e.g. rainfall) and its associated stochastic components, the simplifications introduced to describe the system under consideration, the variability and heterogeneity of the physical media, the difficulty in describing complex situations quantitatively, and even delayed effects (Rossi et al., 1992). This is a common situation when dealing with nature.

From a hydrologic point of view, the first step for a correct water management is to evaluate past, present, and future practices with respect to the basin water balance. Basin water balance refers to the equilibrium between the inflow of water as precipitation and the outflow of water as evapo-transpiration, groundwater discharge, and stream flow (Fig. 1 and 3). Basically, water balance is an accounting tool to keep track of the hydrologic cycle of a watershed over time (Simmers, 1988; 1997; Freeze and Cherry, 1979). When the water balance concept is used in conjunction with probability analysis one can evaluate the hydrologic, economic, and ecological feasibility of past, present, and potential activities on the catchments area. Water balance equation is

$$P = ET + SF + GWD \pm SMC \pm GWS$$

P = Precipitation (inflow)
ET = Evapo-transpiration (outflow)
SF = Stream flow (outflow)
GWD = Groundwater discharge (outflow)
SMC = Soil moisture content (inflow or outflow)
GWS = Groundwater storage (inflow or outflow)

As above mentioned, the difficulty lies in the accuracy and the precision at which we can measure or predict the components of the equation. Accuracy refers to how close a measurement or estimate is to the "true" value. Precision refers to how exact or fine our measuring device might be (Simmers, 1988; 1997; Freeze and Cherry, 1979).

From the above equation it clearly appears that water resources can only be understood within the context of the dynamics of the water cycle. These resources are renewable, except for some groundwater, but only within clear limits, as in most cases water flows through catchments that are more or less self-contained. Water resources are also variable, over both space and time, with huge differences in availability in different parts of the world and wide variations in seasonal and annual precipitation in many places (Foster and Chilton, 2003; UNESCO-WATER PORTAL www.unesco.org/water/). This variability of water availability is one of the most essential characteristics of water resource management. Although progress has been made most notably with respect to land-atmosphere exchanges at both continental and global scales, substantial uncertainties exist with regard to stocks, fluxes, and the inherent nature of interactions among key hydrologic elements. Nonetheless, several opportunities exist for analyzing the global status of the hydrological cycle and associated water resources. Chemical and physical techniques, and the straightforward access to remote sensing imagery provide an incomparable occasion to monitor the hydrological cycle everywhere and every time. In addition, appropriate models can be used with the information obtained to improve the understanding of the spatial and temporal aspects of global water resources (UNESCO-WATER PORTAL www.unesco.org/water/).

It is broadly accepted that groundwater moves very slowly from recharge areas to discharge areas (springs, seepages to watercourses, wetlands and coastal zones). Although recharge may be variable through the year or from year to year, the large storage available in aquifers provides a buffer and gives a more constant discharge regime.

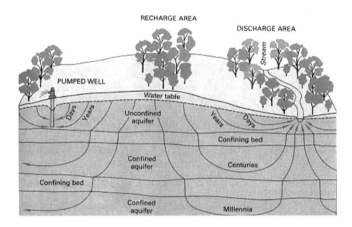

Figure 3. Schematic section of typical ground-water travel times as function of depth and distance of flow (source: U.S. Geological Survey Circular 1139).

Rates of recharge vary because of:

- changes in land use and vegetation cover, notably introduction of irrigated agriculture with imported surface water, but also with natural vegetation clearance, soil compaction, etc;
- changes in surface water regime, especially diversion of river flow;
- lowering of the water-table, by groundwater abstraction and/or land drainage, leading to increased infiltration;

- longer term climatic cycles.

These variations mean that estimating the groundwater resource from rates of recharge has to be done with caution. Aquifer recharge does not only depend on the points above mentioned, but on the extent of the surface area, which is often unclear, especially when there are lateral inflows, and vertical flows from other aquifers. However, long-term monitoring, through more accurate calculation and mathematical modelling of aquifer behaviour under a given set of conditions generally helps to refine the estimated recharge value, provided the exploitation rate and pattern, and hydraulic circumstances of the aquifer do not change significantly in the evaluation period (Custodio, 2002).

It is also important to distinguish between shallow groundwater replenished by modern recharge and deeper groundwater that may not receive any direct contribution from rainfall. As a result of the very large storage of groundwater systems, aquifer residence times can often be counted in decades or centuries and sometimes in millennia.

Thus, the option of continuing to increase the usable quantities of water is no longer possible in most countries, as only limited quantities remain available for mobilization and given the prohibitive costs required for their mobilization.

3. CLIMATE CHANGE AND GROUNDWATER RESOURCES

It has long been known that natural climate variability and climate change both affect groundwater resources. Today, it is widely accepted that recent warming is largely a product of enhanced greenhouse gas concentrations in the atmosphere derived from post-industrial combustion of fossil fuels and biomass energy sources. However, great uncertainty remains regarding the causal relationships between specific parameters and climate phenomena, and regarding the impacts of climate change on the earth's water cycle. Although widespread environmental problems, such as melting sea and glacier ice, higher surface air temperatures, intensification of weather patterns, ecosystem disruption, and rising sea levels are expected to impact freshwater resources in a warmer climate, the magnitude of these changes is more difficult to predict (Gibson and Aggarwal, 2001).

Thus, as an important part of the hydrologic cycle, groundwater will be affected by climate change in relation to the nature of recharge, the kinds of interactions between the groundwater and surface water systems, and changes in water use (Fig. 1 and 3). We expect that changes in temperature and precipitation will alter recharge to groundwater aquifers, causing shifts in water table levels in unconfined aquifers as a first response. Decreases in groundwater recharge will not only affect water supply, but may also lead to reduced water quality. From a regional or continental perspective, our understanding of climate variability and change impacts on groundwater resources – related to availability, vulnerability and sustainability of freshwater – remains limited (Rivera *et al.*, 2004). In fact most parameters of the water balance equation can not be directly measured with meaningful precision (Custodio, 2002). Rates of infiltration and evapo-transpiration can not be measured directly, nor can runoff, and hence, recharge at each location can not be established directly, especially in relation to confined aquifers. Similarly, discharge is not visible in many systems, especially in arid and inland regions, and the origin of the scarce springs that do occur is often equivocal. Hence, discharge in many instances is assumed and not demonstrated (Gibson and Aggarwal, 2001; Rivera *et al.*, 2004). All ecologically fragile areas such as arid and semi arid zones, are characterized by low and irregular rainfall, high temperatures and evaporation, and notable drought periods. Under such conditions, the distribution of annual precipitation and the partitioning between surface water runoff, evapotranspiration loss and recharge to groundwater varies from year to year (de Vries and Simmers, 2002). However the conjunctive use of interstitial water profiles and regional shallow groundwater chemistry provide an inexpensive technique for recharge estimation which can be widely applied in sedimentary terrains under arid and semiarid climate (Edmunds and Gaye, 1994; Ma *et al.*, 2005). The origin of dissolved ions has to be deduced from several possible sources, e.g. sea-derived airborne (atmospheric) transport, water-rock interactions, and intermixing with remnant seawater or brines. Two important factors serve to complicate and limit our understanding and ability to measure these potential impacts (Mazor, 1995; Rivera *et al.*, 2004).

Timing of recharge: Climate change challenges existing water resources management practices by adding uncertainty. Integrated water resources management will enhance the potential for adaptation to change. While surface waters typically see rapid response to climate variability, the response of groundwater systems is often difficult to detect because the magnitude of the response is lower and delayed. In other words, the crucial aspect of hydrology is the dissimulated nature of groundwater. Large systems, and particularly confined aquifers, can be studied only indirectly by a phenomenological approach. In this frame, the study of real cases is based on detailed measurements, leading to hydrological models which are then evaluated in the light of basic principles of physics, hydrology, lithology, structural geology and chemistry (Fig. 3). Longer-term variations in climate are often well preserved in aquifers. Thus, the magnitude and timing of the impact of climate variability and change on aquifers, as reflected in water levels, are difficult to recognize and quantify. This is because of the difference in time frame that exists between climate variations and the aquifer's response to them.

Aquifer Characteristics: Different types of aquifers respond differently to surface stresses.

Shallow aquifers are more responsive to stresses imposed at the ground surface compared to deeper aquifers. Climate variability, being of relatively short term compared to climate change, will have greater impact on these shallow aquifer systems as indicated by large fluctuations of water levels. Similarly, shallow aquifers are strongly affected by anthropogenic stress: over-depletion and pollution. Conversely water levels in deeper aquifers are affected essentially by regional changes more than local climate changes. Thus, entrapped groundwaters are renewed at geological time scale, and their exploitation is equivalent to mining, which may be regarded as a disadvantage. However, their basic properties more than compensate for this shortcoming, provided their exploitation is reassessed in the light of very specific needs of mankind (Mazor, 1995). As pressurized systems have no discharge losses, they are most suitable as water reserves. Pressurized trapped groundwater is shielded from anthropogenic pollution. Hence, trapped groundwaters of good quality are precious and should be used, or preserved, for domestic consumption only, and should be kept for use in times of major pollution accidents.

4. WATER PROBLEMS FROM THE PAST TO THE FUTURE

Despite their importance, there is still not enough concern about protecting quantity and quality groundwater resources. Their being "out of the public sight" has caused them also to be "out of the political mind" and they are too often abandoned to chance (Foster and Chilton, 2003; Morris et al., 2003; Foster et al., 2005). Throughout the world, also in regions that have large water resources and sustainable groundwater balance, withdrawal is visible by the day. Three features dominate groundwater use: *depletion* due to overexploitation; *salinization* and water logging due mostly to inadequate drainage and insufficient conjunctive use or to seawater intrusion linked to the costal aquifer overdraft; and *pollution* due to agricultural, industrial and other human activities.

These three points summarize three key areas: water scarcity, water quality and water-related disasters, linked reciprocally, that often create serious and irreversible problems to underground resources (Abramovitz, 2001). For instance, under some geological conditions, the falling groundwater level induces compaction of underground strata and serious subsidence of the land surface, causing costly damage to urban infrastructure and increasing the risk of flooding (Holzer, 1981). On many coasts and small islands, over-abstraction is leading to the intrusion of saline water inland, causing effectively irreversible deterioration of groundwater resources (UNESCO-WATER PORTAL www.unesco.org/water/).

4.1. Aquifer overexploitation and reserves depletion

Groundwater is now being abstracted at unsustainable rates in many areas, seriously depleting reserves. This happens when uncontrolled drilling of wells causes the overall rates of withdrawal from aquifers greatly to exceed their replenishment from rainfall and other sources over decades or more (Custodio, 2002; Foster and Chilton, 2003; Llamas, 2004). This "overabstraction" causes many serious problems. Often the yield of wells is reduced and the cost of pumping increased. In extreme cases, this may lead to the wells being abandoned, with premature loss of infrastructure investment. Thus groundwater resources are coming under increasing pressure essentially from a rapidly growing human population – both through an ever-increasing demand and through a contaminant load on the land surface which is steadily growing in volume and chemical complexity, especially in Southeast Asia and Latin America (Foster and Chilton, 2003; Simmers et al., 1997).

Reserves depletion and aquifer overexploitation are terms that are becoming common in water-resources management. Hydrologists, managers and journalists use them when talking about stressed aquifers or some groundwater conflict. Overexploitation may be defined as the situation in which, for some years, average aquifer abstraction rate is greater than, or close to the average recharge rate. But rate and extent of recharge areas are often very uncertain. Besides, they may be modified by human activities and aquifer development (Custodio and Gurgui, 1989; Custodio, 2002).

In practice, however, an aquifer is often considered as overexploited when some persistent negative results of aquifer development are felt or perceived, such as a continuous water-level drawdown, progressive water-quality deterioration, increase of abstraction cost, or ecological damage. But negative results do not necessarily imply that abstraction is greater than recharge. They may be simply due to well interferences and the long transient period that follow changes in the aquifer water balance. Groundwater storage is depleted to some extent during the transient period after abstraction is increased. Its duration depends on aquifer size, specific storage and permeability. Which level of "aquifer overexploitation" is advisable or bearable, depends on the detailed and updated consideration of aquifer-development effects and the measures implemented for correction. This should not be the result of applying general rules based on some indirect data. Monitoring, sound aquifer knowledge, and calculation or modelling of behaviour are needed in the framework of a set of objectives and policies. They should be established by a management institution, with the involvement of groundwater stakeholders, and take into account the environmental and social constraints. Aquifer overexploitation, which often is perceived to be associated with something ethically bad, is not necessarily detrimental if it is not permanent (Foster and Chilton, 2003; Llamas, 2004). It may be a step towards sustainable development. Actually, the term aquifer overexploitation is mostly a qualifier that intends to point to a concern about the evolution of the aquifer flow system in some specific, restricted points of view, but without a precise hydrodynamic meaning. Implementing groundwater management and protection measures needs quantitative appraisal of aquifer evolution and effects based on detailed multidisciplinary studies, which have to be supported by reliable data (Simmers *et al.*1997; Dijon and Custodio, 1992; Margat, 1993; Custodio, 2002; Foster and Chilton, 2003; Llamas, 2004).

4.2. Groundwater decline: salinization

Steady increase in the salinity of most of the major aquifers being used for water supply in the arid and semi-arid regions of South Europe, Africa, Asia and West Asia provides evidence of water quality deterioration. Salinization processes are often due to depletion in regards to heavy withdrawals from coastal aquifers and the subsequent sea water intrusion on the one hand and to the mobilization of saline formation waters, due to exploitation even in inland areas on the other hand. The extensive irrigation and the use of fertilisers and other pesticides contribute to soil and groundwater salinization inside the continent (Araguás y Araguás *et al.*, 2005).

With respect to actual measures, priority should be given to projects that are beneficial to presently existing problems in coastal areas. The lowlands along the world's seas will be the areas most vulnerable to impact. They include the deltaic, barrier island, atoll, and marshy coastlines. Increased storm-induced flooding represents the major danger in developing countries because of loss of life. In western countries, beach erosion will be a primary concern, requiring substantial expenditure of public funds to maintain existing recreational beaches. Marshlands will probably be left to their own destiny, which signals a marked decline in most places.

Aquifers in low plain and in coastal areas constitute the most important source of renewable freshwater around the globe (Vellinga and Leatherman, 1989). Therefore, mainly in arid and semiarid regions abstraction rates exceed by far natural replenishment rates. The tremendous increase in population along with the associated increase in human activities has imposed additional demands on groundwater resources in these coastal aquifers (Diersch and Kolditz, 1989; Sherif and Singh, 1996; Sherif and Hamza, 2001). The dynamic balance between the freshwater and the seawater bodies has been disturbed causing additional seawater intrusion. A three to five percent mixing of seawater with the freshwater resource would render it unsuitable for human consumption (Sherif and Singh, 1996; Sherif and Hamza, 2001). Several fields in coastal regions were abandoned due to the high salinity of the pumped water. In such cases, the only practical near-term solution may be to relocate the well field at a site further inland (Sherif and Hamza, 2001; Milnes and Renard, 2004).

The shape and degree of seawater intrusion in a coastal aquifer depend on several factors, among others: the type of aquifer (confined, phreatic, leaky, or multi-layer) and its geology, water table and/or piezometric head, seawater concentration and density, natural rate of flow, capacity and duration of water withdrawal or recharge, rainfall intensities and frequencies, physical and geometric characteristics of the aquifer, land use, geometric and hydraulic boundaries, tidal effects, variations in barometric pressure, earth tides, earthquakes, and water wave actions. Some of these factors are natural and related to the hydraulic and geometric characteristics of the hydrogeological system, while others are artificial and related to human activities (Ghassemi *et al.*, 1995; Sherif and Singh, 1996; Sherif and Hamza, 2001; Milnes and Renard, 2004). Seawater intrusion is an inevitable process that occurs over a long period of time due to over exploitation of coastal aquifers. Undisturbed coastal systems are less vulnerable to this phenomenon. Pumping activities accelerate intrusion rates. Unfortunately, most of the time, the problem is not recognized at the time it occurs. One of the natural factors must be emphasized: the effect of seawater rise due to climatic changes on the intrusion processes. Accelerated sea-level rise and the effects on coastal areas represent one of the most important impacts of global climate warming (Vellinga and Leatherman, 1989).

All these processes have developed as a result of: rising groundwater table, associated with the introduction of inefficient irrigation with imported surface water in areas of inadequate natural drainage; natural salinity having been mobilized from the landscape, consequent upon vegetation clearing for farming development with increased rates of groundwater recharge; and excessive disturbance of natural groundwater salinity stratification in the ground through uncontrolled well construction and pumping. All salinization is likely to prove costly to remediate (Ghassemi *et al.*, 1995; Foster and Chilton, 2003). Aquifer remediation by groundwater drainage to try to lower regional water tables has been underway since the 1960s, at very high investment and operating costs. "Natural inversions" of the classical salinity-depth profile (with a good freshwater aquifer beneath a body of overlying brackish groundwater) can occur in certain, but quite widely distributed, hydrogeological settings. Such situations are especially susceptible to hydraulic disturbance during groundwater abstraction and require careful diagnosis and management (Ghassemi *et al.*, 1995; Sherif and Singh, 1996; Sherif and Hamza, 2001; Foster and Chilton, 2003). Salinization of land and water is brought about by physical and chemical processes that increase concentrations of salt in soil and water. The processes responsible for the development of saline land and water are complex and intimately related to the transport of dissolved mass in groundwater flow systems. The redistribution of soluble salts accumulated in a catchment is evident mainly in topographically lower areas by terminal salt lakes, dry salinas, and areas of saline seeps. The countries affected by salinization are mainly located in arid and semi-arid regions and include areas in North and South America, Australia, China, India, regions in the Mediterranean and Middle East, and Southeast Asia (Ghassemi *et al.*, 1995). Secondary salinization induced by human activity occurs in irrigated and non-irrigated dryland salinity areas (Salama *et al.*, 1999). The following four mechanisms of water uptake and salt concentration are generally recognized by authors: evaporation, evapotranspiration, hydrolysis, and leakage (Freeze and Cherry, 1979). Salt accumulates when mineralized groundwater at or near the ground surface continually evaporates and causes minerals to precipitate; by evapotranspiration where infiltrating recharge water is continually taken up by plants and salt is concentrated in the unsaturated zone; by hydrolysis where water is taken up in the formation of new minerals in the weathering processes; and by leakage between aquifers through confining beds (Edmunds and Gaye, 1994). First-, second-, and third-order catchments refer to catchments with first-, second-, and third-order streams, respectively (Salama *et al.*, 1999).

4.3. Groundwater decline: pollution

Evidence is also accumulating that groundwater is becoming increasingly polluted. Nitrate pollution of groundwaters by agrochemicals poses a serious threat to groundwater resources everywhere in the world. In areas of intensive agriculture, aquifer recharge can be contaminated with nitrate at concentrations well above requirements for safe drinking water, presenting problems for present and future aquifer management. Nitrate salinity, soluble organic compounds (including synthetic toxic species) and, in certain conditions, some faecal pathogens in an aquifer may be dispersed by mixing with older, unpolluted groundwater. Moreover, the aquifer mineralogical matrix can eliminate or attenuate water pollutants by natural physical, chemical and biological processes either in Unsaturated or in Saturated Zones. In the case of nitrate, bacterial nitrate reduction provides a potential natural remediation mechanism in anoxic zones of an aquifer. However, this natural capacity does not extend to all types of water pollutants and varies widely in effectiveness under different hydrogeological conditions, being rather limited in the more vulnerable areas (Moncaster *et al.*, 2000). Serious pollution of groundwater occurs when contaminants are discharged to, deposited on, or leached from the land surface, at rates significantly exceeding the natural attenuation capacity. This is occurring widely as a result of both the indiscriminate disposal of liquid effluents and solid wastes from urban development with inadequate sanitation arrangements, and of uncontrolled effluent disposal and leakage of stored chemicals into the ground from industrial activity (Moncaster et al., 2000; Foster and Chilton, 2003; Morris et al., 2003). Intensification of agricultural cultivation can also lead – and has led – to significant and widespread deterioration in groundwater quality in some conditions. The principal problems are the leaching of nutrients and pesticides, and increasing salinity in the more arid environments (Matson et al., 1997; Lloyd, 1998). Groundwater pollution is insidious and expensive; insidious because it takes many years to show its full effect in the quality of water pumped from deep wells; expensive because, by this time, the cost of remediating polluted aquifers will be extremely high. Indeed, restoration to drinking water standards is often practically impossible (Salman, 1999; Foster et al., 2000).

5. ACTION NEEDED FOR GROUNDWATER MAINTENANCE AND PROTECTION

The growing demands on freshwater resources create an urgent need to link research with improved water management. Better monitoring, assessment, and forecasting of water resources will help to allocate water more efficiently among competing needs. Without protection there is a serious risk of irreversible decline of water on an increasingly general basis. Under the pressure of the need to rapidly develop new water supplies, there is rarely adequate attention to, and investment in, the maintenance, protection and longer-term sustainability of groundwater.

Water resources can only be understood within the context of the dynamics of the water cycle. These resources are renewable (except for some groundwater), but only within clear limits, as in most cases water flows through catchments that are more or less self-contained (UNESCO-WATER PORTAL www.unesco.org/water/).

Water resources are also variable, over both space and time, with huge differences in availability in different parts of the world and wide variations in seasonal and annual precipitation in many places. This discrepancy on water availability is one of the most essential characteristics of water resource management. Most efforts are intended to overcome the variability and to reduce the unpredictability of water resource flows.

Both the availability and use of water are changing. The reasons for concern over the world's water resources can be summarized within three key areas: water scarcity, water quality and water-related disasters.

5.1. Water scarcity

The precipitation that falls on land surfaces is the predominant source of water required for water balance equation. Consequently water resources are strictly controlled by climate change. Morcover, inflowing water is either to be 'taken up' by plants and soil and then eventually returned to the atmosphere by evapotranspiration, or to drain from the land into the sea via rivers, lakes and wetlands. Unsustainable levels of extraction of water diminish the total amount available. This will inevitably lead to the further disturbance and degradation of 'natural' systems and will have profound impacts upon the future availability of water resources. Actions to ensure that the needs of the environment are taken into account as a central part of water management are critical if present trends are to be reversed (Wichelns, 2004; Zhou and Tol, 2005). This situation is aggravated by the fact that many water resources are shared by two or more countries. In the majority of cases, the institutional arrangements needed to regulate equity of resource use are weak or missing (UNESCO-WATER PORTAL www.unesco.org/water/).

5.2. Water quality

Even where there is enough water to meet current needs water resources are becoming increasingly polluted. It has been estimated that half of the population of the developing countries is exposed to polluted sources of water that increase disease incidence. The situation worsens because institutional and structural arrangements for the treatment of municipal, industrial and agricultural waste are poor (UNESCO-WATER PORTAL www.unesco.org/water/).

5.3. Water-related disasters

Growing concentrations of people and increased infrastructure in vulnerable areas such as coasts and floodplains and on marginal lands mean that more people are at risk (Abramovitz, 2001). Poor countries are more vulnerable: in every country poor population is especially hard hit during and after disasters. For example, the destructive consequences of tsunami, typhoons, hurricanes and recursive floods made disaster related institutions determine that they should not meet a similar threat unprepared again. In other words, the significance of disasters as a driver of water resource management should not be underestimated. What is important is thus not just the specific impact of disasters, but the way in which they interact with other aspects of water management, and the ways in which vulnerable people adjust their resource management to take account of the risks. Thus, rapid surveys of the state of groundwater exploitation, aquifer pollution vulnerability, subsurface contaminant load and exposure to natural disaster are needed (UNESCO-WATER PORTAL www.unesco.org/water/). The risk of pollution and the susceptibility of aquifers to the effects of "overabstraction" can then be assessed and protection measures can be prioritized and initiated. In some cases aquifer exploitation must be better controlled, through a combination of regulation, pricing and incentives. Similarly the risk of pollution and the threat of natural catastrophe must be reduced by incorporating groundwater vulnerability as a factor in land-use planning and environmental controls (Foster *et al.*, 2000).

6. CONCLUSIONS

It appears evident that our generation is responsible for protecting part of this hidden resource and precious commodity for future generations. It is necessary that international organizations (every agency of UN family is involved on water problems), governments and scientific communities understand the need for co-ordinated scientific programmes focusing on water. Research to support integrated water resources management should thus be integrated. This implies more inter- and multi-disciplinary approaches (geophysical, geochemical, biological and subsequently socio-economical). Hence there is need for a close investigation of water science and policy. Phenomena and possible changes are to be studied in quantity and quality, prior to addressing science and policy, and water and civilization aspects (UNESCO-WATER PORTAL www.unesco.org/water/).

Water resources are often one of the primary limiting factors for harmonious development in many regions and countries of the world. The incorporation of the social dimension underlines the need for improved, more efficient management of water resources and more accurate knowledge of the hydrological cycle for better water resources assessment. Consequently, water has been and is managed only under the economic aspect, and thus in a fragmented way. Surface water and groundwater are considered separately in development activities without due recognition of their interdependence. Quantity is generally managed separately from quality, as are water science and water policy. This fragmentation of approach also impedes coherent hydrological analyses at regional, continental and global scales. In other words management of aquifers suffers from the myriad of hydroinstitutional settings. Moreover, water resources in many places are still not managed in conjunction with land resources and land development. Water supply schemes, eventually generating large amounts of waste water in consumer areas, are normally designed and built, especially in developing countries, without the required matching drainage networks and waste water treatment facilities. In fact, much more important than volume, is the depth from which water is pumped (cost of the action) and the definition of groundwater replenishment (controls and uncertainties of the recharge rate). The pathway to a therapy focuses on governmental, intergovernmental and international agencies serving as "guardians of groundwater" by partnering with local stakeholders in groundwater administration, protection and monitoring, and broader water-resource planning and management strategies. Many countries need better focused monitoring of groundwater levels and quality so that a clearer picture can be painted of the actual state of resources and of what must be done to use them more effectively – and to preserve them for future generations (UNESCO-WATER PORTAL www.unesco.org/water/).

Giovanni Maria Zuppi
Environmental Sciences Department,
Ca' Foscari University of Venice, Italy

REFERENCES

Abramovitz J., 2001. Unnatural Disasters. Worldwatch Paper 158, Worldwatch Institute, Washington D.C.
Araguás y Araguás L., Custodio E. and Manzano M.S., (eds.), 2005. Groundwater and saline intrusion. Selected papers from the 18th Salt Water Intrusion Meeting. Instituto Geológico y Minero de España, Madrid, Spain, Hidrogeología y Aguas Subterráneas Series, 15, p. 726.
Aureli A. and Ganoulis J., 2005. The Unesco project on internationally aquifer resources management (UNESCO-ISARM). Overview and recent developments UNESCO-ISARM-MED. Report of a consultative meeting: "Key issues for sustainable management of transboundary aquifers in the Mediterranean and in South Eastern Europe (SEE)", Thessaloniki, Greece, $21^{st} - 23^{rd}$ October 2004.
Campana M., 2005. Foreword: Transboundary Groundwater. Groundwater, 43, 5, pp. 646-650.
Chilton P.J. and Foster S.S.D., 1995. Hydrogeological characterisation and water-supply potential of basement aquifers in Tropical Africa. Hydrogeology Journal, 3, 1, 36-49, DOI 10.1007/s100400050061.
Custodio E., 2002. Aquifer overexploitation: what does it mean? Hydrogeology Journal, 10, 2, 254-277, DOI 10.1007/s10040-002-0188-6.
Custodio E. and Gurgui A. (eds.), 1989. Groundwater Economics. Selected Paper from a UN Symposium Held in Barcelona, Spain. Elsevier. Amsterdam, p. 625.
Diersch H-J. G. and Kolditz O., 2002. Variable-density flow and transport in porous media: approaches and challenges. Advances in Water Resources, 25, 8-12, pp. 899-944.
Dijon R. and Custodio E., 1992. Groundwater overexploitation in developing countries: report of an interregional workshop (Las Palmas de Gran Canaria, Canary Islands, Spain). United Nations Department of Technical Cooperation for Development, New York. Doc UN INT/90/R43, p. 109.
Edmunds M. W. and Gaye C. B., 1994. Estimating the Spatial Variability of Groundwater Recharge in the Sahel Using Chloride. Journal of Hydrology, 156, pp. 47-59.
Foster S.S.D. and Chilton P.J., 2003. Groundwater: the processes and global significance of aquifer degradation, Philosophical Transactions of the Royal Society B: Biological Sciences, 358, 1440, 1957 - 1972, DOI 10.1098/rstb.2003.1380.
Foster S.S.D., Chilton J., Moencg M., Cardy F. and Schiffler M., 2000. Groundwater in rural development. World Bank Technical Paper NO. 463, World Bank, Washington D.C., p. 101.
Foster S., Garduno H., Evans R., Olson D., Tian Y., Zhang W. and Han Z., 2004. Quaternary Aquifer of the North China Plain – assessing and achieving groundwater resource sustainability. Hydrogeology Journal, 12, 81–93, 10.1007/s10040-003-0300-6.
Freeze A.R. and Cherry J.A., 1979. Groundwater. Prentice Hall, p. 609.
Ghassemi F., Jakeman A.J. and Nix H.A., 1995. Salinization of Land and Water Resources: Human Causes, Extent, Management and Case Studies. New South Wales Press, Sydney, Australia, p. 520.
Gibson J. and Aggarwal P., 2001. Revisiting climate changes. Isotopes studies open scientific windows to the past. IAEA Bulletin, IAEA, Vienna, 43, 2, pp. 2-5.
Hayton R.D., 1982. The Law of International Aquifers. Natural Resources Journal, 22, 1, pp. 71-94.
Hayton R. and Utton A.E., 1989. Transboundary groundwaters: The Bellagio draft treaty. Natural Resources Journal 29, pp. 663-722.
Holzer T.L., 1981. Preconsolidation Stress of Aquifer Systems in Areas of Induced Land Subsidence. Water Resources Research, 17, 3, pp. 693-704.
IWMI – International Water Management Institute, 2000. Strategic Plan 2000-2005. Improving Water and Land Resources Management for Food, Livelihoods and Nature, p. 30.
Jarvis T., Giordano M., Puri S., Matsumoto K. and Wolf A., 2005. International Borders, Groundwater Flow, Hydroschizophrenia. Groundwater, 43, 5, pp. 764-770.
Jones I.C. and Banner J.L., 2003. Estimating recharge thresholds in tropical karst island aquifers: Barbados, Puerto Rico and Guam. Journal of Hydrology, 278, 1-4, pp. 131-143.
Llamas R., 2004. Water and ethics: Use of groundwater. Series on Water and Ethics, Essay 7, Unesco, Paris, p. 34.
Lloyd J.W., 1994. Groundwater Management problems in the Developing World, Hydrogeology Journal, 2, 4, 35-48, DOI 10.1007/s100400050042.
Lloyd J.W., 1998. A changing approach to arid-zone groundwater resources in developing countries? In: Gambling with groundwater: physical, chemical and biological aspects of aquifer-stream relations. van Brahana et al. (eds.). International Association of Hydrogeologists, Las Vegas, pp. 7-12.

Ma J.Z., Wang X.S. and Edmunds W M., 2005. The characteristics of groundwater resources and their changes under the impacts of human activity in the arid North-West China – A case study of the Shiyang river basin. J. Arid Environments, 61, pp. 277-295.

Margat J., 1993. The overexploitation of aquifers. In: Selected Papers on Aquifer Overexploitation, Simmers et al. (eds.). International Association of Hydrogeologists, Heise, Hannover, Vol. 3, pp. 29-40.

Matson P.A., Parton W.J., Power A.G. and Swift M.J., 1997. Agricultural Intensification and Ecosystem Properties. Science, 277, 5325, 504-509, DOI: 10.1126/science.277.5325.504.

Mazor E., 1995. Stagnant aquifer concept Part 1. Large-scale artesian systems – Great Artesian Basin, Australia. Journal of Hydrology, 173, 1-4, pp. 219-240.

Milnes E. and Renard P., 2004. The problem of salt recycling and seawater intrusion in coastal irrigated plains: an example from the Kiti aquifer (Southern Cyprus). Journal of Hydrology, 288, 3-4, pp. 327-343.

Moncaster S.J., Bottrell S.H., Tellam J.H., Lloyd J.W. and Konhauser K.O., 2000. Migration and attenuation of agrochemical pollutants: insights from isotopic analysis of groundwater sulphate. Journal of Contaminant Hydrology, 43, 2, pp. 147-163.

Morris B.L., Lawrence A.R.L., Chilton P.J., Adams B., Calow R.C. and Klinck B.A., 2003. Groundwater and its Susceptibility to Degradation: A Global Assessment of the Problem and Options for Management. Early Warning and Assessment. Report Series RS.03–3. United Nations Environment Programme, Nairobi, Kenya, p. 138.

Puri S. and Aureli A., 2005. Transboundary Aquifers: A Global Program to Assess, Evaluate, and Develop Policy. Groundwater, 43, 5, pp. 661-668.

Rivera A., Allen D.A. and Maathuis H., 2004. Climate variability and change-Groundwater, Chapter 10 in Environment Canada, Threats to the Availability of Water in Canada, Report No. 3, National Water Research Institute, Burlington, Ontario, 2004, pp. 89-95.

Rossi G., Benedini M., Tsakiris G. and Giakoumakis S., 1992. On regional drought estimation and analysis, Water Resources Management, 6, 4, 249-277, DOI 10.1007/BF00872280.

Salama R.B., Otto C.J. and Fitzpatrick R.W., 1999. Contributions of groundwater conditions to soil and water salinization, Hydrogeology Journal, 7, 1, 46-64, DOI 10.1007/s100400050179.

Salman S. M. A., 1999. Groundwater: Legal and Policy Perspectives. World Bank Technical Paper No. 456, World Bank, Washington D.C., p. 260.

Sherif M.M. and Hamza K.I., 2001. Mitigation of Seawater Intrusion by Pumping Brackish Water, Transport in Porous Media, 43, 1, 29-44, http://dx.doi.org/10.1023/A:1010601208708, DOI 10.1023/A:1010601208708.

Sherif M.M. and Singh V. P., 1996. Saltwater Intrusion, in "Hydrology of Disasters", Book Series: Water Science and Technology Library, 18, 269-319. Kluwer Academic Publishers, The Netherlands.

Simmers I. (ed.), 1988. Estimation of Natural Groundwater Recharge NATO ASI Series. Series C: Mathematical and Physical Sciences, D. Reidel Publishing Company, Dordrecht, The Netherlands, p. 510.

Simmers I. (ed.), 1997. Recharge of phreatic aquifers in (semi-)arid areas. IAH Int. Contrib. Hydrogeol. 19, AA Balkema, Rotterdam, p. 277.

UNESCO-ISARM, 2004. Managing Shared Aquifer Resources in Africa, Bo Appelgren (ed.), Paris, UNESCO, IHP-VI, Series in Groundwater, No. 8.

UNESCO-WATER PORTAL, www.unesco.org/water/

U.S. GEOLOGICAL SURVEY (USGS), 1998. Groundwater and surface water a single resource. U.S. Geological Survey Circular 1139. Denver, Colorado, USA, p. 77.

Vellinga P. and Leatherman S.P., 1989. Sea level rise, consequences and policies, Climatic Change, 15, 1, 175-189, DOI 10.1007/BF00138851.

De Vries J. and Simmers I., 2002. Groundwater recharge: an overview of processes and challenges, Hydrogeology Journal, 10, 1, 5-7, DOI 10.1007/s10040-001-0171-7.

Wichelns D., 2004. The policy of virtual water can be enhanced by considering comparative advantages. Agricultural Water Management, 66, pp. 49-63.

Zhou Y. and Tol R.S.J., 2005. Evaluating the costs of desalination and water transport. Water Resource Research, 41, 3, 10.1029-10.1035.

G. MURARO

WATER SERVICES AND WATER POLICY IN ITALY

Abstract. Water supply in Italy covers 96% of the population, while the sewage system covers 84% and the wastewater treatment covers 75% of the "equivalent inhabitants". A radical reform was introduced in 1994, with the goal to exploit the economies of scale and of scope existing in the sector. It prescribed that supply, collection, and treatment are unified in an "integrated water service", to be managed on a large territorial basis by a unique firm. Consumers must bear all the costs. Because of the relevant investment needed, the average tariff is forecasted to increase in the next fifteen years, but the negative social impact should be avoided through an appropriate tariff discrimination. The reform was strongly opposed by local governments, which were before responsible for water services. Now it covers the majority of the territory but some difficulties remain, since local governments are against the choice of the managing firm by auction and they prefer to directly assign the service to a public firm.

1. INTRODUCTION

Water policy deals with institutions, criteria and procedures. Each aspect is relevant and all of them must fit in a consistent system. The aim of this paper is exactly to investigate such problems in the field of drinking water.

Water services will then mean in this paper some or all of the following items: water acquisition, either by extraction of underground waters or by derivation of surface waters; its transport from the source to the areas of consumption, either towns or productive zones; its distribution to the final consumers, either families (house users) or collective users (added to the house users to form the civil users), or firms, either agricultural or industrial or commercial firms (productive users); the collection of wastewater; the treatment of wastewaters; their final destination, either discharge in the environment or recycling. Integrated water services will then mean the whole of such services.

Water policy may now be specified as the complex of institutions, criteria and procedures that feature the above mentioned drinking water services in one country. The paper will put special emphasis on the choices implied in water policy, namely:

- the role of central, regional and local Governments,
- the role of the public authorities vs. the role of the market,
- the financing of current operations and investments,
- the structure of the tariff.

The analysis of the Italian water policy, after a look to the state of water services (par. 2) and a brief outline of the history of the rules (par. 3), is focused on the reform introduced in 1994. Par. 4 describes the aims, the reasons and goals of the reform, and its prescriptions as far as the institutions, rules and procedures are concerned. Par. 5 will describe the degree of implementation of the reform. Par. 6

will try to explain the troubles found in implementing the reform and the reasons of the successive legislative evolution. Par. 7 will finally attempt to draw some lessons from the Italian case study.

2. THE STATE OF WATER SERVICES

For a better understanding of the situation of water services, Tab. 1 offers some preliminary data on the structure of the country.

Table 1. Italy's Main Indicators

Territorial Surface (km^2)	301,336
Regions	20
Provinces	103
Municipalities	8,101
Resident population (Jan. 1, 2005)	58,462,375
North	26,469,091
Centre	11,245,959
South	20,747,325
Resident foreigners (approximately)	2,000,000
Family members	2.5
Population density (inhab/km^2)	194

Source: ISTAT (2005).

A recent survey by the Italian Supervising Committee on the Use of Water Resources (Committee, from now on) offers a picture of the supply of water services in the country that may be summarized as follows: very good coverage of the interested population as far as water supply by the network is concerned and abundant availability of water on the average, but with persistent irregularities of supply in summer periods that explain the diffused customers' complaints in certain areas; not very good coverage as far as the collection of wastewaters is concerned; remarkable gap still to be filled as far as the wastewater treatment is concerned. The following tables show the main data about the water services on a national basis (avoiding one northern region, namely Trentino-Alto Adige, which has special autonomy).

2.1. The supply of drinking water

With reference to the resident population[1] water supply coverage is generally found to be higher than 90%, with an average weighted value of 96%. Improvements are always possible, but the population non-connected to the network is very probably living in "spread houses" in rural land, and it is not granted that the connection is worthwhile. The per capita supplied water is on the average very high, being equal to 286 litres per day. Domestic use accounts for less than the total, namely 74.8%;

but the absolute figure (214 l/inhabitant) remains very satisfactory. It is in the main case (82.7%) good underground water, though the dependence from surface water is remarkable in some areas of the North and in the islands (Tab. 2).

Table 2. *Main indicators of drinking water supply in Italy (2005)*

Population served	96.0%
Per capita supplied water (l/inhabitant)	286.0
Source of supplied water:	
- underground water	82.7%
- surface water	17.3%
Invoiced water/network available water	59.9%
Total leaks (% of network availability)	40.1%
Composition of the invoiced water uses :	
- domestic use	74.8%
- non domestic civil use	11.7%
- civil use (sum)	86.5%
- industrial use	13.5%

Sources: *Committee (2005, pp. 38-43). For the composition of water uses, data are taken from ISTAT – Istituto Nazionale di Statistica (2003).*

The incidence of leaks is impressive everywhere, but especially in the South, where they almost constantly exceed 50% of the network available water. However, leaks represent the sum – that with the present available information cannot be broken down further – of physical leaks and economic losses linked to unaccounted public uses (for instance, water used on roads and municipal gardens) or to unauthorized users; indeed, the physical leaks themselves could derive from network skims that in areas with great abundance of water might be rational. It is therefore a general opinion that the supply network calls for remarkable expenses in better maintenance and investments; but the problem is not that huge as the figure of leaks would suggest.

As for territorial data, in general North is alike the Centre of Italy and both are in a better situation than the South: more water availability, higher percentage of underground water, inferior level of leaks. The differences are however not very relevant. Southern Italy is more exposed to irregularities of supply in summer, and it can suffer from extensive periods of draughts. The last emergency situation from this point of view happened in 2002 in four Regions (Basilicata, Apulia, Sardinia, and Sicily), reaching an almost dramatic level in Sicily.

As for the quality of the service, the situation is not everywhere optimal indeed, at least as far as the perceived quality is concerned. A survey by the National Institute for Statistics (Istituto Nazionale di Statistica – ISTAT) shows that in 2002 still 14.7% of the population complained about irregularities of supply, but the dispersion of the phenomenon was very high: the percentage in almost all the northern regions is irrelevant, while it is over 20% in almost all the southern ones, with Calabria and Sicily reaching the top with 37% and 39%. An analogous territorial pattern, but with much higher data, is shown by the "mistrust on the drinking water quality" (40.1% of the population, but over 50% in Calabria, Sicily and Sardinia)[2].

2.2. Collection and treatment of wastewater

The sewage system covers in Italy 83.6% of the relevant population, defined as the total number of "equivalent inhabitants" that includes the non-domestic discharges (allowed to use the network because compatible with the domestic ones from the point of view of the necessary treatment). It overcomes 90% only in three regions, while the minimum, equals to 70.1%, is amazingly found in the rich North East of the country, namely in the Regions Veneto and Friuli Venezia Giulia. The collection is based mainly on a unique "mixed" network (for 70.9% of the covered population), collecting both wastewater and rain. The use of separate facilities (29.1% of the covered population) generally dates back to the first half of the last century.

The level of coverage of the purification system as against the equivalent population reaches an average of 74.8%, with considerable variations and with the further disadvantage given by the presence of small plants, which make up 80% of the units. For these small plants it is impossible to guarantee a level of service in line with the current legislation.

In conclusion, the supply of drinking water is the most felt need, so that any irregularity, not to speak of the draught, immediately becomes a highly discussed problem. But the really unsatisfactory situation concerns the not full coverage of the relevant population by the sewage system and by water purification services. The gap is made even more serious by the European legislation, namely by the Water Framework Directive 2000/60, which compels to improve remarkably the quality of water basins in the next ten years.

2.3. Investments and tariffs

The cause of such a gap is explained by two factors: the general fall in the last twenty years of investments in public works for water services, mainly due to the need to recover from the huge public debt accumulated between 1970 and 1990; and the lack of political visibility of the problem of collecting and treating wastewaters, that induced to sacrifice mostly the investments in that specific field.

Needless to say, this means that Italy has a big gap to fill in water related investments. There are already many official programmes approved by local authorities. Assembling those programmes and extrapolating their results on a national basis (ProAqua and ANEA, 2005, pp. 43-45), it comes out that in the next 26 years Italy must invest the total sum of 55,127 million Euros, which is equal to an annual per capita investment of € 37.7, with the following composition:

i) purpose:
- for water supply   44.3%

- for sewage system 35.7%
- for treatment plants 20.0%

ii) yearly per capita investment (€):
- North Italy 39.3
- Centre Italy 28.2
- South Italy 42.8

Such programs are heavy, though not much heavier than those implemented by the country in the years 1980-85. This time, however, the new rules adopted in Italy oblige the consumers to bear directly all the burden of the investment through increasing tariffs, while the help of public finance, through national and European funds, will be restricted to the South of the country.

Tariffs are then forecasted to increase almost 50% in the next fifteen years: from the present level of € 0.94 for cubic meter, it will reach the maximum level, in real terms, of € 1.37; and in Southern Italy tariff development will have a sharper increase than in the rest of the country. The low starting level of the tariffs, when compared to other tariffs in Europe, assures that on the average the burden is bearable. Care is however needed in order not to hurt poor families, which are frequently the most numerous ones. A tariff carefully structured is therefore necessary. A system of differentiated tariffs should ensure both the coverage of total costs as well as the sustainability of individual costs by the weaker users.

Observers strongly recommend that the tariff dynamics, however, should not delay the reform. On the contrary, it should accelerate it, for two reasons: because, as investments are indispensable, especially for the commitments deriving from the European legislation, it is better to tackle them with the best possible efficiency granted by the industrial management of the service; and because the South of Italy shall be able to count on the considerable European contribution foreseen in the "Community Support Plan" only for a few more years.

3. THE EVOLUTION OF THE RULES

Passing now to the organization of water services in Italy, the present features, deriving from the radical legislative innovations introduced in 1994, can be better understood by recalling briefly the preceding history as well as the successive evolution.

For more than one century, water services as almost all local services have been in the power of municipalities. Because of the population growth and the process of industrialization and modernization of the productive apparatus, many local governments started already in the second half of the 19th century to constitute municipal enterprises. The first experiences concentrated in towns where the industrial development was bigger. In fact, between 1845 and 1870 only Genova, Vercelli, Brescia, Rovigo, La Spezia and Cesena had a public gas service. After 1880, other sectors were involved such as waterworks (Spoleto, Udine, Terni, Vercelli), urban transports (1895 Genova), and power-stations (1893 Brescia, 1898 Verona).

In this context, in 1903, the so-called Municipalization Law was elaborated (Law 29 March 1903 No. 103, or Giolitti Law, from the name of the President of the Government). This law confirmed the power to municipalize the services to local governments, according to specific conditions, and listed the categories of services which could be municipalized. Such an approach was confirmed, during the fascist regime, by the Consolidation Act of 1925 (R.D. 15th October 1925, No. 2578).

No important innovation as far as the legislation is concerned occurred until the Second World War, when social problems called for a strict regulation of the dynamics of the tariffs. The tariff aspects were then regulated by the D.L.Lgt. 19th October 1944, No. 347 which set up the Interdepartmental Price Committee (Comitato Interministeriale Prezzi – CIP) and the Provincial Price Committee (Comitato Provinciale Prezzi – CPP). With reference to water services, the relevant legislative and administrative measures were issued in 1974-75 (CIP 45/1974; Resolution 26/1975). Those rules developed in a context of great inflation (near 20%), in which the tariff should inevitably be incorporated together with the investment and operating costs. Much care was then given to the structure of the tariff, in order to protect the weakest consumers. As a result, there was a tariff of growing blocks, with a lower tariff for the first stage of consumption, a standard tariff, and a high tariff for the consumption over a definite threshold.

From the second half of the seventies, there was a development of the debate on water services, mainly in relation with the growing attention to environmental problems. As a result, the important Law 10th May 1976, No. 319 (Merli Law) was issued. However, the theme of poor efficiency in water service management was much debated as well: the excessive fragmentation of the single services was considered no longer tolerable, also in consideration of the English reform which established water authorities. Such a problem was faced up in Law 8th June 1990, No. 142 which consented to manage local public services in general, and water services in particular, through joint-stock companies. A further development of that "managerial approach" was offered in year 2000 by the Consolidation Act on Local Governments (D.lgs. 18th August 2000, No. 267). More recent changes concern the rules for the assignment of local public services, and they will be considered later on with reference to water services.

For the specific field of water services, a vast reform was introduced in 1994. A successive important step was the D.lgs. 11th May 1999 No. 152 that fully translated the European Directive 91/271/CE on the treatment of urban wastewater in the national legislation. The water sector is now facing the implementation of the European Directive 2000/60/CE, which imposes more ambitious targets for water quality protection. Finally, the recently issued "Environmental Law" (D.lgs. 3rd April 2006, No. 152) has modified the governance of water service by creating an independent Authority with competence on water services and waste disposal services.

4. THE 1994 REFORM

The Law 5th January 1994, No. 36 (often labelled as the reform of water services, by its contents, or the Galli Law, by the name of its proponent) remains the main reference point of the present Italian policies concerning the water sector, since its implementation is still underway and is accompanied by various difficulties.

4.1. Reasons and targets of the law

It might be useful, first of all, to recall its motivations and contents. As already said, the starting point was the wide fragmentation of the sector, over and above the extreme backwardness of investments. When the law was passed, there were probably more than 8,000 firms in the sector that, according to estimates by ISTAT, decreased to 7,848 in 1999. About 82% of these were municipalities acting on an internal providing basis. Therefore the structure of the sector looked completely unfitted to face the challenges of the new and costly water policy, especially as far as the protection of the environment was concerned.

It was a common belief, based on some scientific researches and on foreign examples, that the sector was featured by strong economies of scale (i.e. decreasing average cost by enlarging the dimension of activity) and strong economies of scope (i.e. decreasing average cost by putting the various water services together). The aim of the law was exactly to exploit those economies in order to decrease the average cost of activity. In addition, the law was inspired by the growing awareness that public production was intrinsically less efficient than market production, and therefore, local governments should preferably not manage directly water services but concentrate in the strictly public functions of planning, control and regulation.

Namely the law disposed:

- the territorial integration, with the definition of the Optimal Territorial Basins (Ambito Territoriale Ottimale – ATO) based on natural borders more than on the administrative ones;
- the functional integration of the various activities of the water cycle – as already said, the acquisition, transport, and distribution of drinking water, and the collection and treatment of waste water – within the integrated water service;
- the rule disposing that tariffs should guarantee the full coverage of the investment and management costs.

It is important to note that those rules are still considered appropriate by many economists in Italy (Robotti 2002; Buratti and Muraro 2002; Muraro and Valbonesi 2003).

4.2. The new governance

As for the governance of the sector, the law prescribed:

- the competence at Government level of the Italian Ministry of the Environment, Land and Sea (IMELS) (initially, until year 2000, the Ministry of Public Works, in agreement with the Ministry for the Environment), however in a context of large autonomy granted to the regional governments;

- the institution of a national structure of control, the Supervising Committee for the Use of Water Resources, composed by seven experts, four designated by the Government (initially by the two competent Ministries and from year 2000, by the Ministry for the Environment) and three by the Conference of the Presidents of the Regions, appointed for a non renewable five years period; the Committee has the task of monitoring the local activities and offering the guidelines for the definition of the tariff; the Committee, however, while fully autonomous in its judgments, is not an independent Authority in strict sense: it cannot impose penalties directly, its tariff rules must be approved by the competent Ministry, and its support structure depends on the IMELS;
- an active role of each regional government in issuing a regional law inspired by the national one and in particular in defining the division of the regional territory in ATOs;
- the institution in each ATO of a governing body (ATO Authority), elected by the assembly of the provinces and municipalities concerned, that should act as the promoter and controller of the reform.

Namely, the ATO Authority should:

- carry out the survey (*"ricognizione"*) of plants and services,
- prepare a plan of investments and improvements of the services,
- contract out the management of the integrated water service to a company and supervise its behaviour with regard to the implementation of the investments and service levels detailed in the plan and the agreement,
- approve the tariffs which shall have to be created according to the so-called "normalized method" drafted by the Supervisory Committee on the Use of Water Resources.

This institutional architecture is still valid, with an important innovation in the near future: as already mentioned, the recently issued "Environmental Law" (D.lgs. 3rd April 2006, No. 152) has modified the water service governance by creating an independent national Authority, that will take the place of the Committee, with more powers to rule and control the water services (together with the waste disposal services).

4.3. The rules on the tariff

As for the tariff, the rules elaborated by the Committee are inspired by a price cap approach. The average tariff, specifically determined by the ATO Authority following the national rules, covers the full cost agreed (operational costs, amortization, and interest at a conventional rate on the investments). The dynamic of the tariff is based on three parameters: the officially forecasted rate of inflation (the so-called "programmed inflation"), which is fully recognized in the tariff; a rate of productivity growth, that is applied *ex ante* to the current costs, thus operating as a decreasing factor of the tariff; a coefficient that for social reasons limits the maximum annual increase of the tariff. The last two parameters are decided by the ATO Authority within definite ranges which are determined at national level by the Committee and are linked with the initial levels of operational costs and tariff (to a lower initial level correspond a lower productivity gain rate and a higher rate of maximum increase to be imposed on the firm).

The Committee has also elaborated (D.M. 1st August 1996) a so-called "normalized method" that offers an econometric assessment of the operational costs for water supply, and wastewater collection and treatment, using various physical and demographic data as independent variables. All the equations are built on the hypotheses of remarkable scale economies and therefore of a decreasing average cost. Applied within each ATO, the method determines the reference levels of those costs: if the ATO Authority wants to recognize some levels of cost over the range allowed by the method, it must justify the reasons and get the authorization from the Committee. The tariff determined remains valid for a three-year period and will then be assessed again by the same method but with up-to-date data. In each three-year period the price cap approach, with its well known incentive to efficiency, is fully applied, in the sense that the firms have the right to keep whatever additional gain obtained thanks to an increase in productivity higher than that implied in the tariff.

A revision of the "normalized method" must be elaborated by the Committee every five years and proposed to the competent Ministry. As a matter of fact, a deep revision of the method has been proposed by the Committee in May 2002, but not yet accepted by the IMELS. One of the main parameter of the method is the interest rate allowed for the capital invested. It is still at the initial level of 7%, determined in 1996, while the proposed revision suggests a two-part rate, with a fixed spread of 3.5% over a variable market rate (namely, the Interest Rate Swap for fifteen years long operations).

The weighted average tariff just discussed is compatible with an articulated structure of the consumers' tariffs. The law 36/1994 prescribes a tariff under the average for the "essential consume" and for low income users, while it allows a penalty rate on non-residents. It allows a differentiated tariff on a territorial basis too, for taking into account the different levels of investment in water services made by the Municipalities before the reform.

As for wastewater collection and treatment, it is worthwhile to note that the specific part of the tariff is applied to the clean water supplied and it must be paid by users which are connected to the sewage network even if the treatment plants are not yet in operation (but in that case, the funds collected must be kept for building the plants).

4.4. The number and size of ATOs

For a correct appreciation of the impact of the reform, note finally that the total number of ATOs created by the regions (with the already mentioned exception of one Northern Region, Trentino-Alto Adige, which is not subjected to law 36/1994) amounts to 91. On the average each ATO has a population of 616,000 inhabitants and covers a territory of 3,162 km^2. They are different in size: at the top extreme is Apulia, where the basin coincides with the whole region, with 4 million inhabitants and a territory of 19,362 km^2; at the bottom there is the ATO Valle del Chiampo in Veneto, with only 54,505 inhabitants and 162 km^2 but with a heavy concentration of polluting industry. Considering the civil population served the distribution of the ATOs in shown in Tab. 3.

Table 3. Distribution of ATOs by population

Number of inhabitants	*Number of ATOs*
1-100,000	2
100,000-250,000	17
250,000-400,000	29
400,000-750,000	22
750,000-1,000,000	8
1,000,000-2,000,000	8
> 2,000,000	5

It is in any case impressive to think that, if the story has a happy ending, there will be less than one hundred big firms in the country, each one caring for the integrated water service in a large territory, instead of the many thousands small firms operating before the reform. It is also worthwhile to note that the current annual cost of the new institution created, the ATO Authority, is very reasonable, being equal on the average to € 1.47 per capita.

5. STATE OF IMPLEMENTATION OF LAW 36/1994, BETWEEN DELAYS AND ACCELERATIONS

However, are we sure about the happy ending? Before trying to reply to the question, let us analyze the situation and the past dynamics.

Table 4. State of advancement of the reform according to Law 36/1994 (December 2004)

	North	Centre	South	**Italy**
Foreseen Optimal Territorial Basins	44	19	28	**91**
ATO Authorities in office	40	19	28	**87**

Carried out Surveys (*)	38	19	28	**85**
Approved Territorial Basin Plans	22	16	28	66
Appointed managing firms	17	17	13	47

(*) ATO Apulia is reviewing the preceding survey prepared by a State agency. In the present Table this survey is however considered available.

The situation is described in Tab. 4. At the end of year 2004 in almost all instances the ATO Authorities were installed in office and carried out the survey on the state of water services. In more than 2/3 of the cases, they approved the plan too, which means that the participating municipalities agreed on the investments to be made and the corresponding tariff. More than half of the Authorities (47 out of 91) made the final step, by assigning the service to a managing firm. Thus, the majority of the population in Italy is already experiencing the new organization of the integrated water service on large area. No doubt therefore that the reform has gone past the point of no return.

At the same time one could argue that such a situation, after ten years since the coming into force of Law 36/94, is by no way satisfactory nor is it assuring about the completion of the reform. A better judgement can however be formed by considering the dynamics of the process so far, as it is depicted in Tab. 5. It becomes then clear that in the first five years there was practically no movement, while the improvement has been rapid in the following period. One becomes then more optimistic about the remaining steps to be done, though some worries remain.

Table 5. The implementation process of the reform

Date (Dec. 31)	ATO set up	Surveys performed	Plans approved	Assignments made
1996	6	0	0	0
1997	14	1	0	0
1998	21	6	0	0
1999	34	12	3	3
2000	46	17	6	4
2001	64	40	11	9
2002	83	58	38	17
2003	86	77	55	33
2004	87	85	66	47

In order to understand the residual difficulties and to make the Italian water policy an interesting case to study, it is worthwhile to explain briefly the dynamics of the reform.

The almost total paralysis of the first 6-7 years, after which only in 3 of the 91 Territorial Basins the reform had been implemented, has many motivations. However, the most important is surely the resistance of the municipalities to the idea of relinquishing a direct competence awarded to them in 1903 with the Giolitti Law. At the basis of the reform there is a clear political compromise that can be described as follows: the national legislator considered that it would be inevitable to respect the municipal authority in the field in order to have the reform approved; therefore, the municipalities were accepted as main players in the new organization, with the obligation, extended also to the provinces, of co-operating (through the institution of agreements or consortia, according to regional preference) and creating together an ATO Authority that would be entrusted with the planning of the water service and the choice and control of the managing firm. The compromise allowed the reform to be launched, but at the same time it hindered its implementation, causing the above mentioned delay (as well as problems in the definition of the Territorial Basins, that was influenced more by the administrative borders than by the hydrographical features of the territory, which the legislator considered more important).

And how to explain the strong acceleration in the last years? A first factor was the stance immediately adopted by the new Government (May 2001) in favour of this law. The second factor was the passing of Law 448/2001 (2002 Finance Act). This law, on the one hand, made it compulsory for the managing firm to be entrusted through public tender, as a standard rule; on the other hand, it left as a transitory solution the possibility of appointing directly, for a few years, a wholly public local company, on condition of having to include, in the short term, a private shareholder chosen by tender. Given the strong opposition of several Municipalities to tenders, in the period of grace after the approval of the Finance Act, in many cases management was quickly entrusted directly to mixed companies, where the majority of capital was held publicly by local authorities, or, provisionally, to public-sector companies[3].

The rule of the compulsory assignment by tender was challenged by some regional governments in front of the Constitutional Court, and it was challenged by the great majority of local governments in the political debate. The central Government was practically forced to introduce a different solution that was passed into law at the end of year 2003 (D.L. 269/2003, turned into Law 326/2003). Precisely the new law allowed the ATO Authorities to choose among three options: contracting out through tender, mixed public company, "in house providing". After only a few months, the 2004 Finance Act (Law 350/2003) art. 4, par. 234, introduced two important rules: the first (par. 234, letter b) saved from early cessation the "appointments already made as at 1st October 2003, of companies listed on the Stock Exchange and of their subsidiaries, as long as they are exclusive managers of the service, and of companies which were formerly wholly owned by the public sector, provided that, as at that date, they had put on the market shares of their capital through public procedures"; the second (par. 234, letter c) allowed directly appointed companies whose appointment expires at the end of 2006 under the new provisions, to participate in the first tender for the services supplied by them.

This meant the end of those strong internal controversies and the further advancement of the reform.

6. THE RESIDUAL CONTROVERSIES ON SERVICE ASSIGNMENT

Why, then, some residual worries about the completion of the process? The main worry is not the high number of ATOs in the North still far from the end (the reform has started there with greater delay, but the performance in the last period is promising); nor the troubles found in some southern ATOs in concluding successfully the tender (probably the tender conditions must be revised in favour of the potential competitors, with some additional public financing). The main worries derive from the feeling that the present peaceful situation as far as the assignment rules are concerned is not stable, because of some deep ambiguities in interpreting the rules.

The assignment rules represent the crucial question of the reform and the most illuminating lesson for other countries. Let's therefore briefly recall the story.

Italy's new political culture, born in 1992 under Mr Amato's Government because of the urgent necessity of decreasing public deficit, aimed at a clear cut separation between the normative and supervisory activities, and the management activities, thus taking the public sector out of the field of industrially relevant goods and services production. In the field under consideration, however, this cultural approach that should have led to the privatization of water services appears attenuated or even lacking. Law 36/1994 refers to Law 142/1990 which, with its subsequent amendments and integrations, lists the following authorized management options[4]: on an internal providing basis, by contracting out, by a municipally-owned enterprise, by an institution, by a mainly public sector-owned company, by a mainly private sector-owned public company[5]. As a matter of fact, only three of these forms had a real appeal; but it remains true that the appointment of a private enterprise, chosen by tender, was just one option.

In brief, in order to reduce hostilities, or even to obtain the approval of the wide political, technical and administrative apparatus present in public sector enterprises, which were widespread and "virtuous" especially in the Centre and North of the country, the legislator accepted the compromise of the optional privatization, thus granting the public sector the opportunity of being involved in the management. And we have already discussed the almost universal preference of the ATO Authorities for direct appointments, rather than appointments by tender.

This approach, however, gained the opposition by the European Community Commission which, with a letter dated 8th November 2000, warned the government then in office that direct appointment violated the non-discrimination and transparency principles. In practice, the Community accepts the two extremes of the range of management options foreseen in Law 142/90 and transferred later in the original art. 113 of the Legislative Decree 267/2000: either a completely public management or a management entrusted to a subject extraneous to public administration chosen according to the principles of transparency and competition.

In the first instance, as declared by the Court of Justice (1999), the presence of a company is possible insofar as it is treated as an office of the shareholding public authorities that shall exercise full control on it, thus remaining within the logic, if not the form, of an in house providing (*in house*).

In the second instance, the external subject has full responsibility for the management, within the boundaries set by the public authority in its planning and supervision, and it must be chosen by tender, with public evidence procedures.

Given this choice, the Government and the Parliament, with art. 35 of the 2002 Finance Act (Law 448/2001), made a choice in favour of the assignment by tender, as the only method for the assignment[6]. However, the diffused hostility against art. 35 forced the Government, in September 2003, to reintroduce the mixed company as a third choice after tender and in house providing. Now, the possibility for the ATO Authorities of choosing among three options – contracting out through tender, mixed public company, "in house providing" – appears sufficient to eliminate future contrasts between ATO Authorities and the Government. The problem is that the European rules are still there; and notwithstanding the common belief that the Government solution has been agreed with the Commission, many observers think that the supposed agreement hides some ambiguity.

Unofficially, it is rumoured that the Commission considers this mixed solution as a variation of the tender process (the tender for the service takes place, and the winning company shall be obliged to take on public shareholders, even majority ones), while it is certain that most ATO Authorities see this formula as a new version of the classic direct appointment of a mixed company.

Further difficulties arise with regard to the *in house* providing. Adhering to the Community's requests, the law foresees that the company must be wholly public sector-owned and that it must also be totally controlled by local authorities (and have a prevailing local activity): it should be, in practice, a very rare exception, not a diffused case. A very restrictive interpretation of those rules has been confirmed by the Italian Antitrust Authority (13th September 2005) and by the European Court of Justice, Section 1 (Sentence 13th October 2005, no.C-458/03). In practice, this means a return to a kind of municipal enterprise, in the form of a company at the service of several local authorities. However, it seems difficult that such authorities and the companies themselves will accept this structure that would mean the negation of all has been done until now to make the management of local public services more efficient. It is easy therefore to predict that a strong pressure shall be brought to bear on the Government, and indirectly on the European Commission, to accept the idea that public ownership of the share capital, added to the mainly local character of the activities, shall be enough to answer to the requirements of an *in house* appointment.

To conclude:

1. the reform has proceeded and has gone past the point of no return;
2. the strong controversies between central and local governments between fall 2001 and fall 2003 should have been solved, in their main points, by the three appointment methods introduced in September 2003;
3. uncertainties with regard to the new rules, however, threaten new controversies.

7. SOME LESSONS FROM THE ITALIAN EXPERIENCE

Italian water policy is surely an interesting case study. Here are the main lessons to learn from it.

The economic reasons of the reform are convincing. It means that there are economies of scale and of scope that justify the rule of unifying water supply and wastewater collection and treatment, operating over a large territory.

The territorial basin will have the borders preferably defined by the hydrographical features of the territory. The reform is valid, however, even if the administrative borders are followed.

One weakness of the reform's institutional architecture is the determinant role of Municipalities. From the point of view of logical consistency, it is clear that the institutional dimension of local bodies would have to be abandoned, given that the municipal geographical dimension had been surpassed. Given also the fact that in the design of territorial basins the Provinces were often broken down, the correct player in the reform would have to be the Region (which, in turn, would probably entrust the Municipalities and Provinces with consulting or decisional competences, though not determinant ones, with regard to the planning and control of water service management)[7].

At the same time, note that the reform has been delayed, not stopped by the power of Municipalities: its implementation simply requires a much stronger political force by the central Government.

The governance of the system is well conceived, but the central level should be reinforced. Therefore, the very recent decision, not yet implemented, to put an Authority with direct power to issue directives on tariffs in place of the existing Supervising Committee is positive.

The tariff, as now conceived in the "normalized method", is a good starting point for regulating the service. Various technical improvements of the discipline are however possible. In addition, for less developed areas it seems unavoidable to assure the public financing of a relevant part of the investments needed.

Gilberto Muraro
Department of Economic Sciences,
University of Padua, Italy

NOTES

[1] The Supervising Committee estimates that the fluctuating population served is 26% of the residential population, so that the total population considered in the statistics is 126% of the residential population.
[2] It is commonly believed that such mistrust explains the peculiar love of Italian families for mineral water: a great business, based on a per capita consumption (more than half a litre per day for more than 75% of the population aged more than 14 years) that is considered the highest in the world. But econometric research shows that by far the most important factor is the per capita income: Italians want to drink high quality water, not only good wine!
[3] As far as the South of Italy is concerned, where in the same years there was a considerable advancement towards implementation of the reform, there are three specific factors to be taken into account: a) the

impulse given by the water emergency in 2002; b) the technical support for the drafting of the surveys and of the territorial basin plans offered by the Treasury owned company Sogesid; c) and, above all, the incentive represented by the access to financing offered by the European Union within the context of the QCS (Quadro Comunitario di Sostegno – Community Frame of Support), access that was conditional to the implementation of the reform. This is an important event that must be underlined: with an efficient system of bonuses and penalties, and with the correct technical support, it is possible to keep to the deadlines set by the law.

[4] "The municipalities and the provinces shall provide the management of the integrated water service through the forms, even compulsory, provided for in Law No. 142 of 8th June 1990, and subsequently integrated by art. 12 of Law No. 498 of 23rd December 1992 art. 9, par. 2, of Law 36/1994". The prescriptions of Law 142/1990 were subsequently integrated, with marginal changes, in art. 113 of Legislative Decree No. 267 of 18th August 2000, heavily amended later by art. 35 of Law 448/2001.

[5] The management on an internal providing basis fits only for small services, and the management by an institution, limited to the exercise of social services without entrepreneurial relevance, must be excluded in practice. Moreover, as far as the "special enterprise" is concerned, the law considers it only for a single public body, in the specific instance it should be a consortium, as foreseen by art. 25 of Law 142/1990.

[6] It has been already recalled the temporary discipline that allowed the solution, widely adopted, of a public firm.

[7] It is enlightening, the profound differences in the starting legal setting notwithstanding, to compare this reform with the previous health care reform of 1978, reviewed and deeply amended by Leg. Decree 502/1992. Of course, in the health care sector, municipal authorities were not as involved as they were in the water service. The legislative and administrative role of the region is stated in the Constitution of 1948. However, in practice, the municipal authorities were deeply involved in the health care sector, at least in the medium-to-large sized municipalities, thanks to several direct responsibilities attributed to the municipalities and through the presence of several charities and similar institutions operating at a municipal level in the hospitals. As soon as the legislator devised the territorial reorganisation of the health care service on the basis of the creation of Local Health Care Units of large dimensions, not bound within the border of the local authorities and administered on a managerial basis, the logical consequence was that the peripheral responsibilities were entrusted exclusively to the Regions, that were appointed to design the service basin of each Local Health Care Unit and to appoint its managers.

REFERENCES

Buratti C. and Muraro G. (eds.), 2002. La riforma dei servizi pubblici locali. *Il diritto della regione*, Padova, Cedam, No. 4/2002.

Comitato, 2005: Comitato per la vigilanza sull'uso delle risorse idriche. *Relazione annuale al Parlamento sullo stato dei servizi idrici, Anno 2004*, Roma. (English translation of the 2003 Report on the web site: www.minambiente.it).

Court of Justice, 1999. *Decision of 18th November 1999 (Lawsuit C-107/98, Teckal)*, Luxembourg.

ISTAT, 2003. *Indagine sui servizi idrici, anno 1999*, Roma.

ISTAT, 2005. *L'Italia in cifre*, Roma.

Muraro G. and Valbonesi P. (eds.), 2003. *I servizi idrici tra mercato e regole*, Roma, Carocci Editore.

Proaqua and ANEA, 2005. *Blue Book, I dati sul Servizio idrico Integrato in Italia*, Roma.

Robotti L. (ed.), 2002. *Competizione e regole nel mercato dei servizi pubblici locali*, Bologna: Il Mulino.

II. WATER MANAGEMENT

2. WATER QUALITY CONTROL

A. BARBANTI

WATER QUALITY CONTROL

Abstract. The aim of WQC is principally to establish weather or not a certain water body is in undisturbed conditions and, if this is not the case, which is the level of such disturbance and whether or not its quality is compatible with its possible or intended uses (i.e. drinking water supply, irrigation, fishing and aquaculture, bathing, etc.). In Europe, quality objectives have been recently redefined by the Water Framework Directive (WFD), combining the chemical and the ecological approach, although particular relevance is given to the second one. Environmental Monitoring is primarily linked to the objectives of WFD. The Quality Elements to be monitored are first of all biological parameters. Among the chemical parameters, those defined as Priority Substances have particular importance and are expected to be eliminated from all the emissions within 20 years. The effectiveness of monitoring programmes and their support to sound management decisions is strongly related to a transparent and efficient data management strategy.

1. INTRODUCTION

The raise of concern in the last decades for water pollution brought to the progressive development of a protective legislation. Legislation on water pollution had a prominent role within the legislation on environmental issues, due to the direct and essential uses of water, first of all water for drinking supply.

Every planned or due action aimed at safeguarding water resources or solving a water pollution issue is based on the knowledge at a reasonable level of the quality of the water body and its main controlling factors. In other words, knowledge is essential for management decisions.

Knowledge is the result of extensive and coordinated programmes for water quality control, coupled with the analysis of all the main factors (driving forces and pressures, according to the European Environmental Agency nomenclature) related to it (e.g. pollutant loadings, water budgets, water uses, etc.).

The concept of "Water Quality Control", which will be further discussed below, is strictly related to the definition of "Water Quality". Among the numerous definitions that can be found in literature, a very recent and official one that comes from the European Water Framework Directive (WFD) (Directive 2000/60/EC) is reported here. The WFD considers "water status", and distinguishes between "chemical status" and "ecological status". The chemical status defines water quality in connection to the concentration of specific pollutants, which must not exceed certain levels. The ecological status is an expression of the quality of the structure and functioning of aquatic ecosystems. It appears quite clear that the use of a wide concept of water quality as "ecological status" according to the WFD has significant impact on the meaning of "water quality control".

The European Union is embracing policies which will require a move from sectorial-based to more ecosystem-based, holistic environmental management.

The issue is so wide and complex that this chapter will focus on providing the reader with some basic concepts and highlighting the main overall tendencies, mainly concerning surface water bodies and marginally groundwaters. Occasionally, specific technical issues will be presented, mostly aiming at clarifying the above general concepts.

2. AIM OF WATER QUALITY CONTROL

The aim of water quality control is to determine weather or not a certain water body is in undisturbed conditions and, if this is not the case, which is the level of such disturbance and whether or not its quality is compatible with its possible or intended uses (i.e. drinking water supply, irrigation, fishing and aquaculture, bathing, etc.). Moreover, the concept of water quality control is about controlling the observance of law prescriptions from responsible public and private subjects: this is the case, for example, of controlling liquid discharges limits or quality standards for drinking waters.

Since the principles that inspire the two objectives are fully different, it seems appropriate to separate them clearly, using the general term of "monitoring" for the first objective and the more strict term of "control" for the second one.

One should say that what distinguishes the two approaches, beyond their specific scope, is the requirement for fast actions under the "control" regime, if exceedances of the prescribed quality criteria are observed.

3. CONCEPT OF "INTEGRATION" IN WATER QUALITY CONTROL

The evolution of the concept of "water quality" from a purely chemical and tabular concept to an ecosystem oriented concept and the consciousness of the need to strictly link the observation of environmental effects to the knowledge of their causes, both determined the progressive application of integrated approaches to water quality control initiatives.

The term "integration" has at this regard several meanings:

- integration of environmental objectives, combining quality, quantity and ecology;
- integration of all water resources (fresh surface water and groundwater bodies, wetlands, coastal water resources at the river basin scale);
- integration of disciplines, analyses and expertise (hydrology, hydraulics, biology, chemistry, toxicology, soil sciences, etc.).

The above statements can appear generic and possibly obvious. As a matter of fact, these principles, if correctly applied, give rise to very precise and effective choices and actions in designing and actuating monitoring programmes and in managing, processing and using their results. Practically, we still see today monitoring programmes and control practices that often are too sectorial and not integrated, with useless duplicated efforts and data underutilized, because not properly accessible to all potential users.

A significant example is the monitoring of river waters or the control of liquid discharges: only recently, after long discussions, these activities are carried out by measuring jointly quality (mainly chemistry) and quantity (flows), although it had been evident for a long time to most technicians and managers that knowing the pollutant load discharged in the receiving water body is equally or even more important than knowing only the concentration of a pollutant in the analyzed water.

It is also evident that the integration concept has great relevance on monitoring and control because the same concept has become more and more the basis for managing water resources and ecosystems.

The management plans at basin or hydrographic district scale depend upon an integrated and constantly updated knowledge, coordinating efforts and actions and unifying competencies and responsibilities. If one subject is responsible for the final result, i.e. a good ecosystem and sustainable uses of the resources, he/she will be interested in having all the data needed for taking the right decisions and he/she will make sure that those data are fully exploited. If several subjects are responsible for distinct sectorial objectives or territories and coordination is lacking, each of them will look first of all at satisfying its needs and defend its role, possibly against the overall efficiency of the system.

4. WATER QUALITY CONTROL IN THE EUROPEAN LEGISLATION

Legislation played and is playing a fundamental role on monitoring and control and therefore on the environmental assessment of water bodies. Hence, it must be admitted that until a law does not prescribe a certain monitoring approach and requires certain measurements, monitoring activities remain scarce, inhomogeneous and incomplete. Their effectiveness is left to the occasional presence of local environmental agencies particularly active or research centers of excellence. Laws often, and as expected, introduce simplifications; having to be applied everywhere, they could not distinguish the peculiarity of situations and territories. However, if correctly designed and predisposed to evolve, they guarantee fixed and homogenous rules.

The reference law on water management is the recent Water Framework Directive (Directive 2000/60/EC).

The Directive establishes a framework for the protection of all waters which: i) prevents further deterioration, protects and enhances the status of water resources through specific measures for the progressive reduction of discharges, emissions and losses of priority substances and the cessation or phasing-out of discharges, emissions and losses of the priority hazardous substances; ii) promotes sustainable use of water based on long-term protection of water resources; iii) ensures the progressive reduction of groundwater pollution and prevents its further pollution; iv) contributes to mitigating the effects of floods and droughts.

This Directive has a very wide and ambitious perspective; its implementation is in fact a challenge in the European Union and in each Member State.

According to the WFD, by 2006 all Member States shall ensure the establishment of programmes for the monitoring of water status within each River Basin District.

Three types of monitoring activities are identified: surveillance monitoring; operational monitoring and investigative monitoring (Annex V of Directive 2000/60/EC).

The surveillance monitoring programme is intended for assessing the long-term changes in the natural conditions and those resulting from significant human activity.

Operational monitoring is conceived as an additional measure to be undertaken by those water bodies identified as being at risk of failing to meet their environmental objectives.

Investigative monitoring shall be performed in individual cases in which the reasons for exceeding environmental quality standards are unknown, or where the surveillance monitoring programme reveals that the objectives set for a body of waters are not likely to be achieved.

Specific monitoring requirements exist for protected areas (Annex V). Protected Areas include bodies of surface water and groundwater used for abstraction of drinking water and habitat and species protection areas identified under the Birds Directive (79/409/EEC) or the Habitats Directive (92/43/EC).

A number of other Directives are connected to WFD and have relevance for water quality control.

Drinking water quality standards and relevant control procedures are defined through Directive 98/83/EC.

Bathing water quality standards are defined through Directive 2006/7/EC.

The Commission proposal for a Groundwater Directive (COM(2003) final of 19/09/2003) suggests further specific criteria for "good groundwater chemical status" and monitoring.

Some Directives on specific pollutants, the so-called daughters of Directive 76/464/EEC, are also still active and define certain values of emission limits and environmental quality standards for the purpose of the WFD:

- Mercury Discharges Directive (82/176/EEC);
- Cadmium Discharges Directive (83/513/EEC);
- Mercury Directive (84/156/EEC);
- Hexachlorcyclohexane Discharges Directive (84/491/EEC);
- Dangerous Substance Discharges Directive (86/280/EEC).

A new Directive is being prepared according to Art. 16 (7) of the WFD, on Environmental Quality Standards applicable to the concentrations of the priority substances in surface water, sediments and biota.

The adoption of the EU Marine Strategy (COM(2002)) and the recent suggestion on the need for an accompanying Marine Framework Directive will apply the integrated ecosystem management philosophies from terrestrial and freshwater areas to estuaries, coasts and open sea.

To conclude, a number of international agreements containing important obligations on the protection of marine waters from pollution, such as the Helsinki Convention, 1992, the Paris Convention, 1992, the Barcelona Convention, 1976 and the Athens Protocol, 1980, have to be mentioned.

5. APPROACHES AND TOOLS FOR WATER QUALITY CONTROL

Classification criteria are needed to transform monitoring into assessment.

Since the WFD establishes general classification criteria, it is straightforward to refer to them when discussing on monitoring and control.

Actually, this is one of the most critical issues of the Directive: on the one hand, for the first time the focus is not on chemistry but on biological quality elements; on the other hand, the classification rules are defined in general terms and are left to be defined in details only during the Common Implementation Strategy (CIS) and on a case-by-case basis.

As reported above, the declared objective of the WFD is to reach a good chemical and ecological status by 2015.

"Good chemical status" is quite easily defined as the concentration of specific pollutants that do not exceed specific levels. In details, the chemical status is evaluated by considering a limited number of substances:

- a list of 33 Priority Substances (Directive 2000/60/EC: Art. 16 (7), Annex X; COM(2001));
- substances with standards established in other EU legislation (Annex IX: Mercury Discharges Directive (82/176/EEC); Cadmium Discharges Directive (83/513/EEC); Mercury Directive (84/156/EEC); Hexachlorcyclohexane Discharges Directive (84/491/EEC); Dangerous Substance Discharges Directive (86/280/EEC)).

The "Ecological status" expresses the quality of the structure and the functioning of aquatic ecosystems; its overall definition is reported in Annex 5, Section 1.4.1:

> ...ii) In order to ensure comparability of such monitoring systems, the results of the systems operated by each Member State shall be expressed as Ecological Quality Ratios (EQR) for the purposes of classification of ecological status. These ratios shall represent the relationship between the values of the biological parameters observed for a given body of surface water and the values for these parameters in the reference conditions applicable to that body. The ratio shall be expressed as a numerical value between zero and one, with high ecological status represented by values close to one and bad ecological status by values close to zero.

The classification is based on biological parameters; chemical and hydro-morphological parameters should be used as support parameters in defining the EQR, according to a predefined hierarchical process (European Commission, 2003). In order to properly classify water bodies, a number of quality elements (QE) are defined and will be presented in details further on.

The classification process can be summarized through the following steps:

- characterization of surface water types, according to one of the two alternative systems (A and B) described in Annex II;
- definition of type-specific Reference Conditions (for all Quality Elements);
- collection of monitoring data available (all Quality Elements, with particular reference to biological parameters);
- classification of EQR for each water body (for all Quality Elements);
- agreement on the definitions contained in the normative: 'slight', 'moderate', 'major' and 'severe' deviations from minimally disturbed conditions, through inter-calibration.

However, the above process is not straightforward, as it faces a number of challenging issues:

- scarcity of data on pressures and impacts, a definition for the significant pressures, relation between pressures and impacts, baseline scenarios before estimating the forecasted impacts, the 2015 objectives to assess the risk of failure;
- data on Reference Conditions (RC) are a prerequisite for assigning ecologically relevant typology;
- need to start monitoring potential RC sites before general monitoring programmes are operational;
- need for monitoring data from inter-calibration sites for calculating EQRs;
- evaluation of the testing and review of guidance was too late for the 2005 reporting of status (Article V reporting);
- typology, reference conditions and class boundaries are not yet available;
- finishing inter-calibration exercise before monitoring programmes are operational.

Based on the above classification criterion, the objective of monitoring within the WFD is to permit the classification of all surface water bodies in one of five classes.

For surface waters, monitoring programmes should cover:

- the volume and level or rate of flow to the extent relevant for ecological and chemical status and ecological potential;
- the ecological and chemical status and ecological potential.

For groundwaters, such programmes should cover monitoring of the chemical and quantitative status.

When defining these general rules, the Directive recognizes the impossibility and inopportuneness to define too stringent rules for the monitoring programme, to be applied always and everywhere; it therefore provides, directly or through the Guidance Documents produced within the Common Implementation Strategy (CIS), a series of guidelines only. Each Member State and especially the technicians operating in the different River Basin Districts are in charge of finding customized optimal solutions. The definition of the level of precision and confidence required from monitoring programmes and status assessments is left to the expert judgement of the operators.

Under the WFD a surveillance monitoring programme is the basis for the classification of a water body. This type of monitoring must include all Quality Elements (biological, hydromorphological and all general and specific physico-chemical quality elements); it must have a minimum duration of 1 year and is expected to strongly support the definition of the first River Basin Management Plan (RBMP). Afterwards, the surveillance monitoring, very demanding and expensive, can be repeated only every 18 years (i.e. every 3 RBMP), when the water body has reached the good status and there is no evidence that impacts on that body have changed.

The operational monitoring should allow following, in the most focused way, the temporal trend of the quality status, which, due to the progressive application of RBMP measures, is expected to be improving. In particular, the operational monitoring should be applied to those water bodies that are at risk of not reaching the good ecological and chemical status by 2015.

The parameters used for operational monitoring should be those indicative of the biological and hydromorphological quality elements most sensitive to the pressures to which the water body is subject, and all priority substances discharged, as well as other substances discharged in significant amount. In other words, operational monitoring must use parameters relevant to the assessment of the effects of the pressures that put the water body at risk.

For example, if organic pollution is a significant pressure on a river, then the benthic invertebrates might be considered as the most sensitive indicator, whereas aquatic flora and fish population may not need to be monitored.

If a water body is not identified as being at risk (because of discharges of priority substances or other pollutants), operational monitoring for these substances is not necessary.

Frequencies of operational monitoring should not exceed those indicated for the different quality elements in Tab. 1. More generally, frequencies shall be selected taking into account the parameters variability due to natural and anthropogenic factors and aiming at reflecting, in particular, those changes due to anthropogenic pressures.

Investigative monitoring should be activated where the mechanisms that determine an unsatisfactory chemical and ecological status are not fully understood; in this case, specific investigation is required.

These situations are more common than one can think, given the complexity of ecosystems and the relationships between the matrices water, sediment and biota.

Therefore, it is foreseen that in many cases surveillance and operational monitoring will be complemented by investigative monitoring activities to better understand ongoing processes and existing cause-effects relationships.

It is not possible here to deeply investigate the techniques to adopt in carrying out monitoring programmes. Nevertheless, it is worth underlining a specific aspect that appears of some importance, particularly on certain applications and environments.

It is the coupling between traditional measuring approaches, (that require the presence of operators and, most of the times, the collection of samples for subsequent laboratory analysis) and approaches based on automatic monitoring techniques.

These types of approach are of particular relevance when the water body to be monitored is characterized by great variability in time and when short-term events may play an important role in determining the overall quality of the environment as well as the comprehension of its functioning.

This could be the case of river flood events and their contribution to the total mass transport to the receiving water body; storm events and their role on the littoral sediment transport; anoxic and dystrophic events and their effect on benthic communities.

Table 1. Frequencies suggested for operational monitoring (Directive 2000/60/EC: Annex V)

Quality Element	River	Lake	Transitional	Coastal
Biological				
Phytoplankton	6 months	6 months	6 months	6 months
Other aquatic flora	3 years	3 years	3 years	3 years
Macroinvertebrates	3 years	3 years	3 years	3 years
Fish	3 years	3 years	3 years	
Hydromorphological				
Continuity	6 years			
Hydrology	Continuous	1 month		
Morphology	6 years	6 years	6 years	6 years
Physico-chemical				
Thermal conditions	3 months	3 months	3 months	3 months
Oxygenation	3 months	3 months	3 months	3 months
Salinity	3 months	3 months	3 months	
Nutrient status	3 months	3 months	3 months	3 months
Acidification status	3 months	3 months		
Other pollutants	3 months	3 months	3 months	3 months
Priority substances	1 month	1 month	1 month	1 month

The technology presently available allows to measure automatically most physical parameters (e.g. levels, flows, waves, currents, etc.), a number of physico-chemical parameters (e.g. salinity, dissolved oxygen, nutrients, etc.), and a few biological parameters (e.g. chlorophyll and phytoplankton, zooplankton, etc.). However, it must be considered that the relatively long time scales typical of the vegetable and animal communities make, in most cases, their monitoring through automatic instruments, less important if compared with hydromorphological and physico-chemical parameters.

Innovative automatic techniques include passive samplers, portable Gas Chromatography/Mass Spectrometer (GC/MS) for organics, in situ biosensors, Biologically Early Warning Systems (BEWS), based on living organisms and the measurement of an end-point related to acute or chronic effects.

More and more, automatic monitoring techniques are being used and integrated to build up complete observatories at local, basin and continental scale.

6. QUALITY ELEMENTS AND PRIORITY POLLUTANTS

The classification of water bodies is based on biological quality elements, while hydromorphological and physico-chemical parameters are used to support the evaluation process, recognizing the fact that biological communities are products of their physical and chemical environment. The use of non-biological indicators to assess the condition of a biological quality element may be a complement to the use of biological indicators, but cannot replace it.

This position, so strongly oriented toward biological and ecological aspects, represents a real revolution if compared to past approaches; it is also a strong distinctive element if compared to the approaches used in other continents, North America *in primis*.

Quality Elements for transitional waters are shown in details in Fig. 1; they are summarized in Tab. 2 when referring to different water categories, as far as biological parameters are concerned.

Biological elements include: composition, abundance and biomass of phytoplankton; composition and abundance of other aquatic flora, benthic inverte-brate fauna and fish fauna.

Hydro-morphological elements supporting the biological elements include: morphological conditions (depth variation; quantity, structure and substrate of the bed; structure of the intertidal zone) and tidal regime (freshwater flow; wave exposure).

Chemical and physico-chemical elements supporting the biological elements include: some general parameters (transparency; thermal conditions; oxygenation conditions; salinity; nutrient conditions), specific synthetic and non-synthetic pollutants (Priority Substances identified as being discharged into the body of water and other substances identified as being discharged in significant quantities into the body of water).

Within the WFD a list of priority chemicals has been defined (WFD: Annex X and COM(2001). Chemicals on this list are of primary concern for European waters; therefore, Member States will be obliged to monitor these compounds in all water bodies. The list includes 33 compounds and compound groups and distinguishes between "Priority Dangerous Substances" and "Priority Hazardous Substances", for which more restrictive actions are needed (Tab. 3).

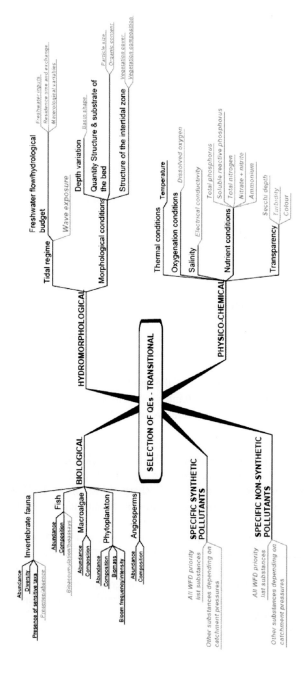

Figure 1. Selection of Quality Elements for transitional waters (European Commission, 2003).

Table 2. Biological quality elements to be considered in different surface water bodies

	Rivers	Lakes	Estuaries	Coastal waters
Phytoplancton		X	X	X
Macrophytes	X	X	X	X
Zoobenthos	X	X	X	X
Fish	X	X	X	

Notable omissions in the list are compounds like polychlorobyphenils and dioxins/furans or emerging chemicals such as pharmaceuticals and personal care products, Methyl Tert Butyl Ether (MTBE) and related compounds, surfactants and their recalcitrant metabolites, endocrine disruptors.

The Priority Substance list will have to be revised 4 years after its entering in force. The first review is expected in correspondence with the Directive defining Environmental Quality Standards.

Priority Substances are the key factor for determining the chemical status of a water body, but this requires the availability of Quality Standards on water, sediment and biota, that presently have not been fixed yet and are under definition by the EU through a specific Directive.

Quality Standards will probably be distinct for inland, transitional and coastal waters and will include "chronic" ("Annual average environmental quality standard" -AA-EQS) and "acute" ("Maximum allowable concentration environmental quality standard"-MAC-EQS) quality standards. At this stage, the Commission envisages to present quality standards for the water phase only, despite the prescriptions included in art.16 (7) and experts' advices on this matter (CSTEE, 2004).

The difficulty of the Commission in providing an official definition of Quality Standards is well comprehensible: defining too restrictive standards based on the precautionary principle could generate *de facto* unreachable or too expensive objectives, while on the other side too high standards could be underprotective and cause an unacceptable number of false negatives. Furthermore, in many cases we deal with substances whose hazard and action mechanisms are not fully understood.

Last but not least, the analytical issue must also be considered: measuring all the present and future Priority Substances at very low concentrations in fresh and saltwater raises the problem of having sampling and analytical methods correspondingly precise and accurate, which could be applied routinely by standard laboratories.

The difficulty of defining absolute Quality Standards to be used for assessing the chemical status of water bodies is exemplified by the mercury case study. For this element, standard values suggested and available in literature, adopted in some national law, vary by about 4 orders of magnitude (from 10 pg/l to 100 ng/l), due to the uncertainties surrounding the bioaccumulation factor, i.e. the transformation of inorganic mercury to methyl mercury and the bioaccumulation of methyl mercury in fish. One could argue then that a unique water quality standard for mercury would not be adopted, being not scientifically defendable. Instead, a maximum residue level tolerable in fish for human methyl mercury exposure should be adopted (CTSEE, 2003).

In Italy, the decree n. 367/03 of the Ministry of the Environment defines Quality Standards of Priority Substances and others substances, for waters and sediments. The decree establishes that if those standards cannot be reached in a water body using all the Best Available Technologies (BAT), the Competent Authorities should indicate lowest reachable values, carry out a risk analysis and define possible restrictions on the use of that water body.

The above discussion brings us to the point that Quality Assurance/Quality Control is a key issue in monitoring programmes.

It is very important that sampling and analytical methods standardized and certified by qualified organizations (EN, ISO, national certification organisms) are used. On the contrary, if non-standardized methods are used, they should at least be documented and qualified in details, so that they can be replicated in future and the data produced can be correctly interpreted and properly used.

The laboratories involved in monitoring programmes should follow a programme of QA/QC (EN ISO 17025) and participate regularly in proficiency testing exercises.

Table 3. Priority Dangerous and Hazardous Substances defined under the Decision 2455/2001/EC (COM, 2001)

Priority hazardous substances	Priority hazardous substances under review	Priority dangerous substances (67/548/CE) not proposed as hazardous substances
Brominated diphenil ethers (penta) Cadmium Mercury C10-C13 Chloroalkanes Hexachlorobenzenes Hexachlorocyclohexane Tributhyltin Hexachlorobutadiene Nonylphenols PAH's Pentachlorobenzene	Anthracene Atrazine Chlorpyrifos Di(2-ethylhexyl)phtalate (DEHP) Endosulfan Lead Naphtalene Octylphenols Pentachlorophenol Trichlorobenzenes Trifuralin	Alachlor Benzene Chlorfenvinphos Dichloromethane 1,2-Dichloroethane Diuron Isoproturon Nickel Simazine Trichloromethane
To be eliminated from the emissions within 20 years.	*Show properties similar to those identified as priority substances; subject to a review for identification as possible priority hazardous substances.*	*Do not fulfil the criteria for being "toxic, persistent and liable to bioaccumulate". Classified as dangerous. Subject to emission control and quality standards.*

The inter-calibration phase of the CIS should also contribute to the definition and development, if necessary, of standard methods and procedures to be used in routine sampling and analysis. Today, the need for standardization of methods and certification of laboratories is particularly strong for the biological sector, whereas the physical and chemical sectors appear more advanced as far as this aspect is concerned.

7. DATA MANAGEMENT AND DISSEMINATION

A huge number of monitoring programmes has been carried out and is active in all European Countries and throughout the world. It is worth asking if this incredible amount of data has been fully exploited from a scientific and operational point of view. I would say that the answer is no, for a relevant number of cases.

This depends above all on the inappropriate management of data, which often remain in the drawers of their producers and/or owners or are disseminated and made available to other users in ineffective and incomplete ways.

The motivations for a better data management and diffusion are several and of a different kind:

- the producer of data accounts for the quality of his work especially if he is forced to fully present his data and results;
- data must be used at best to support management decisions, rewarding the motivations that generated their collection;
- data may have different uses, often not foreseen at their collection and coming up with time, that must be safeguarded;
- copyright must be guaranteed, but it cannot be claimed, as it often happens, as a reason for not allowing the use of data other than by their owner;
- duplications and waste of resources due to inaccessibility or simply difficulties in existing data mining and recovery should be avoided;
- when operating on transnational basins and coastal areas, influenced often by environmental driving forces and pressures located very far and spread, it is fundamental that environmental agencies of different countries share their data and, if possible, manage and elaborate them in an integrated way (see, for example, the experience of the HELCOM and OSPAR Conferences);
- in many cases the chance for citizens of controlling decision-makers and environmental managers is based on data transparency;
- a scientifically correct but popular dissemination of data and related environmental information is essential for a growing, informed and diffused environmental culture.

To satisfy the above listed needs, concerted actions on several fronts are required; from technical ones, through the use of the available information technologies (Databases, Geographic Information Systems, Decision Support Systems, the web, etc.) and the definition of standardized formats and procedures, to the ones that could be defined as "political", which are often the real critical element.

In fact, data are often unavailable because the producer/owner, that can be a researcher, a stakeholder, an agency or a state, has some reasons for keeping them reserved or restricted.

At this regard, we could say that this tendency and attitude is more frequent in Europe than in the United States, where probably more value is given to the capacity of using data than to data themselves. However, in the last years, several steps forward have been made in Europe, possibly due also to the reinforcement of the Union and of its Institutions and Agencies. Examples of these steps forward are some national and international centers created in the last years for collecting, standardizing, archiving environmental data, both at wide spectrum and on specific topics (e.g.: EIONET – EU, BODC – UK, SINAnet – IT, etc.).

The WFD attributes great importance to public information and consultation and to the active involvement of all the parties directly or indirectly interested in the production, review and updating of the RBMP (art. 14).

In particular, Member States should make the following documents available, allowing at least six months for comments:

- a timetable and work programme for the production of the RBMP, including a statement of the consultation measures to be taken;
- an overview of the significant water management issues identified in the river basin;
- draft copies of the RBMP.

The WFD requires that processed data produced by monitoring programmes being part of the RBMP (Annex VII), must be transmitted to the Commission at their first emission and at their updates every 3 years.

As far as monitoring is concerned, the following must be at least reported in the RBMP:

- maps of the monitoring networks;
- maps of water status;
- estimates of the confidence and precision attained by the monitoring systems.

Andrea Barbanti
ICRAM, Rome, Italy

REFERENCES

COM, 2001. Decision n.2455/2001/EC of the European Parliament and of the Council of 20 November 2001 establishing the list of priority substances in the field of water policy and amending Directive 2000/60/EC. *Official Journal of the European Communities* L 331, 15/12/2001, pp. 1-5.

COM, 2002. Communication from the Commission to the Council and the European Parliament: Towards a strategy to protect and conserve the marine environment. Brussels: Commission of the European Communities. *Report* nr. COM(2002) 539 final.

CSTEE, 2004. Opinion of the Scientific Committee on Toxicity, Ecotoxicity and the Environment on "The setting of Environmental Quality Standards for the Priority Substances in Annex X of Directive

2000/60/EC in accordance with Article 16 thereof". Brussels, C7/GF/csteeop/WFD/280504 D(04), p. 32.
Directive 76/464/EC. Council Directive 76/464/EEC of 4 May 1976 on pollution caused by certain dangerous substances discharged into the aquatic environment of the Community. *Official Journal of the European Communities* L 129, 18/05/1976, pp. 23-29.
Directive 79/409/EC. Council Directive 79/409/EC of 2 April 1979 on the conservation of the wild birds. *Official Journal of the European Communities* L 103, 25/4/1979, pp. 1-18.
Directive 82/176/EEC. Council Directive 82/176/EEC of 22 March 1982 on limit values and quality objectives for mercury discharges by the chlor-alkali electrolysis industry. *Official Journal of the European Communities* L 81, 27/03/1982, pp. 29-34.
Directive 83/513/EEC. Council Directive 83/513/EEC of 26 September 1983 on limit values and quality objectives for cadmium discharges. *Official Journal of the European Communities* L 291, 24/10/1983, pp. 1-8.
Directive 84/156/EEC. Council Directive 84/156/EEC of 8 March 1984 on limit values and quality objectives for mercury discharges other than the chlor-alkali electrolysis industry. *Official Journal of the European Communities* L 74, 17/03/1984, pp. 49-54.
Directive 84/491/EEC. Council Directive 84/491/EEC of 9 October 1984 on limit values and quality objectives for discharges of hexachlorocyclohexane. *Official Journal of the European Communities* L 274, 17/10/1984, pp. 11-17.
Directive 86/280/EEC. Council Directive 86/280/EEC of 12 June 1986 on limit values and quality objectives for discharges of certain dangerous substances included in List I of the Annex to Directive 76/464/EEC as amended by Council Directive 88/347/EEC and Council Directive 90/415/EEC. *Official Journal of the European Communities* L 181, 04/07/1986, pp. 16-27.
Directive 92/43/EC. Council Directive 92/43/EEC of 21 May 1992 on the conservation of natural habitats and of wild fauna and flora. *Official Journal of the European Communities* L 206, 22.7.1992, pp. 7-50.
Directive 98/83/EC. Council Directive 98/83/EC of 3 November 1998 on the quality of water intended for human consumption. *Official Journal of the European Communities* L 330, 05/12/1998, pp. 32-54.
Directive 2000/60/EC. Directive 2000/60/EC of the European Parliament and of the Council of 23 October 2000 establishing a framework for Community action in the field of water policy. *Official Journal of the European Communities* L 327, 22/12/2000, pp. 1-73.
Directive 2006/7/EC. Directive 2006/7/EC of the European Parliament and of the Council of 15 February 2006 concerning the management of bathing water quality and repealing Directive 76/160/EEC. *Official Journal of the European Communities* L 64, 04/03/2006, pp. 37-51.
European Commission, 2003. Monitoring under the Water Framework Directive. *Guidance Document* No. 7, p. 153.

F. CECCHI, P. BATTISTONI AND P. PAVAN

NEW APPROACHES TO WASTEWATER TREATMENT PLANTS IN ITALY

Abstract. The chapter deals with some new experienced processes in the fields of civil wastewater treatments and organic wastes. These processes were studied by the authors and designed for the full scale application by Ingegneria Ambiente s.r.l. The following topics are reported: the integrated treatment of organic fraction of municipal solid waste (OFMSW) and wastewater, the reclamation of phosphorous as struvite salt (anaerobic fermentation, biological nutrient removal struvite crystallization process: AF, BNR, SCP), a simple and cheap on line signal (dissolved oxygen and oxidation reduction potential: DO, ORP) for the automatic and remote management of nitrification/denitrification processes, the Alternate Cycles Process (ACP) joint to the ultrafiltration membrane (MBR) to reach levels of water quality high enough for the reuse purpose. Furthermore the same process approach is illustrated as a pre-treatment for the micro-pollutants removal. All the situations taken into consideration are completed by some figures on the investment and management costs.

1. INTRODUCTION

Wastewater treatment facilities have gone through a massive development from the early 1900s until now and they are still subject to continuous improvement and optimization in order to respond to new treatment objectives and law requirements. In particular, the new regulations (EC 91/271, 91/676) require more stringent limits about carbon, nutrients, suspended solids and pathogenic organisms. More attention is paid to the control of xenobiotic compounds, the safe guard of water bodies and to the water reuse in general. Tab. 1 reports a brief summary of the system evolution in the last years in Italy.

Table 1. Law evolution, pollutant control objectives and type of wastewater treatment in Italy

Period	Law	Objectives	Treatment
1900-1970s	-	Suspended materials, carbon and pathogens	Activated sludge process
1970s to early 1990s	319/76 First important national law	Suspended solids, carbon, nutrients and pathogens	Activated sludge, nitrification/ denitrification
1990s	152/99	Stringent limits for nutrients	Biological nutrient removal
2003	185/2003	Reclaim and reuse	Advanced technologies (membrane bioreactors)
2003	367/2003	Micropollutants control	

C. Clini et al. (eds.), Sustainable Development and Environmental Management: Experiences and Case Studies, 99-111.
© 2008 *Springer.*

According to these concerns, the first aim of this chapter is to summarily explain how to promote nutrients biological removal from wastewaters using the interactions between wastewater treatment and fermentation of the organic fraction of municipal solid waste (Cecchi *et al.*, 1994; Llabrés *et al.*, 1999; Pavan *et al.*, 2000). That is, the diversion of the refuse fraction from landfill and its use as energy source for biogas production and cogeneration, and, finally, the recovery of a non-renewable resource: phosphorus itself (Battistoni *et al.*, 1997). In the frame of the resources reclamation a low energy cost process for wastewater treatment is then presented. This goal can be reached only when the nitrogen removal goes up to 80-90%; that is, a nearly complete recovery of the oxygen bound to nitrogen in nitrates. This target is linked to adequate plant and a reliable process control. The use of the continuously fed alternate process (the so called Alternate Cycles Process ACP) allows an effective nitrogen removal thanks to two different cycles: the first, the aerobic one, achieves the complete ammonia oxidation, whereas the second, the anoxic one, allows the nearly total nitrates denitrification. The automatic control of the process may be performed according to typical profiles of the dissolved oxygen and the oxidation-reduction potential within the process cycles. This approach is illustrated in the second section of the chapter.

Increasing needs for wastewater reclamation and reuse call for further developments of conventional activated sludge processes in the field of wastewater treatment. Among the available technologies, the membrane biological reactors (MBR) are supposedly the best choice for the improvement of conventional processes. It allows the enhancement of biochemical processes and the production of a permeate free of suspended solids; this results in the removal of micropollutants adsorbed on particulate matter and the direct reuse of the treated water for non potable purposes. Although several advantages come from the adoption of the MBR technology compared to the conventional activated sludge processes (CASPs), the MBR technology is characterized by increased energy consumptions. Hence the choice of an "energy saving" process, such as the alternate cycles®, is of particular importance since the biological process operating in the reactor drastically influences both the treatment performances and the managing costs. This will be the topic of the third section in this chapter, underlining the effectiveness of the MBR in the micropollutants control.

2. THE PROCESSES: BASIC BALANCES AND DESIGN

2.1. The AF-BNR-SCP processes

The integrated wastewater and OFMSW treatment system proposed by Cecchi *et al.* (1994) involves a combination of anaerobic fermentation (AF), biological nutrients removal (BNR) and struvite crystallization processes (SCP) (AF-BNR-SCP) as shown in Fig. 1.a. The process can also operate without the fermentation step; that is with the only co-digestion process of OFMSW and sewage sludge after the sorting line of the refuses. In this case the energy recovery is maximized (see Fig. 1.b).

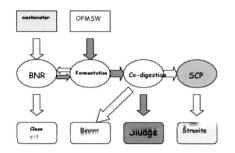

Figure 1a. Scheme of the integrated wastewater treatment system using OFMSW fermentate as BNR promoter (AF-BNR-SCP) (Cecchi et al., 1994).

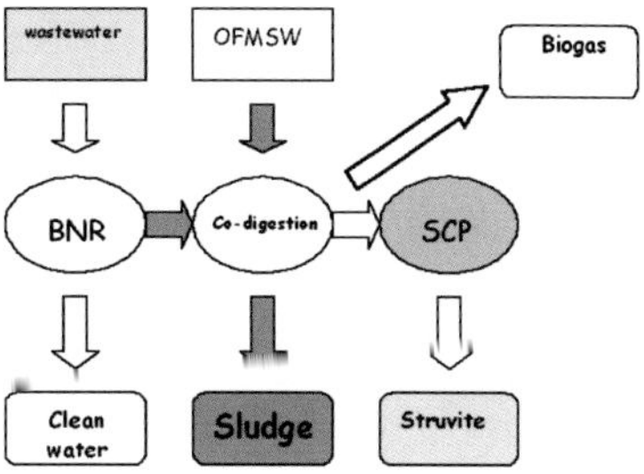

Figure 1b. Scheme of the integrated wastewater treatment system maximizing energy recovery (BNR-CoD-SCP).

The single processes characterizing the whole approach are:

a) anaerobic acid fermentation (AF) of the organic fraction of municipal solid waste (OFMSW) in mesophilic conditions (T = 35-37°C);
b) anaerobic co-digestion of OFMSW and waste activated sludge (WAS) under mesophilic (T = 35°C) or thermophilic conditions (T = 55°C);
c) biological nutrient removal (BNR) wastewater treatment for carbon, nitrogen and phosphorus control;
d) phosphorus and ammonia recovery from supernatants by struvite and/or hydroxyapatite production by crystallization (SCP).

In synthesis, the goals of the integrated process are:

- to reduce and stabilize the organic matter in the OFMSW together with the waste activated sludge in wastewater treatment plant. Since the OFMSW is mainly composed by water (more than 80%), it seems to be the simplest and most logical approach;
- to allow high performances in BNR process by using easily biodegradable carbon source from fermentation of OFMSW;
- to perform the energy production from renewable sources: the OFMSW;
- to perform the recovery of a non-renewable resource like phosphorus into a usable form.

2.2. The mass balance of the AF-BNR-SCP processes

According to the two schemes, OFMSW fermentation followed by anaerobic co-digestion of solid phase (AF-BNR-SCP) and OFMSW co-digestion alone (AF-CoD-

SCP), the mass balance process, can be evaluated as follows, on a wet weight basis and considering 250 l of wastewater discharged and 0.3 kg of OFMSW produced per capita per day:

AF-BNR-SCP:
- 20-30% rejects from sorting line
- 35-40% solid phase to co-digestion
- 35-40% liquid phase to denitrification/Phosporous removal

AF-CoD-SCP:
- 20-30% rejects from sorting line
- 70-80% to co-digestion

In Tab. 2 the mass balance of the BNR section in terms of carbon and nutrient flow mass is reported. This is based on literature data (Battistoni *et al.*, 1998) for medium strength untreated domestic wastewater characteristics. Tab. 3 shows the mass balance of co-digestion of residual solid from OFMSW fermentation and waste activated sludge (WAS).

Table 2. Mass balance of the BNR section on a Carbon-Nitrogen-Phosphorous base

	Flow	C, %	N, %	P, %
IN	Sewage inlet	88	98	83
	OFMSW fermentation effluent	12	2	17
OUT	Gases	60	59	0
	Waste sludge	34	39	67
	Water outlet	6	2	33

Table 3. Mass balance of the anaerobic co-digestion step of the AF-BNR-SCP on a TVS basis

	Flow	TVS, %
IN	OFMW fermentation residue	27
	Waste activated sludge from BNR plant	73
OUT	Biogas from sludge	8
	Biogas from OFMSW	22
	Sludge	70

When the only co-digestion approach is considered the mass balance is that represented in Tab. 4.

Table 4. Mass balance of the anaerobic co-digestion step of the BNR-CoD-SCP process on a TVS basis

	Flow	TVS, %
IN	OFMW	51
	Waste activated sludge from BNR plant	49
OUT	Biogas from sludge	5
	Biogas from OFMSW	40
	Sludge	55

A N,P mass balance can be evaluated also for struvite crystallization (SCP): the results are reported in Tab. 5.

Table 5. Mass balance of the Struvite Crystallization Process (SCP) on a Carbon-Nitrogen-Phosphorous base

		Flow	N, %	P, %
IN	Supernatant influent		100	100
OUT	Stripped		3	0
	Struvite		10-12	70-80
	Sludge		85-87	20-30

2.3. Economics for the integrated process

An economical comparison of a conventional C-N process and the AF-BNR-SCP, considering both configurations (co-digestion only and fermentated addition in denitrification step) is shown in Tab. 6. The comparison is given only on the basis of the non common costs for the processes, assuming that the other ones are the same. The costs are on a daily basis, considering 100.000 AE basin, 250 l/AE per day and 300 g OFMSW/AE per day. The amortization costs are evaluated on a 3 year basis, and considering the opportunity of the 'green certificates' given by the Italian Government for the energy produced using renewable sources (108.92 Euro/MWh in 2005).

Table 6. Economic comparison between C-N and AF-BNR-SCP approaches (costs in Euro/d)

	C-N	BNR-CoD-SCP	AF-BNR-SCP
OFMSW disposal	-	-1680	-1680
Chemicals	486	-	-
Heat production	38	-684	-280
Electricity production	-	-1206	-661
Sludge disposal	2064	2964	3144
Additional personnel	-	219	219
Amortization	-	2024	2024
Total	2587	1796	2766
From 4th year	2587	-227	742

The advantages derived from the adoption of these approaches are clear: after the amortization period of 3 years, about 2800 Euro/d can be recovered in respect to the C-N process by using the BNR-CoD-SCP approach, and about 1800 Euro/d by using the AF-BNR-SCP approach.

2.4. Case study: Treviso city wastewater treatment plant

The process scheme of the new wastewater treatment plant of Treviso city is reported in Fig. 2: it is possible to see the general scheme of AF-BNR-SCP in the application of the concept from a straight – line logic (solid/liquid waste → final disposal) to a circular logic (solid/liquid waste → treatments integration → resources' recovery → final disposal).

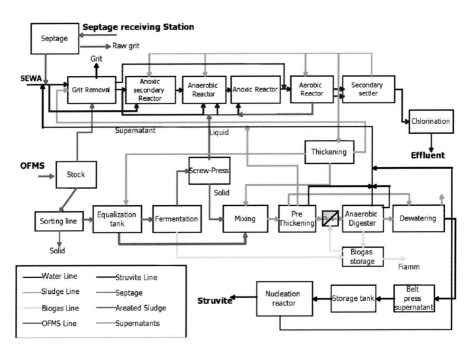

Figure 2. Flow-sheet of the new plant of Treviso city.

2.4.1. Results of two years of management

A complete analysis of the performance of the sorting section and of the whole process was carried out, also considering the fractioning of each effluent stream from the line. The sorting line performance is briefly summarized in Tab. 7. The operational conditions adopted and the yields obtained in the Treviso plant are reported in Tab. 8. The main cost results are reported in Tab. 9 and 10.

Table 7. Efficiency of the sorting line adopted in the Treviso plant (Patent RN20044A000003)

Fraction	Removal
Plastics	90%
Metals	99%
Organics (as capture percentage)	76%

Table 8. Digester operational conditions and yields in co-digestion with the pre-treated OFMSW obtained in the Treviso plant

Parameter	Unit	Sludge only	Co-digestion
T	°C	36.5	36.3
HRT	d	27	22.0
Total OLR	kgTVS/m^{3}d	0.53	0.78
GP	m^3/month	4500	19500
CH$_4$	%	65.6	64.7

Table 9. Management cost fractions of the whole Treviso plant

Plant section	Management cost, %
Wastewater line	47
Sludge line	17
Septage treatment	21
OFMSW treatment	14
SCP	1

Table 10. Management cost fractions of OFMSW treatment area

Fraction	Management cost, %
Sludge disposal	25
Sorting line refuses	23
Personnel	30
Maintenance	6
Energy	7
General	9

It was also observed that, when the capacity of 50 tons/week is reached, the final disposal cost of OFMSW is about 50 Euros/ton. This cost can be lowered under 40 Euros/ton. Furthermore, an economic income of 90-95 Euros/ton of OFMSW treated is recoverable from the government support for energy production from renewable sources. Investments costs can be evaluated in 20-30 Euros/EI.

3. THE ACP

The alternate cycles process (ACP) was studied by the authors and applied by Ingegneria Ambiente s.r.l. in a network of small and medium size Italian plants for municipal wastewater treatment. This section deals with the ACP process ranging from the theoretical fundamentals to the more practical aspects of the full scale applications.

3.1. Theoretical fundamentals and real applications

Besides the more conventional biological processes performed in separated tanks (Fig. 3a), a time succession of nitrification (aerobic) and denitrification (anoxic) phases can be adopted to obtain high C and N removals in the same bioreactor (Fig. 3b).

Figure 3a. Conventional biological process.   *Figure 3b. Succession of nitrification and denitrification phases.*

The behaviour of the nitrogen formed in the bioreactor can be used to determine the optimal time lengths of the phases. In particular, the ammonia disappearing during the aerobic phase is identified by a flex point on the pattern of the dissolved oxygen (DO) versus time or the oxidation reduction potential (ORP) profile. The nitrates disappearing (breakpoint) in the anoxic phase are identified by a flex point in the ORP profile versus time at the end of the denitrification phase.

In the alternate cycles process the time sequence of aerobic and anoxic phases in the same continuously fed reactor is managed by a patented expert control device (patent N. RM99A0000182.6.99,1999) which operates an on-line analysis of DO and ORP signals detecting the above mentioned bending points (Battistoni *et al.*, 2003; 2004).

A simplified mathematical model has been used to predict both nitrates and ammonia variations inside the reactor or in the effluent when the ACP is applied. The assumptions of this model are:

- zero order kinetics for nitrification and denitrification rates;
- negligible influent nitrates;
- total influent nitrogen approximated as ammonia;
- autothrophs and air supply able to nitrify all the nitrifiable loadings.

Therefore a simplified solution for ammonia and nitrates mass balance during oxic and anoxic phases can be found (Fig. 4a). Since the nitrification and denitrification kinetic values remain constant, the time cycle depends only on the volume of the reactor and on the physical chemical characteristics of the influent.

Moreover the model mainly confirms that the ORP and DO signals are good tools to control the process since the trend of nitrogen compounds is similar to the one observed in sequencing batch reactors. The automatic on-line control of both nitrification and denitrification phases involves the best process flexibility with respect to the influent loadings (Battistoni *et al.*, 2003). In fact, the control device is able to change the anoxic and oxic phases time lengths depending on the influent carbon and nitrogen hourly fluctuations (Fig. 4b) or on the dry and wet weather periods (Fatone *et al.*, 2005).

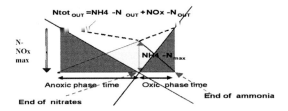

Figure 4a. Theoretic profiles of N-NOx and NH$_4$-N.

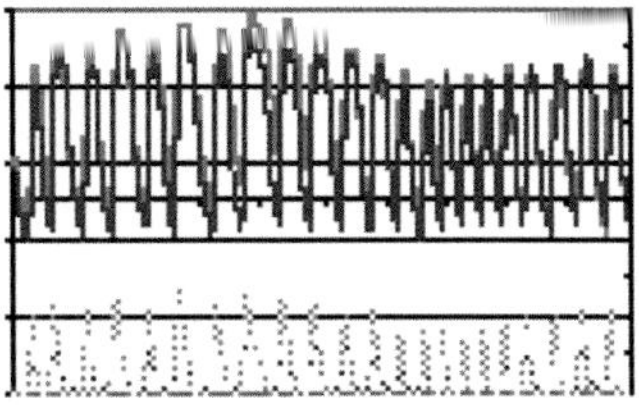

Figure 4b. Real trends of DO and ORP in the ACP process.

3.2. Mass balance and energy savings

According to the process fundamentals and the reliability of the automatic control device, the mass balances of the process in terms of carbon and nitrogen removal performances are always optimized (Tab. 11). The high denitrification efficiency of the ACP allows for the best exploitation of the nitrates bound oxygen for the carbon oxidation. Consequently relevant savings for the air supply are obtained.

Besides the optimization of the air supply, ACP also involves lower flows to be recycled due to the absence of mixed liquor recycle. This aspect avoids to pump back a flowrate 2-3 times influent. As overall result, the ACP energy consumptions are 20-30% lower than the Conventional Activated Sludge Processes (CASPs).

Table 11. Mass balance of ACP on C,N,P basis

	Flow	C, %	N, %	P, %
IN	Sewage inlet	100	100	100
OUT	Gases	62	80	0
	Waste sludge	35	19	50
	Water outlet	3	1	50

3.2.1. Managing costs in full scale applications

At the moment, the ACP is operating in a number of Waste Water Treatment Plants WWTPs with potentiality ranging from 700 to 30,000 EI, while other plants up to 100,000 EI have been designed.

Besides the discussed advantages deriving from the high performances and low energy consumptions, if compared to the conventional activated sludge process (CASP), the ACP allows also cost saving thanks to:

- lower specific reactor volumes required (100-130 versus 150-200 $l_{reactor}$/EI);
- low manpower for small and decentralized plants thanks to a cheap local and remote control of the process;
- low investments in case of plant upgrading thanks to the best recovery of the existing structures.

Tab. 12 shows the managing costs concerning a real plant that was operating a conventional process firstly and then was upgraded to the ACP.

Table 12. Managing costs comparison: conventional extended aeration Vs ACP ([a]Two skilled workers, each employed 2.5 hours per day and 5 days per week; [b]Two skilled workers, each employed 3 hours per week; [c]Lab Analysis, maintenance, chemicals)

	Conventional		ACP	
	€ year^{-1}	€ m^{-3}	€ year^{-1}	€ m^{-3}
Manpower	26780[a]	0.19[a]	6500[b]	0.03[b]
Ordinary maintenance[c]	6320	0.05	4750	0.03
Energy Consumptions	21560	0.15	20820	0.09
Waste sludge disposal in landfill	9400	0.07	11000	0.05
Total Managing Costs	64060	0.46	41470	0.20
	Conventional		ACP	
	€ year^{-1}	€ m^{-3}	€ year^{-1}	€ m^{-3}
Manpower	26780[a]	0.19[a]	6500[b]	0.03[b]
Ordinary maintenance[c]	6320	0.05	4750	0.03
Energy Consumptions	21560	0.15	20820	0.09
Waste sludge disposal in landfill	9400	0.07	11000	0.05
Total Managing Costs	64060	0.46	41470	0.20

The yearly managing costs decreased of some 30% while, considering the values per treated water, the cost saving was about 60%. These results came mainly from the remote process control (involving less manpower) and the energy savings.

4. THE ACP-MBR

The xenobiotics (heavy metals and persistent organic compounds) control is the next challenge in the field of wastewater treatment. Since the adsorption on particulate fractions is the main xenobiotics vehicle, the membrane technology can optimize the removal of both conventional pollutants and xenobiotics, especially when coupled with a very energetically effective biological process. Therefore the ACP was firstly applied by the authors to a demonstrative bioreactor (MBR) located in Treviso. Subsequently, Ingegneria Ambiente s.r.l. designed and monitored the first operating year of a full scale ACP-MBR located in the Viareggio WWTP (Battistoni et al., 2006).

4.1. The ACP-MBR: a case study

Since the Viareggio municipality called for wastewater reuse for agricultural purposes, the existing municipal WWTP was upgraded to produce reusable water up to 6000 m^3/day^{-1}. Realizing the best recovery of the existing structures, the water line was modified as follows:

- converting primary clarifier to ACP;
- coupling ultra-filtration with ACP section.

After one year experience, the plant confirmed the high removal efficiencies already pointed out by the demonstrative application on both conventional pollutants and xenobiotics. Tab. 13 shows these performances also in terms of larger abatements with respect to the conventional process.

Table 13. ACP-MBR macro-pollutants removals

	Removal (%)		Removal (%)	(MBR-CASP) (%)
COD	94-96	Arsenic	40-45	+20-30
N	85-90	Other metals	74-97	+10-50
P^d	52-58	Organic micropollutants	66-96	+10-40
SS	> 99			

^{d}obtained without any chemical/physical treatment

The energy consumptions of the ultrafiltration section depended upon the solids concentration operated in the bioreactor (Tab. 14).

Table 14. Energy Specific Consumptions of the ACP-MBR

Treatment Unit	kWh m^{-3} (case A)	kWh m^{-3}(case B)
First Pumping and Pre-treatments	0.13	0.13
AC Reactor	0.13	0.12
UF Section	0.20	0.16
Total ACP/MBR	0.46	0.41
CASP	0.36^e	

$^{case\ A}$ MLSS$_{ACPtank}$ = 8 kg m^{-3} $^{case\ B}$ MLSS$_{ACPtank}$ = 6 kg m^{-3} ^{e} Burton (1996)

Furthermore the specific energy consumptions were only 15÷30% higher than the CASP thanks to the coupling of the ultrafiltration with the ACP energy saving process.

5. CONCLUSIONS

As a conclusion, considering the approaches presented and the data collected, the main benefits of the AF-BNR-SCP process are:

- the recovery of renewable energy by methanization of the carbonaceous substrates (solid fraction from the OFMSW fermentation and waste activated sludge with or without primary sludge);
- the very efficient removal of nitrogen to gaseous N_2 through assisted denitrification step using RBCOD from OFMSW fermentation;
- the reclamation of phosphorus and ammonia for agricultural purposes;
- the environmental-friendly disposal of a significant and otherwise difficult to dispose fraction of the MSW;
- the economical reliability of the integrated process thanks to the biogas production that reduces the OFMSW disposal costs;
- the reduction of waste sludge production and the abatement of released nutrients;
- the limited investment cost of the OFMSW treatment in existing digesters. It can be evaluated in 20-30 Euro/EI.

Considering the Alternate Cycle Process alone or coupled with membrane technology, the main benefits can be summarized as follows:

- the recovery of the existing structures through a plant upgrading which implies a low capital investment;
- high carbon and total nitrogen removals either in conventional or membrane technology plants;
- the energy saving up to 20-35%;
- saving of manpower especially for small and decentralized communities;
- with the application of the ACP-MBR process water reuse at sustainable costs especially in highly urbanized areas;
- ACP-MBR reached impressive macropollutants removals also with a specific tank volume of 81 l_{tank} PE^{-1};
- Acceptable specific energy consumptions with ACP-MBR (from 0.41 to 0.46 kWh m^{-3} depending on the operating biomass concentration).

Franco Cecchi
University of Verona, Italy
Ingegneria Ambiente s.r.l., Italy

Paolo Battistoni
University of Ancona, Italy
Ingegneria Ambiente s.r.l., Italy

Paolo Pavan
Ca' Foscari University of Venice, Italy
Ingegneria Ambiente, Venice, Italy

REFERENCES

Battistoni P., Fava G., Pavan P., Musacco A. and Cecchi F., 1997. Phosphate removal in anaerobic liquors by struvite crystallization without addition of chemicals. Preliminary results. *Water Research*, 11, 2959-2929.
Battistoni P., Pavan P., Cecchi F. and Mata J., 1998. Phosphorous removal in real anaerobic supernatants: modelling and performance of a fluidized bed reactor, *Water Science & Technology*, 38, 1, pp. 275-283.
Battistoni P., De Angelis A., Boccadoro R. and Bolzonella D., 2003. An automatically controlled alternate oxic-anoxic process. A feasible way to perform high nitrogen biological removal also during wet weather periods. *Industrial and Engineering Chemistry Research*, 42(3), pp. 509-515.
Battistoni P., Boccadoro R., Bolzonella D. and Marinelli M., 2004. An alternate oxic-anoxic process automatically controlled. Theory and practice in a real treatment plant network. *Water Science and Technology*. 48(11-12), pp. 337-344.
Battistoni P., Fatone F., Bolzonella D. and Pavan P., 2006. Full scale application of the coupled alternate cycles-membrane bioreactor (ACP-MBR) process for wastewater reclamation and reuse. *Accepted for long platform presentation at the IWA World Water Congress. Beijing (China)* 10-14 September 2006.
Burton F.L., 1996. *Water and Wastewater Industries: Characteristics and Energy Management Opportunities*, CR-10691, Electric Power Research Institute, St. Louis, MO.
Cecchi F., Battistoni P., Pavan P., Fava G. and Mata-Alvarez J., 1994. Anaerobic digestion of OFMSW and BNR processes: a possible integration. Preliminary results. *Wat.Sci.Tech.*, 30, 8, pp. 65-72.
Fatone F., Bolzonella D., Battistoni P. and Cecchi F., 2005. Removal of nutrients and micropollutants treating low loaded wastewaters in a membrane bioreactor operating the automatic alternate cycles process. *Desalination*. 183(1-3), pp. 395-405.
Llabrés P., Pavan P., Battistioni P., Cecchi F. and Mata Alvarez J., 1999. The use of organic fraction of municipal solid waste hydrolysis products for biological nutrient removal in wastewater treatment plants. *Water Research*, 33(1), pp. 214-222.
Pavan P., Battistoni P., Bolzonella D., Innocenti L., Traverso P. and Cecchi F., 2000. Integration of wastewater and OFMSW treatment cycles: from the pilot scale experiment to the industrial realisation. The new full scale plant of Treviso (Italy). *Water Science & Technology*, 41(12), International Water Association Publishing, London (UK), pp. 165-173.

P. GARDIN AND A. MARCHINI

INTEGRATED POLICY APPROACH TO WATER SUPPLY MANAGEMENT: A VENETIAN PROJECT

Abstract. Water demand for different applications is increasing year by year, while on the contrary, its availability, due to various reasons, tends to decrease. Also, districts that are historically rich in such resource have to deal with this new situation and give great attention and care to water management. Good examples of an integrated approach to water management in the Veneto Region are the "Modello strutturale degli Acquedotti del Veneto" and the "Progetto integrato Fusina". The first project aims at managing water resources of a district with 4,500,000 inhabitants for the next 30 years, while the second one intends to realise an integrated plant capable of purifying wastewaters produced by a great industrial and urban area of 600,000 inhabitants and to supply purified waters instead of clean fresh water for industrial needs.

1. INTEGRATION NEEDS AMONG DIFFERENT USES IN WATER RESOURCES MANAGEMENT

Integration of different governmental policies for environmental sustainability has been stressed in Muraro (2007). Such issue refers to the particular case of water resources management of a specific regional or national territory which can be considered under different points of view. At the physical level, with reference to the quantity and quality of water resources in a territory, management objectives targeted to the optimization of the final use of the available resources must be achieved through their efficient distribution, by curbing dissipation and water losses in the pipes and through treatment and a possible reuse. In other words, the policy target is to promote water saving through regulations or incentives to restrain the waste of this resource which appears to be more and more limited and precious, and to safeguard its quality being its pollution the first prejudice to its use.

It is worth noticing that also in temperate regions, characterized by relative frequency of rainfall, water supply is becoming more and more problematic essentially because of the continuous increase of water demand and consumption for urban, industrial and agricultural uses. Such explosion of water demand comparing to the past has affected both the quality and quantity of its availability. Water consumption especially for industrial production processes causes not only a large quantity to be used, but sometimes also a considerable modification of its quality since residual discharges of water deriving from such processes can be very harmful for the environment if released without a previous and adequate depuration treatment. Moreover, the whole economy and the cost of water supply itself are burdened by the cost of such treatment.

The interrelationship between the quantitative and qualitative aspects and also, between the uses characterized by different quality criteria of water demand calls for an integrated management; in other words, a tight integration among the different phases of the water use cycle is necessary: from the supply of ground or surface water to the pipe distribution system; from the use in different sectors to the treatment of water effluents and sewages; from recycling to reuse in a saving process.

The trend of the recent governmental policy for quality protection of water resources and for their optimization is based on different regulations applied for each type of use: drinking water and water for food preparation, water for hygienic and sanitary consumption, fishing water, bathing and swimming water and for other leisure, for landscapes, etc. Every different type of demand calls for specific corresponding quality criteria in order to allow the best use of water resources, defending human health first and the whole environment, where water is a fundamental and vital element. In fact, beyond a certain level of pollution, the environment is not able to ensure a natural capacity to recover the original quality. Therefore, public institutions must intervene to regulate and govern the matter by preventing by law any possible damage, by prescribing the right behaviour from users and by requiring the adoption of the best available technologies to reach the quality criteria established by each sectoral regulation.

2. INTEGRATION OF LOW ENVIRONMENTAL IMPACT TECHNOLOGY UPSTREAM IN PRODUCTIVE PROCESSES AND DOWNSTREAM WATER DISCHARGE TREATMENT

It is known that the most adopted technologies in water treatment are the ones for the depuration of effluents and civil and industrial wastewater downstream. There are different types of plants and processes, functioning with advanced technologies, which are generally well known; they are referred to the treatment (mechanical, chemical, biological) of residual water and discharges by way of bringing them to satisfy the parameter values set by the law and consistent with their intake in the natural environment without polluting it. Wastewater treatment, even by integrative methods and techniques, is a delicate and complex matter difficult to manage, often submitted to the control of supervisory authorities and environmental control police, both for civil and industrial plants. There are many laws and regulations concerning this sector enacted at European, national and regional levels.

As mentioned earlier, the technology in this field is well known and has been tested by worldwide experience over many decades. However, technologies that have been introduced in the productive cycle for this purpose can have an increasing role. They are targeted to use materials, chemical products and substances, production and logistic processes which allow a less negative impact on the quantity and quality of supplied water, mainly through an intensive use of recycling processes of used waters.

"Clean" or "environmental sound" technologies in this sector are probably destined to grow and expand their presence in the next future, especially in water intensive consumption of polluting processes and industries. The increasing adoption of such technologies is in any case due not only to a stringent environmental prescription but also to an increasing cost of the treatment downstream of discharges and of water supply itself.

3. INTEGRATION OF WATER MANAGEMENT AT REGIONAL LEVEL

The management of water resources in its different and integrated uses, as described above, cannot disregard the regional area in which it naturally belongs. For such a reason, any water policy is necessarily referred to the area of regional scope generally corresponding to a water basin. Although this concept is evident, the organization of water resources management according to this principle has never been easily achievable, for in the past, the political and administrative competences and borders of municipalities and provinces have always prevailed. We also must remember that water (which in this paper is considered only as a vital and fundamental resource for life) can become, in specific meteorological situations, a factor of danger and destruction, a natural element capable of causing serious damages and responsible for calamitous events affecting territories, infrastructures and populations when rivers overflow due to hydrological instability. Defence from water and use of water are parts of a sole issue which must be dealt with through integrated policies including hydraulic regulation of rivers, protection of water-table from withdrawal and environment protection rules.

For this reason, in Italy as in other developed countries, the trend is to operate water management policies based on territorial units corresponding to river basin scopes, overcoming the fragmentation of the past management.

In particular, in the past, the organization of this sector was characterized by a wide fragmentation of public water services at municipal level (more than 8000 local utilities). The first result of such situation was a low scale economy and the incapability to ensure investments levels which would modernize the sector. In 1994, a law introduced a national reform that is still in progress. The Law 36/1994 subdivided the national territory in 91 Territorial Basins (called ATO – Ambito Territoriale Omogeneo, homogeneous territorial unit): this law aimed at increasing the cooperation among the municipalities involved in each ATO in order to create an ATO authority that would be entrusted with the planning of water services, the choice of a manager and the supervision of the management.

Despite the controversies and obstacles that arose in the municipalities, the enactment of Law 36/94 led to the strategy of integrating and consolidating the national system of water services for implementing investments and controlling tariff increase. The option was setting up a sole management unit for each ATO and choosing a technological private partner to be selected through a public tender. This strategy started a system of joint partnerships among the main public and private companies in Italy. Foreign operators were also involved in the take-over of shares in Italian companies.

4. VERTICAL INTEGRATION AND MULTI-UTILITIES APPROACH

Along with the "horizontal" integration between water management operators located in different and possibly contiguous regions or municipalities, we must also consider a possible "vertical" integration among local utilities operating in different segments of the larger sector of public urban services, e.g. energy, waste and other residential utilities. Due to the fact that citizens or residents are also final consumers, the concessionary is often the same (the municipality) and the business organization is very similar (tariff system, billing, call centre, customer satisfaction and customer relationship management, etc.). A joint management of various but contiguous local utilities can bring significant scale economies and a recovery of efficiency. The legislator has promoted this kind of integration mainly for the "hydro-environmental" sector (water distribution and waste collection) by possibly overlapping the waste management territorial optimal unit with the scope of the ATO.

The deriving restraint of operational costs becomes a fundamental condition to improve profits against the multiplicity of local bodies which do not reach a financial equilibrium and find increasing limitations to access public financing. Moreover the cost reduction has been imposed by a new regulation framework through the tariff system based on the "price cap" approach.

Now, the tariff structure adopted for example in Italy (where the average price is 1.0-1.2 euro per cubic meter) includes total investments coverage, operational and depletion costs and the annual adaptation on the basis of the "price cap" mechanism. This method ensures an increasing tariff corresponding to the inflation rate, diminished by a factor X that expresses the increase of productivity that must be guaranteed by the utility operator. This method stimulates the operator to contain costs and improve efficiency to reach profit.

The so-called "multi-utility approach" requires an enterprise organization with a central directive management for planning activities, control, finance and administration, billing and commercial supply, and several productive units for each business area. In major multi-utilities, created from the merger of utility operators, a holding which controls several one-business companies is set up. Non-strategic activities are outsourced.

5. GENERAL FIGURES ON WATER RESOURCES MANAGEMENT IN ITALY

Nowadays in Italy, due to the recent law favouring competition and market liberalization, important mergers of relevant utility companies are being carried out. It is a complex issue since often such companies belong to municipalities which also own the water infrastructure network and plants. These assets must be concentrated in a public company and a possible private partner must be selected through a public tender.

In order to show the extent of such a process, it is important to notice that the number of mergers involving water utilities – which were only 9 in year 2000, has increased up to 36 in 2001, 51 in 2002 and 69 in 2003.

However, the merger trend has only relatively reduced the elevated fragmentation of operators in Italy, that include about 8,000 municipal drinkable water distribution mains, almost the same number of sewerage and more than 15,000 treatment plants. Almost 8 billion cubic meters of drinkable water are pumped yearly (2000) in the water network but 29% is lost in the pipes; therefore, the supplied water amounts to 5.7 billion cubic meters: 4.9 billion cubic meters (87%) are destined to civil customers, 0.8 billion cubic meters go to industrial and agricultural activities (but the direct supply from water table or surface water – non drinkable – is about 20 billion cubic meters for agriculture, 8 billion cubic meters for industry and 6 billion cubic meters for energy production). The per capita consumption of civil customers is about 267 litres/day.

As for the sewage system, almost 99% of the Italian population is served by it, but 36% of total discharges flow directly into rivers or into the sea without any treatment. 85% of municipalities have their wastewaters depurated by treatment plants even though it must be specified that 80% of the plants are small-sized, serving towns or urban districts with less than 2,000 inhabitants.

As for water management operators (still more than 7,000), at the end of the 90s, 30% of them managed the entire integrated water cycle (withdrawal, distribution, discharge treatment), while most of them (generally the same municipalities) operated the first two phases only. At present, more than ten years after the enactment of Law 36/94 which has introduced the above-mentioned 91 territorial units, 87 of them have been set up, while 68 have approved the ATO Plan as established by the law and only 42 (representing 33 million users) are operative since they have appointed the integrated service to a single operator through a direct assignment to public or mixed public-private companies. In fact, the original law encountered many obstacles on its path especially because of the controversies arisen between central and local governments, concerning mainly the services' appointment procedures. Local authorities clearly preferred a direct appointment; this approach however violates the European Union's principle of market competitiveness and transparency. The last years were characterized by certain confusion and by a search for compromise that would emend the first law in order to produce other acts that would introduce more options and a transitory period for adjusting the most controversial cases. However, the reform of 1994 can be considered under way and it strives for the integration and consolidation of water resources management in Italy.

6. APPLICATIONS IN THE VENETO REGION OF WATER SAVING THROUGH WATER RECYCLING

Because of the reduction of water resources, water recycling has become more and more important in water supply management. As explained earlier, at the physical level, with reference to the quantity and quality of water resources on the territory, the management objectives aiming at optimizing the final use of the available resources must be achieved through an efficient distribution, by curbing dissipation and water losses in the pipes, and through water treatment and its possible reuse.

A good example of this strategy application is evident in the MOSAV project (transport optimization and interconnected distribution of large volumes of drinking water in the Veneto Region) and in the PIF project (an integrated project for the wastewater treatment in Fusina, Venice). Both these projects will be realized before 2010.

6.1. The MOSAV project

The most important aim of the MOSAV project is to grant full and sure availability, together with high quality, of an indispensable resource such as drinking water to an area having the following characteristics:

- High concentration of people (4,500,000 inhabitants).
- Presence of important and large industrial plants on the territory.
- Qualified agriculture.

Several significant actions have been planned to reach this objective. A new connected system of high capacity water pipes will be realized to ensure drinking water convoyed from rich to poor areas. New sources will be identified and dedicated only to drinking utilization and, where it will be possible, drawing from underground water or from good quality rivers with a stable flow will be preferred to any other sources. High capacity reservoirs will be realized to ensure water supply. Sewage and purification systems will be improved introducing advanced and flexible technologies. The reuse of purified water will be promoted for process, cooling and irrigation waters and for any application where the use of potable water is not necessary (e.g. street cleaning, car washing, etc.).

Some features of the MOSAV project:

- Population involved — 4,526,000 inhabitants
- Estimated drinking water needs in 2015 — 34,224 litres/second
- Total length of the new high capacity interconnected pipes — 241,000 meters
- Total distribution capacity — 47,500 litres/second

6.2. The PIF project

The PIF project responds to the necessity of reducing pollution in the Venice Lagoon and the Adriatic Sea.

To reach this target, all wastewaters produced in the territory around the Venice Lagoon, both from human and industrial origins will be collected and transported to a new high capacity treatment plant.

Here, using the most advanced and appropriate technologies, it will be possible not only to purify wastewaters so as to enable the discharging into the Adriatic Sea without any environmental risk and according to very strict law limits, but also to save a large volume of good quality river water that will be available for drinking purpose in the MOSAV project (Fig. 1).

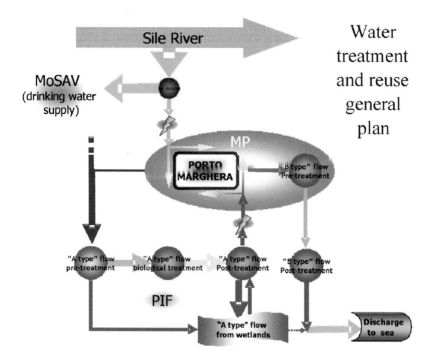

Figure 1. Scheme of the PIF project in connection with the MOSAV project. By courtesy of the Joint Venture "S.I.F.A." (Sistema Integrato Fusina Ambiente) and the Engineering Consulting "Porto Marghera Servizi Ingegneria".

In this project, pollution prevention and water saving are the most important aspects; water is reused more than once in "cascade" according to the required standards (Fig. 2).

In the plant, new technologies on large scale will be adopted and an advanced policy of the territory and water will be introduced:

- 50% of the total wastewater treated in the plant will be reused in local industries in substitution to good quality water coming from the same source that is supplying the drinking water network of the city of Venice. Water saved will be sufficient as drinking water for 250,000 inhabitants.
- The process includes a wetland system that consists of a large phytoremediation area with very low running costs. This area, considered one of the largest in Europe, will be enjoyable as a natural park, with exhibition areas, conference and visit centres, a book shop and a rest area.

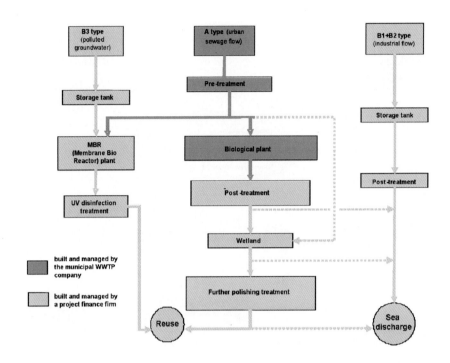

Figure 2. Lay-out of the PIF process. By courtesy of the Joint Venture "S.I.F.A." (Sistema Integrato Fusina Ambiente) and the Engineering Consulting "Porto Marghera Servizi Ingegneria".

The wastewater volumes treated in the new plant, as shown in Fig. 2, are the following:

- Industrial wastewater (B1+B2+B3 type)   940   l/s
- Urban wastewater (A type)   1215   l/s
- Expected volume of available reuse water   870   l/s

The reuse of depurated human wastewaters (Tab. 1) will produce several advantages:

- better control on the complete cycle of industrial water;
- cost reduction of the treatment plant by selling treated water for reuse;
- cost reduction of investments and plant management due to the fact that large volumes of reused treated water allow the realization of a smaller sea discharge unit (treated waters that are not reused will be discharged into the Adriatic Sea at 10,000 meters from the coast by a special pipe. The cost of this unit represents about 20% of the project's total cost).

Another important aspect of this process is that, due to the "open lay-out", reused waters can be utilized only once. In fact, industrial reused waters will return to the PIF treatment plant as industrial wastewaters and will be discharged into the sea after treatment.

Reused water is obtained only by treating human wastewater and a limited volume of polluted groundwater; reused water is discharged and not retreated. This approach avoids the risk of increasing concentrations of pollutants or other effects (such as salt concentration) that occur with the progressive reuse of water.

Table 1. Reused water properties obtained at the wastewater treatment plant of Fusina compared with water drawn from the Sile River. By courtesy of the Joint Venture "S.I.F.A." (Sistema Integrato Fusina Ambiente) and the Engineering Consulting "Porto Marghera Servizi Ingegneria"

Parameters	Unit	P.I.F. reused water	Sile River
Temperature	°C	14 – 15	14.01
PH		7 – 8	7.81
Conductivity	µS/cm		486
Turbidity	NTU	< 1	4.68
Suspended solids	mg/l	< 3	12
COD		10 – 12	10.9
BOD 5		2 – 5	2.5
Ammonium	mg/l NH_4	0.1	0.23
Hardness	°F	30 – 32	25.26
Oxygen	mg/l O_2		8.89
Chloride	mg/l Cl	130 – 150	7.90
Total Nitrogen	mg/N	4 – 7	
Nitrates	mgN/l	0.9 – 4	3.31
Total phosphor	mgP/l	0.6	
Phosphates	mgP/l	0.2	0.08
Sulphates	mg/l SO_4	120	45.51
Aluminium	µg/l Al	50 – 150	104.90
Cadmium	µg/l Cd	0.1 – 0.3	0.00
Total chrome	µg/l Cr	1 – 2	0.69
Iron	µg/l Fe	50 – 150	79.72
Lead	µg/l Pb	1 – 3	0.48
Manganese	µg/l Mn	10 – 20	7.97
Mercury	µg/l Hg	0.2 – 0.3	0.00

Nickel	μg/l Ni	**10 – 20**	1.80
Copper	μg/l Cu	**5 – 7**	5.35
Zinc	μg/l Zn	**50 – 150**	12.81
PAHs	μg/l	**< 0.1**	0.00
PCBs	μg/l	**< 1**	
Total phenols	μg/l C_6H_5OH	**< 30**	1.88

By comparing the properties of water destined to industrial reuse obtained at the PIF plants with those of water drawn for the same application from the Sile River, we can say that they are quite similar.

Only Sulphates and Chlorides contents are higher than those of the river water. Also, the values of Nickel and Zinc are higher but acceptable for standard industrial applications.

To reduce Sulphates and Chlorides contents, it would be necessary to introduce in the process an inverse osmosis unit. This treatment is very expensive and complex and will be introduced only if necessary.

From a financial point of view, the PIF project will be realized through project financing. The Joint Venture (J.V.) "Sistema Integrato Fusina Ambiente" (S.I.F.A.) between public and private subjects has been established for realizing the structure and its technical and economic features have been approved by the authority. The J.V. has been authorized to realize and manage the structure for 25 years. The expected total cost is about 200,000,000 € and will be covered as for operating costs by charges paid by industrial companies to purify wastewaters and to buy reused water. The cost of one cubic meter of reused water will be equivalent to that of the river water currently used (0.07 €/cubic meter).

Paolo Gardin
Former Vesta S.p.A., Venice, Italy

Adriano Marchini
Department of Industrial Development and Special Projects,
Vesta S.p.A., Venice, Italy

REFERENCES

Costantini A., Pinato T., Zanette D. and Del Rizzo S., 1998. *Modello strutturale degli acquedotti del Veneto (27/03/1998)",* Regione Veneto.
Engineering consulting "Porto Marghera Servizi Ingegneria", (17/03/2005). *Progetto Integrato di Fusina/ Proposal of Project Financing/Preliminary Project.*
Regione Veneto. Direzione Tutela dell'Ambiente. *Piano per la prevenzione dell'inquinamento e il risanamento delle acque del bacino idrografico immediatamente sversante nella Laguna Venezia/Piano Direttore 2000.*

III. AIR QUALITY CONTROL

I. ALLEGRINI AND F. COSTABILE

MONITORING AIR QUALITY IN URBAN AREAS

Experiences in China

Abstract. Air quality improvement in rapidly developing countries like China requires to generate valuable data with high-level of representativeness to be transformed into useful information about air pollution. However, the problem of monitoring representativeness has been overlooked. In this respect, a comprehensive methodology to monitor air quality has been developed by this Institute, engaged since 2001 in several projects in China dealing with air quality. The new approach requires rearranging the existing monitoring strategies. In short, saturation monitoring is employed to preliminary assess air pollution and provide spatial resolution over the study area. Automatic monitoring is reduced to a minimum to make time-related data available. The information so-measured with high degree of both spatial and temporal resolution, are completed by statistical and mathematical modelling to generate data interpretation schemes. This methodology extensively tested in China during these years has been found to be very useful for macro/micro-designing emission and air quality monitoring networks, apportioning emission sources, assessing spatial representativeness of measurements, and evaluating spatial frequency distributions of air pollutants. Case-studies of Suzhou, Lanzhou and Beijing are presented to discuss these findings.

1. INTRODUCTION

Fuel combustion, diffuse primary industrial activities and automotive traffic bear the main responsibility for the degradation of air quality in urban areas. In rapidly developing countries like China the processes toward the reduction of atmospheric emissions, resulting in lower pollution levels, will rapidly move into governing processes which need to be efficient, relatively inexpensive to the society and very effective in the health protection of a high number of citizens potentially exposed to such pollutants. This is especially true in large conurbations where pollution levels and human exposures, caused by both industrial and civil emissions, are expected to be very high and to substantially increase in the next few years.

China has long suffered from serious air pollution characterized by high ambient concentrations of SO_2 and particles. The main reason is that more than 75% of primary energy comes from coal combustion. Since 1980, great efforts have been put to abate SO_2 and particulate matter emissions from coal combustion. As a result, ambient pollutant concentrations, especially particulate, have dropped gradually (World Bank, 1997). However, in the process of economic development during the last decades, the number of vehicles has increased dramatically and the local pollution problem due to automotive traffic is becoming severe in big Chinese cities (Zhou et al., 2001). Pollution related with traffic exhaust first gained attention in the early 1980s (Sun, 1982). In 1995 the total amount of vehicles in China was more than 20 million with an average annual increase rate of 15%. Vehicle emission factors developed by the China National Mobile Source Emission Laboratory were found to be much higher than those in Europe and the United States. Since 1990, NO_x emissions from stationary sources in China have levelled off somewhat, but vehicular NO_x emissions have increased. Also hydrocarbons are significant components in urban air mainly because of combustion, solvent use, fuel evaporation, and tank leakage. Among them, Benzene, Toluene and Xylene (BTX) are listed as toxic or potential toxic contaminants while other Volatile Organic Compounds (VOC) are certainly contributing to tropospheric Ozone formation. However, only a few studies have investigated these issues in China (e.g.: Wang et al., 2002; Barletta et al., 2005).

In the design of cost effective abatement strategies, it must be realized that the relations between emissions and resulting concentrations are not simple as generally thought. A lot of research has been done to investigate this issue. Most of the emission factors used by the emission inventories to quantify traffic contribution upon total emissions are derived from laboratory measurements carried out according to specific protocols. On the opposite, very often, pollutant concentrations are measured on the road to infer emission profiles and rates (e.g.: Shi et al., 1999); thus, they account for a mix of vehicles, other than air temperature and humidity. Additionally, the concentrations are strongly dependent upon wind speed and direction and, consequently, on the temporal variation of atmospheric dilution rate. In particular, the microscale pollutant dispersion from road traffic is strongly influenced by the wind flow field around urban street-building configurations. This affects overall dilution and creates localized spatial variations of pollutant concentrations (Scaperdas and Colvile, 1999). Classic examples of microscale dispersion effects in urban areas are those associated with *street canyons*, the common urban street-building geometry of a street bounded by buildings on both sides (e.g.: Vardoulakis et al., 2003). As a result, air quality measurements made at different urban monitoring sites may also be strongly influenced by the way the street-building geometry surrounding the monitoring site interacts with flow, as well as by the road traffic emissions dispersion at a localized scale. In order to investigate this point, several studies mainly based upon three different approaches have been done. They are concerned with i) field monitoring in street canyon (e.g.: Kukkonen et al., 2001), ii) wind tunnel simulation (e.g.: Pavageau and Schcatzmann, 1999), and iii) computational fluid-dynamic modeling (e.g.: Zhou and Sperling, 2001). Although measurements are still the basis of our understanding, the integration with mathematical and physical modeling is of increasing importance in urban air pollution management (Fenger, 1999). This is true on local scale (through the use of CFD, Computational Fluid Dynamics), as well as on urban scale, to build-up full decision support systems. However, a further improvement of the integration of field measurements and models is needed. Besides the numerous issues rising with model application, the continuous measurements (monitoring) are also constrained by several factors, ranging from the cost and bulk of the monitoring equipment, to the need of power supply and maintenance requirements. The number of monitoring stations is, therefore, reduced as much as possible resulting in the representativeness of air quality monitoring network to become a key-issue for the evaluation and management of air pollution. However, the problem of representativeness and data interpretation has been overlooked all around the world and there isn't a quantitative methodology to describe it.

In this respect, this paper reports the main findings of the work carried out by our Institute in China since 2001 in a number of projects dealing with air quality. The objective is to present the methodology that has been developed in order to monitor air quality and apportion emission sources with the aim to generate dataset highly representative of the main urban air pollutants, in terms of both time and space changes.

2. METHODOLOGY

The proposed methodological approach requires a rearrangement of the existing air quality monitoring strategies and emission monitoring requirements. A preliminary assessment of air quality is the first step of each monitoring intervention. The obtained data are integrated with statistical data modelling to infer source contributions and spatial distribution of atmospheric pollutants, then frequency distributions are calculated and maps of air pollution generated. The results are used to design the network of monitoring stations (macrositing) and the type of monitoring equipment to be installed at each station (micrositing), as well as to evaluate their representativeness. In such a way, automatic monitoring may be reduced to a minimum to make time-related air quality data available and develop emission control strategies. The spatial measurement of air pollutants is resolved by the so-called saturation monitoring. In order to perform it, diffusive sampling techniques are applied as a supplementary method to the manual, active sampling methods and automated ones. All together, these three groups of methods may form an air pollution monitoring system that is capable of providing data of high quality and very good resolution in terms of time and space, and representativeness of the area to be monitored as well (UNEP/WHO, 1994; European Commission, 1998). The data are used as input for advanced mathematical models to evaluate the source-receptor relationship: integrated systems of measurements and modelling may be used to evaluate the impact of emission sources on urban-scale; additionally, typical urban micro-environments, such as street canyons, enclosed yards, etc., may be conveniently simulated by CFD.

2.1. Preliminary assessment

The preliminary assessment of atmospheric pollution is a very important step to gain important information about air quality, such as, for instance, the definition of locations where to deploy monitoring stations. Analyst® diffusive monitors (Marbaglass, Rome, Italy) may be conveniently used for this work. Sampling is carried out on a monthly basis throughout one year (12 monthly samples) by mounting the Analyst samplers at heights of approximately 2.5 meters above ground level under a protective shelter. SO_2, NO_x, NO_2, O_3, NH_3, BTX (benzene, toluene and xylene), VOC and H_2S concentrations may be assessed. The sampling protocol is broadly based on methods and approaches described in the EC document on guidance on preliminary assessment in relation to the Framework Directive (1998). Essentially, a grid over the area under investigation is identified and for each cell of the grid, a location as much as possible representative of the pollution level in the grid is chosen. At the end of each period of exposure, samplers are analyzed by ion-chromatography for SO_2, NO_2 and NO_x, NH_3, O_3, H_2S and by gas-chromatography for benzene, toluene, xylene and VOC. The accuracy of the Analyst® samplers compared to the reference techniques expressed as percent relative error is around ± 20%. This is in compliance with the Directive 1999/30/EC requirement for uncertainty (precision and bias) < 25% for NO_2, SO_2 and NO_x and within the recommended accuracy for diffusive sampling indicated in the guidance on diffusive sampling as provided for NH_3 and BTX by EC (1998). Several stations (typically several tens) are sited and set up throughout the urban area to provide a spatial scheme that represents the extent of local variability within the investigated area.

After the sampling campaigns, the results are complemented by statistical studies. Geostatistical Analysis is used to map air pollutants (e.g.: Atkins and Lee, 1995). The sampler sites are geo-referenced and registered to a base-map, then formatted and imported into a GIS. Thus, the measurement of pollutant concentrations carried out at different sites, are used to create, by interpolation, a continuous surface with the aim to forecast a value for the considered pollutant in each point of the map. Inverse-distance-weighting and Kriging algorithms are used to find the most suitable model to represent that phenomenon. Frequency diagrams and histograms are calculated. They have usually irregular trends since they are based on real samples. However, the bigger the number of measurements, the more efficient is the data analysis. The comparison between the measured distributions and the expected theoretical distributions, such as Lognormal, Weibull, pseudo-lognormal and Type V Pearson and Normal is conveniently used to infer source contributions and pollutant spatial distributions (e.g.: Mage and Ott, 1984; Morel et al., 1999). In this way, a relatively simple assessment technique is used to provide extensive and detailed results.

2.2. Network design of representative monitoring stations

The process of air quality recovery and improvement requires to generate valuable data and to provide tools to transform these data into useful information about air pollution, to be integrated into technical and social measures. It is worth to mention that the comparison between pollution data taken at different sites is often not significant, as well as the emission source apportionment or data modelling are often difficult to be achieved by using those data. The United States Environmental Protection Agency defined five categories of spatial scales – micro, middle, neighbourhood, urban, and regional scales – in its guidelines for siting State, Local and National Air Monitoring Stations. The European legislation suggests having stations for the monitoring of the population exposure in heavy traffic sites (Italian type C: traffic exposure) or in urban background (Italian type A: urban background exposure), but also considering the need to monitor sites where the exposure of population may be a problem (Italian type B: residential population exposure); in addition, it suggests that every Member State should set-up a proper background monitoring network for regional or national background evaluation of atmospheric pollution (Italian type D: non-urban background station).

The issue of site representativeness may be considered as a spatial time-dependent problem due to the fact that site representativeness may be determined by spatial variations in measurements around monitoring sites over a period of time and that air measurements are reported as time-averaged concentrations. Atmospheric chemistry and source/receptor distance, meteorology, and topographic conditions determine how each emission source is partitioned between the different pollutants. Knowing the emission profile, it is possible to previously evaluate the distance of the monitoring points from the main emission sources, especially when primary pollutants are considered.

The case of traffic related NO_x is presented to better discuss this issue. A useful way of considering the significant aspects of NO_x-NO_2 chemistry in urban centres is to plot the observed concentration of NO_x against that of NO_2 [e.g.: Carslaw and Carslaw, 2001] giving a convenient measure of the extent and completeness of $NO \rightarrow NO_2$ oxidation processes. The non-linear nature of the relationship between the concentrations of NO_x and NO_2 observed by Carslaw and Carslaw (2001) may be looked at. From approximately 0 to 80 ppbv NO_x, the NO_2/NO_x curve rises steeply and corresponds to conditions where there is generally enough O_3 present to convert NO to NO_2; therefore, concentrations of NO_2 rise as NO_x concentrations increase. From about 100 to 600 ppbv NO_x, NO_2 concentrations increase slowly with increasing NO_x concentrations. In this regime there is very little or no O_3 present to convert NO to NO_2 and the gradient of 0.1 tends to approximate the proportion of NO_2 found in exhaust gases of vehicles [Shi and Harrison, 1997]. Throughout the year, NO_2/NO_x values measured at the different districts may be compared to the corresponding NO_2/NO_x ratios for motor vehicle emissions: these emission ratios are typically in the range 0.05-0.1, although somewhat higher emission ratios up to 0.3 are possible from warm idling car engines in cold weather conditions. A high degree of NO oxidation is consistent with a relative average distancing from emission sources; in contrast, measured NO_2/NO_x ratios are on average markedly lower throughout the year at the traffic-exposed measurement sites, clearly showing values of the corresponding NO_2/NO_x ratio lower than the ones at urban background stations. Therefore, the NO_x/NO_2 values interpolated among all the measured points by using Kriging algorithm are used to create the relative distribution maps. The areas where NO_x/NO_2 ratios are expected to be higher according to these maps may be assumed to be more impacted from traffic and considered as possible sites where to establish traffic-oriented (type C) fixed stations. On the contrary, the areas where NO_x/NO_2 ratios are expected to be lower may be considered suitable for both residential and background fixed stations. The same procedure may be followed for other sources using suitable pollutants that may be considered as tracers for those sources.

2.3. Integration of measurements and modelling

The framework for integrating measurements and modelling on urban areas is based on the development of a methodology for the analysis of the source-receptor relationship related to air pollution. In particular, in urban areas, it is very useful to analyze traffic-related air pollution. The data of air quality, and public and private traffic flows measured by representative air quality monitoring stations (AQMS), and traffic measurements are used to input mathematical and physical models.

On micro-scale, Computational fluid dynamic (CFD) is used to explore the transfer and diffusion characteristics of single emission and the constraints associated with significant monitoring locations such as street canyons, courtyards and enclosed spaces, and the conditions propitious to pollutant diffusion and air dispersion. CFD solves the mass and momentum conservation equations (the Navier-Stokes equations) for general incompressible and compressible fluid flows. The terms representing the additional Reynolds stresses due to turbulent motion are linked to the mean velocity field via the turbulence models. The well-known k-ε turbulence model may be conveniently used, comprising transport equations for the turbulence kinetic energy k and its dissipation rate ε. In this model, k and ε are chosen as typical turbulent velocity scale and length scale, respectively. The success or failure of a fluid simulation depends not only on the code capabilities, but obviously also upon the input data, such as: geometry of the flow domain, fluid properties, boundary conditions, solution control parameters. For a simulation to have any chance of success, such information should be physically realistic and correctly presented to the analysis code. Thus, only high accuracy and temporal resolution measurements are suitable to model the data.

On urban scale, where local emission measurements (e.g.: traffic counts and flows) are available, the AQMS core-module may provide the input data to a system where a first model suite (e.g.: transportation) calculates the emissions generated into the grid by each source (e.g.: traffic emissions) and a dispersion model calculates the concentrations all over the interested areas. It is worth mentioning that only high quality and time, and space representative data measured by the AQMS can guarantee the reliability of air pollutant concentrations modelling. If the system determines into the grid pollution levels exceeding target values, long-term emission-oriented interventions and policies may be planned (e.g.: public transport fleet may be upgraded; sensible areas of the city may be identified, banned to pollutant vehicles and transformed in low emission zones with extra public transportation; industrial point sources may be moved out of the city; heating emission contribution may be better understood; etc.).

3. RESULTS AND DISCUSSION

It was found that this methodology extensively tested in China and Europe during the past years can provide valuable (and often necessary) tools to manage urban air pollution. Several targets were found to be reached: macro/micro-design of emission and air quality monitoring networks, apportioning of emission sources, assessment of spatial representativeness of measurements and zones where to extend the monitoring, measurement of spatial frequency distributions of air pollutants, pollution modelling on urban and micro scale, integration of measurements and modelling for policy planning and interventions. A more detailed discussion follows; the focus is on case-studies of pilot projects carried out in Suzhou, Lanzhou and Beijing in the framework of the Sino-Italian Cooperation Program for Environmental Protection, held by the Italian Ministry of the Environment Land and Sea and the Chinese State Environment Protection Agency.

3.1. The preliminary assessment of Lanzhou

A project for the establishment of an Air Quality Monitoring System and Greenhouse Gas emission Inventory in the city of Lanzhou has been carried out since 2004. Analyst® passive samplers for BTX and SO_2, NO_x, NO_2, NH_3, O_3 have been placed in 40 locations and statistical analysis of raw data performed (Costabile et al., 2006a). Quality assurance and Quality Control have been taken into account by duplicating 8 sampling sites among the 40 locations. The correlation coefficient r of the duplicates samples showed a good correspondence between the co-located samples. The angular coefficients of the BTX regression lines (Fig. 1) were close to 1 as well as the biases: that indicates the low errors of sampling found in this work suggesting a good accuracy of measurements.

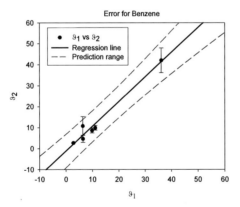

Figure 1. Scatter plot with error bars and regression of Benzene duplicated sampler ϑ. The dotted line indicates the prediction range.

The focus on BTX data is particularly valuable considering that only a little work has already been done in China concerning these pollutants. Benzene values are low dispersed and centered on 10 µg/m³, corresponding to the limit value set by the European Union. More dispersed values were found for Xylene and Toluene with the 75th percentiles always less than 20 µg/m³. BTX spatial data distributions were found to be normal (Costabile *et al.*, 2006a). Thus, concentration mean and median values (quite similar) can be correctly used to represent the concentration values of these pollutants all over the city (for air quality management purposes and calculation of air pollution indexes API) only if the data distribution is demonstrated to be normal. In fact, in this case, the mean equals the median and, therefore, it is representative of the air quality across the city.

3.2. Network design and representativeness in Suzhou

An extensive air pollution assessment was carried out in 2003 during the course of the AQMS project started in 2002 for the establishment of an Air Quality Monitoring System in the City of Suzhou. A broad preliminary assessment of air pollution was carried out by passive sampling technique throughout 100 measurement sites and three sampling campaigns with the aim of network designing the AQMS (Allegrini *et al.*, 2004). In particular, the representativeness of urban zones where to site the traffic-oriented air monitoring stations was studied (Costabile *et al.*, 2006b). The sampling sites were geo-referenced and reported on the map of the city, together with the NO_x/NO_2 values to evaluate the impact of primary emission sources (Fig. 2).

The distribution trend of NO and NO_2 was used to identify the heavy traffic polluted zones where to extend the measurement of the other pollutants by means of automatic monitoring technique. The NO_x/NO_2 ratio was calculated during summer and winter campaigns, respectively for the district-average values and for all the measured sites. The values measured during the summer period were generally lower than the ones measured during the winter period, showing a superimposition between the highest summer values and the lowest winter values. Considering traffic emissions reasonably constant throughout the year, other emissions (mainly from power stations) rise during winter (in Suzhou and all over the South of China there is no heating either in summer or in winter seasons). Winter conditions are also associated with increased atmospheric stability, reduced mixing depths and consequently poorer dispersion of air pollutants. Taken together, these factors resulted in higher urban NO_x level during winter. The regression lines had the same slope for both periods indicating a quite similar relationship between NO_2 and NO_x. The NO_2/NO_x ratios measured at Suzhou's districts were calculated. The high degree of NO oxidation observed at the measurement sites in Pinjiang, Huqiu and Wuzhong districts is consistent with their relative average distances from emission sources. In contrast, measured NO_2/NO_x ratios are on average markedly lower throughout the year at the traffic-exposed measurement sites, in Jinchang, Canlang and New district, clearly showing values of the corresponding NO_2/NO_x ratio lower than the ones at urban background stations.

The distribution of NO_2/NO_x values during winter around 0.8 was consistent with that of primary emissions. At these traffic-exposed sites (in Jinchang, Canlang and New districts), during the summer months, NO_2/NO_x values were higher, demonstrating substantial oxidation to have occurred. Enhanced NO→NO_2 oxidation due to increased ozone levels, together with higher temperature and insulation rates, resulted in average summer NO_2/NO_x values higher than in winter. On the contrary, it should be noted that the summer NO_2/NO_x values at background urban sites (in Pinjiang, Huqiu, Wuzhong districts) were substantially lower than the corresponding winter values: higher photolysis rates during the summer may have likely acted against the oxidation processes described above tending to result in a photo-stationary equilibrated state between NO, NO_2 and O_3.

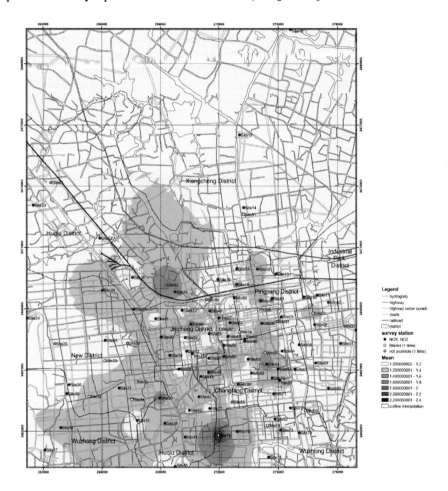

Figure 2. Map of NO_x/NO_2 ratio in Suzhou throughout 100 measurement points in 2003.

The final network design was established by the Suzhou Environmental Protection Bureau matching these results with the administrative constraints due to Chinese environmental regulations. Five stations were sited in the urban area: Station A (Urban background) at Jingchang district; Stations C1 (Traffic oriented) and C2 at Canglang district; Station B1 (Urban exposure) at New districts; Station

B2 at Pinjiang district. Three stations were sited in the suburban area: Station B3 at Wuzhong district; Station B4 at Pinjiang/Xiangcheng district; Station C3 at Pinjiang/Industrial park district. Finally the station D (Regional background) was sited in Huqiu district out of the urban area as regional background station. All the stations were equipped with automatic NO_x analyzers.

The frequency distributions of summer/winter-average NO_x, SO_2, O_3 and NO_2 values measured by passive sampling were calculated. A consistent variation of NO_x district by district was observed. Since NO_x is directly emitted from various mobile sources, its simultaneous measurements at two different locations are influenced by emission sources at individual locations during the monitoring periods. The same was assumed for CO and HC. Therefore, a lower correlation in measurements over time between two different locations for these pollutants was expected. Accordingly, the stations at Jinchang, Canlang and New districts (more influenced by direct traffic emissions) were also equipped with automatic CO analyzers and HC analyzers (especially BTX analyzer). By contrast, a lower variation over space of O_3, SO_2, NO_2 was found, since they are less influenced by direct emissions from local sources. Instead, they are either formed by chemical reactions or contributed by various sources in larger areas surrounding the station. Accordingly, higher correlation in the measurement over time between two different locations for these pollutants was expected. The same was assumed for PM_{10}. For this reason, automatic O_3 analyzers were only put at the stations of Jinchang, Canlang and New districts (in order to investigate the O_3 pollution variation and compare its values with NO_x/NO_2, HC data) and one at the regional background station. Since the measurement of SO_2 and PM_{10} throughout the network is compulsory in China, all the stations were equipped with automatic analyzers for these two pollutants.

3.3. Micro-scale modeling of traffic emissions in Suzhou

A study to evaluate the micro-scale spatial distribution of traffic pollutants inside a canyon of the urban area of Suzhou was performed in 2005 (Costabile *et al.*, 2006c, d). A monitoring station belonging to the AQMS established in 2004 was selected: the area includes five buildings southward with a width/height (W/H) ratio of 1.8-3, and four buildings northward with W/H ratio of 3-3.5. Considering it a typical open street-canyon lined by middle-low buildings with W/H ratio greater than 1.8, this station was selected as the first case-site to conduct field monitoring and investigate the vertical distribution of traffic air pollution. The fixed monitoring station is located on the roof of a building (site A), sampling inlet is 13 meters high. The mobile unit belonging to the AQMS system was located into the yard (site B), sampling inlet 3 meters high. During a one week campaign, the pollutant concentration profiles of CO, NO_x, NO_2, NO measured by the fixed station and the mobile unit, both equipped with the same automatic analyzers, were measured. At the same time, meteorological parameters were recorded as well (particularly vertical wind speed). The data were logged and transferred by a continuous data collection and analysis system into the data quality control centre of the AQMS.

Furthermore, automatic counting of total traffic vehicles per 30 minutes was carried out by the Communication Agency of Suzhou.

Modelling of the fluid domain allowed for the clear identification of both the entrapment zone into the yard and the up to down mass transfer (Fig. 3). The case of wind speed entering the canyon greater than 3 m/s is showed here in order to better understand the ground and roof NO_x concentration trends measured. The results modelled by CFD fit well with the field measurements: the wind velocity modelled at site B, inside the yard, ranged from 0 to 1 m/s; wind direction modelled at site A, at the top of the building, ranged from 2.5 to 3.5 m/s. The vertical component of the wind vector inside the street-canyon was always down-ward causing the impact of polluted air masses from the street to be more significant upon site B than site A. The higher concentrations measured at site B were clearly shown by comparing the velocity vectors based on the K-ε turbulence model along the section perpendicular to the street and surrounding the two measurement points. The negative vertical component of the velocity measured at site A was also modelled. Finally, the strong decrease of the horizontal component of velocity vector at site B was also identified.

As a common finding, the vertical wind speed was found to strongly affect the fluctuation of all the pollutant concentrations. The daily average of NO_x, NO_2 and NO concentrations, together with their NO_2/NO_x ratios, were found to have different relative average daily trends alongside the vertical: oxidation processes and photochemical activity were the main reasons governing these phenomena (Costabile and Allegrini, 2007a).

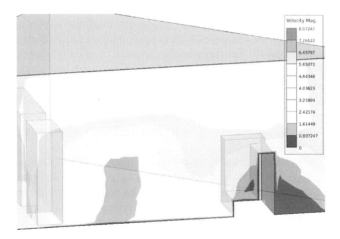

Figure 3. CFD modelling of an open street-canyon in Suzhou: section of the fluid domain perpendicular to the traffic street and forming a 45° with the prevailing wind direction.

The daily CO trend tracked closely with both NO_x and Benzene measured values, while their ratios were different if compared with other published works (e.g.: Kukkonen *et al.*, 2001). Different emission factors and driving conditions, together with other types of emission sources (especially for Benzene), were considered to explain this point. Therefore the representativeness of any sampling site is the key element to be taken into account to drive environmental information from concentration data.

3.4. The integrated system of Beijing

A feasibility study of a system design and planning for an Intelligent Transport System driven by a Traffic Air Pollution network was performed for the mega-city of Beijing in 2004 (Costabile and Allegrini, 2007b). In Beijing air quality is in transition from coal burning to traffic exhaust related pollution. Vehicle emissions are projected to double within the next two decades unless drastic strategies to decrease emissions are enforced. In this respect, a study was performed to analyze traffic and air pollution relationship and to develop a more flexible framework to allow communication between traffic emissions and air pollutant concentrations. This would suggest a methodological tool to mitigate environmental problems and show an example of its application in a real situation (Costabile et al., 2004a, b).

The system is intended to provide high time/space resolution measurements of air pollutant concentrations and traffic emissions, as well as real-time transportation and dispersion modelling of those data. The key advantage would be the runtime integration of measurements and modelling. Road transport is distinguished from other sources of air pollution in that the emissions are released in very close proximity to human receptors (Colvile et al., 2001). This reduces the opportunity for the atmosphere to dilute the emissions which would render them less likely to damage human health. Thus, it seems necessary to directly monitor pollutant emissions on the road to be compared with the pollutant concentrations measured by the AQMS sited relatively far from traffic sources. Critically, concentrations of pollutants from non-road traffic sources tend to exhibit much less spatial variability than those from road networks. Detailed information about traffic emissions and sources may be investigated by this system in order to determine which combination of policy interventions (e.g.: fuel type, engine type and end-of-pipe emissions abatement technology, etc.) is likely to be most effective at reducing impacts on human health.

The aim of the study was also to provide policy-makers with lessons learned regarding transportation programs which promote economic development while reducing pollution impacts. The following are the initial policy control measures suggested on the basis of the feasibility study in Beijing: i) establishment of a Low Emission Zone (LEZ) where to run the system; ii) management of private car traffic on the basis of air pollution; iii) imposition of penalties on those vehicles which do not meet emission standards (yellow tags); iv) adoption of special policies to invest money taken from penalties in transportation and environmental improvement actions; v) long-term and infrastructural interventions to control transport-related air pollution focused on public transport development. The implementation of the project is currently under way in Beijing. The framework showed in this paper uses a large amount of data and time. The initial results will need to be evaluated and then further optimization may be required.

4. CONCLUSIONS

The implementation of the methodology here described in several Chinese and European urban environments has produced a number of findings.

The integration of conventional measurements with supplementary diffusive and active sampling techniques has given the possibility to build-up air quality datasets with high time and space resolution and representativeness of values. The measurement campaigns performed to preliminary assess the main air pollutants in Suzhou and Lanzhou gave comprehensive multi-pollutant, multi-seasonal assessments of air pollution. In such a way it was possible to understand reasons and causes of measured concentration values, as well as investigate the source-receptor relationship of these pollutants.

It was demonstrated that the passive sampling method can be used in large-scale projects for the measurements of BTX, Nitrogen Oxides, and SO_2 spatial variability at extremely low cost. Errors found in measurement technique were extremely low and valuable for the purposes of this work. Moreover, many of the measurement sites could not be operated using conventional methods due to the difficulties in transportation, power supply, location, etc. Thus this technique allowed mapping large areas difficult to be reached with heavy or large instrumentation.

Additionally, NH_3, O_3, VOC, H_2S were also measured to provide a first assessment of these pollutants. The considerable number of samples taken with the technique proposed allowed for the statistical treatment of data along the space coordinates, that is very useful in air quality management and control, due to the possibility to infer the statistical distribution of data and compare its parameters with reference values, such as mean, limit or standard values, etc.; additionally, emission source contributions were also preliminary deduced.

The representativeness of high polluted zones where to site emission source-oriented air monitoring stations (traffic, industry, etc.) was assessed in Suzhou by means of diffusive sampling technique giving the necessary findings to design the network of AQMS established in the city. The main focus was on the generation of meaningful data for further interpretation. The distribution trend of NO and NO_2 was used to identify the heavy traffic polluted zones: NO_2/NO_x ratios calculated and mapped by these measurements clearly demonstrated the different extent and completeness of NO$\rightarrow$$NO_2$ oxidation processes by time and districts. The high resolution air quality dataset obtained by field monitoring in Suzhou, representative of a Chinese open street canyon, was used to simulate the traffic-related pollutant distribution by means of CFD.

The results modelled were analyzed to better understand the values of gaseous pollutants measured by means of automatic analyzers along the same street-canyon at two different heights. The vortex recirculation pattern around the building geometry studied was correctly simulated, and it was very useful in assessing the micro-scale pollution trends. Finally, the development framework for an integrated system of measurement and modelling founded on the control of air pollution was presented. The results from the case-study of Beijing showed that the integration of modelling, to interpret the data measured, with measurements, to validate the data modelled, can offer useful tools to assess in real-time traffic-related air pollution.

However, there remains a need to continue research to improve our understanding of the mechanisms leading to impacts of air pollution emissions from transport, to reduce uncertainty in our ability to quantify relationships between all emissions and all impacts. For urban air pollutants from road traffic there are still some doubts concerning both the existence and the mechanisms of cause and effect, especially for NO_2 and particles air pollution, which are currently two of the pollutants causing most concern. Integrated assessment modelling has not yet been widely applied to the comparison of such markedly different impacts as on urban air quality and upper troposphere chemistry, but its application to future integrated systems has the potential to be extremely valuable and would therefore be an intellectually challenging and worthwhile development to pursue.

In conclusion, it seems that the new approach pursued in China on monitoring atmospheric pollutants by means of passive and active sampling and combining the results of the two approaches with models, may represent an important step for the advancement of pollution control strategies in developing countries to protect the local population from the effects which are today very impressive even in developed countries.

Ivo Allegrini and Francesca Costabile
Institute for Atmospheric Pollution,
National Research Council, Rome, Italy

REFERENCES

Allegrini I., Paternò C., Biscotto M., Hong W., Liu F., Yin Z., and Costabile F., 2004. According to the Framework Directive 96/62/EC, preliminary assessment as a tool for Air Quality Monitoring Network Design in a Chinese City. Air Pollution XII, series *Advances in Air Pollution*, edited by C.A. Brebbia, 14, pp. 414-424.

Atkins D.H.F. and Lee D.S., 1995. Spatial and temporal variation of rural nitrogen dioxide concentrations across the United Kingdom. *Atmospheric Environment* 29, pp. 223-239.

Barletta B., Meinardi S., Sherwood Rowland F., Chan C., Wang X., Zoud S., Chan L. and Blake D.R., 2005. Volatile organic compounds in 43 Chinese cities. *Atmospheric Environment* 39, pp. 5979-5990.

Carslaw N. and Carslaw D., 2001. The gas-phase chemistry of urban atmospheres. *Surveys in Geophysics* 22: pp. 31-53.

Colvile R.N., Hutchinson E.J., Mindell J.S. and Warren R.F., 2001. The transport sector as a source of air pollution. *Atmospheric environment* 35, pp. 1537-1565.

Costabile F., Fiorani L., Picchi S. and Allegrini I., 2004a. Intelligent Transport Systems for Traffic Air Pollution Management: Beijing and Vehicle emissions control. *Proceedings of the 13th World Conf. On Clean Air and Environmental Protection*, London, United Kingdom, 22 – 27 August 2004.

Costabile F. and Allegrini I., 2004b. An Intelligent Transport System based on Traffic Air Pollution Control. Air Pollution XII, series *Advances in Air Pollution*, edited by C.A. Brebbia, 14, pp. 541-550.

Costabile F., Wang Q., Bertoni, Ciuchini, Wang F. and Allegrini I, 2006a. A preliminary assessment of BTX pollution in urban air in the city of Lanzhou. *Air Pollution* XIV edited by C.A. Brebbia.

Costabile F., Desantis F., Wang F., Hong W., Liu F. and Allegrini I., 2006b. Representativeness of Urban Highest Polluted Zones for Sitting Traffic-Oriented Air Monitoring Stations in a Chinese City. *JSME International Journal*-B, 49 (1), pp. 35-41.

Costabile F., Wang F., Hong W., Liu F. and Allegrini I., 2006c. Spatial Distribution of Traffic Air Pollution and Evaluation of Transport Vehicle Emission Dispersion in Ambient Air in Urban Areas. *JSME International Journal*-B, 49 (1), pp. 27-34.

Costabile F., Wang, Hong, Liu and Allegrini I., 2006d. CFD modelling of traffic-related air pollutants around urban street-canyon in Suzhou. *Air Pollution* XIV edited by C.A. Brebbia.

Costabile F., Bertoni G., Desantis F., Wang F., Hong W., Liu F., Allegrini, I., 2006e. A preliminary assessment of major air pollutants in the city of Suzhou, China. *Atmospheric Environment* 40(33), pp. 6380-6395.

Costabile F., Allegrini I., 2007a. Measurements and analysis of nitrogen oxides and ozone in the yard and on the roof of a street-canyon in Suzhou. *Atmospheric Environment* 41(31), pp. 6637-6647 doi:10.1016/j.atmosenv.2007.04.018

Costabile F., Allegrini I., 2007b. A new approach to link transport emissions and air quality: an intelligent transport system based on the control of traffic air pollution. *Environmental Modelling & Software*, doi: 10.1016/J.envsoft.2007.03.001 (in press)

European Commission, 1998. Guidance Report on Preliminary Assessment under EC Air Quality Directives of January 1998. Project Manager Gabriel Kielland, European Environment Agency.

Fenger J., 1999. Urban air quality. *Atmospheric environment* 33, pp. 4877-4900.

Kukkonen J., Valkonen E., Walden J., Koskentalo T., Aarnio P., Karppinem A., Berkowicz R. and Kartastenpaa R., 2001. A measurement campaign in a street canyon in Helsinki and comparison of results with predictions of the OSPM model. *Atmospheric Environment* 35, pp. 231-243.

Mage D.T. and Ott W.R., 1984. An evaluation of the method of fractiles, moments and maximum likelihood for estimating parameters when sampling air quality data from a stationary lognormal distribution. *Atmospheric environment* 18, pp. 163-171.

Morel B., Yen S. and Cifuentes L., 1999. Statistical distribution for air pollutant applied to the study of the particulate problem in Santiago. *Atmospheric Environment* 33, pp. 2575-2585.

Pavageau M. and Schcatzmann M., 1999. Wind tunnel measurements of concentration fluctuations in an urban street canyon. *Atmospheric Environment* 33, pp. 3961-3971.

Scaperdas A. and Colvile R.N., 1999. Assessing the representativeness of monitoring data from an urban intersection site in central London, UK. *Atmospheric Environment* 33, pp. 661-674.

Shi J.P., Khan A.A. and Harrison R.M., 1999. Measurements of ultrafine particle concentration and size distribution in the urban atmosphere. *The Science of Total Environment*. 235, pp. 51-64.

Shi J.P. and Harrison R.M., 1997, Rapid NO_2 formation in diluted petrol-fuelled engine exhaust – a source of NO_2 in winter smog episodes, *Atmospheric Environment* 31, pp. 3857-3866.

Sun J. 1982. Intersection carbon monoxide pollution investigation in Beijing. *Research report of Beijing Institute of environment protection science*, Beijing, pp. 1-98.

UNEP/WHO: 1994, Passive and Active Sampling Methodologies for Measurement of Air Quality.

Vardoulakis S., Fisher B., Pericleous K. and Gonzalez-Flesca N., 2003. Modelling air quality in street canyons: a review. *Atmospheric Environment* 37, pp. 155-182.

Wang X., Shenga G., Fua J., Chan C., Lee S., Chan L. and Wang Z., 2002. Urban roadside aromatic hydrocarbons in three cities of the Pearl River Delta, People's Republic of China. *Atmospheric Environment* 36, pp. 5141-5148.

World Bank, 1997. Clean air and blue skies. The World Bank, Washington, DC, USA.

Zhou H. and Sperling D., 2001. Traffic emission pollution sampling and analysis on urban streets with high-rising buildings. *Transportation research* part D6, pp. 269-181.

M. MAZZON

SUSTAINABLE MOBILITY: MITIGATION OF TRAFFIC ORIGINATED POLLUTION

Abstract. This contribution will recall the European framework at the basis of urban sustainable mobility actions and will introduce how ITS (Intelligent Transport Systems) may help reaching sustainable mobility goals in cities. Basically, the objectives are to improve mobility and reduce the impact of traffic originated emissions over the air quality conditions. ITS technology may contribute through a better control of mobility (redesign of traffic strategy, traffic signal priority, traffic monitoring and control, etc.) and through a better management of public transport, freight delivery and urban logistics, garbage collection and the like. A practical example implemented in China will be discussed.

1. THE EUROPEAN FRAMEWORK TO MANAGE AIR QUALITY AND URBAN DEVELOPMENT

The Directives issued by the European Commission (EC) are acknowledged by the Member States and incorporated into the national legal frameworks.

A great impulse to a better planning of traffic was provided by the Directives on air quality, which established criteria related to acceptable limits and the pollution sources. While originally their rationale was to protect population health, recently also the principle of environmental protection was stated, hence the principle of sustainability and transport.

1.1. Air quality Directives

In the following the relevant Directives on Air Quality are summarized:

- Directive 96/62/CE issued in 1996 provides regulation for the evaluation and management of air quality. The most important factors determining air quality in the national environments were identified as Heating, Industry, Traffic, Cattle and Agriculture. In European cities, traffic is recognized as the most important pollution source.
- Directive 99/30/CE defines the admissible air quality limit values for SO_2, NO_2, NO_x, PM10 and Pb. It is established that air quality in all Europe shall shall respect those limits and that those values shall be respected within 2010, while a transitory regime is allowed for the period 2000-2010. All member states are requested to send to the Commission the national action plans to respect the stated limits and the measurements related to maximum values recorded yearly. Member States classify the respective territories into areas, related to the amount of population and other parameters, and mitigation programs are normally finalized to those areas, taking into

account their respective characteristics (sources of pollution, effects, cities, local conditions and the like).
- Directive 2000/69/CE establishes the limit values for benzene and carbon monoxide, as well as the allowed criteria for measurement and admissible control thresholds.
- Directive 2001/81/CE establishes National Emission Ceilings for SO_x, NO_x, NH_3, ozone precursors and other pollutants having acid eutrophication effects. Each Member State shall respect such ceilings by 2010 and decide autonomously how, provided that a National Program is issued and sent to the EC by 2006 and updated regularly.
- Directive 2002/03/CE establishes the target values and the threshold values for information to the public related to Ozone concentration. It grants that all Member States use uniform methods and criteria for ozone concentration to safeguard and improve air quality. Moreover it establishes that information related to concentration levels are available to the population and promotes the cooperation among Member States related to crossborder areas.

1.2. The European Strategy on Urban Environment

Besides air quality, also cities and sustainable urban development are given attention by the EU. On 11 January 2006, the European Commission issued a Communication to the European Council and to the European Parliament relevant to the "Thematic Strategy on Urban Environment" – SEC(2006) 16.

This Strategy is one of seven foreseen under the 6th Environmental Action Program. Its goal is to facilitate better implementation of EU environmental policies and legislation at the local level through the exchange of experience and good practice between Europe's local authorities. Four out of five European citizens now live in towns and cities and their quality of life is directly influenced by the state of the urban environment.

Among the issues of the Strategy Document, the central role of cities in Europe related to environmental degradation problems is recognized, and the requirement that the Regional and City Governments issue Integrated Development Plans considering Sustainable Mobility is established as well.

1.2.1. The Strategy's Approach
The Strategy premise is that Europe's urban areas face a number of environmental challenges including poor air quality, high levels of traffic and congestion, urban sprawl, greenhouse gas emissions and generation of waste and waste water. These can cause environmental damage and affect human health.

Local authorities have a decisive role to play in implementing environmental legislation and improving the environmental performance of a city. The best performing cities have developed integrated approaches to urban management where daily decisions are guided by a strategic vision and objectives. These can improve quality of life and the city's economic performance, which in turn can attract new residents and businesses.

While local action is essential, it is stated that public authorities at regional, national and European level also need to be proactive. To this respect, it is said that the EU can provide support by promoting Europe's good practices, and that it can do so best by encouraging effective networking and exchange of experience between cities. Many solutions already exist in cities but they are not sufficiently disseminated or implemented.

The Commission has coordinated the objectives of the Strategy with other Community programs and offers support for investments, research and demonstration projects on key urban environment issues such as investment in urban transport and reuse of derelict land, or training in urban management. Member States are suggested to exploit the opportunities offered at the European level to make improvements in the environmental performance of their cities.

1.2.2. Proposed Measures

The Strategy Document relies on the responsibility of the Local Governments (Regions, Municipalities) to manage their territory. In support to these responsibilities, the main actions under the strategy are:

- Guidance on Integrated Environmental Management and on Sustainable Urban Transport Plans. The guidance will be based on cities' experiences, expert views and research, and will help ensure full implementation of EU legislation. It will provide sources of further information to help prepare and implement action plans.
- Training. A number of Community programs will provide opportunities for training and capacity-building for local authorities to develop the skills needed for managing the urban environment. Moreover, support will be offered for local authorities to work together and learn from each other.
- Support for EU wide exchange of best practices. Consideration will be given for the establishment of a new European program to exchange knowledge and experience on urban issues under the new Cohesion Policy.
- Commission internet portal for local authorities. The feasibility of creating a new internet portal for local authorities on the Europe website will be explored to provide better access to the latest information.

The Guidance on Integrated Environmental Management and on Sustainable Urban Transport Plans is expected to be issued in 2006. As to the Integrated Environmental Management, indications will be provided about its function within urban politics, the minimum contents and flexibility they need to ensure, decisions for the implementation, selection of targets and goals, indicators for monitoring, evaluation of environmental performances. The importance will be underlined of considering not only the administrative urban boundaries, but the whole urban unit, as composed of different areas linked in a functional continuum (functional area).

The Guidance will also analyze Environmental Management Systems as instruments to put in practice the strategy of the Environmental Management Plans, show ways and criteria to define implementation procedures and indicate some preliminary conditions for the implementation.

1.3. The European Clean Air Strategy

The European Commission proposed on 21 September 2005 an ambitious strategy for achieving further significant improvements in air quality across Europe. The Thematic Strategy on air pollution aims by 2020 at cutting the annual number of premature deaths from air pollution-related diseases by almost 40% from the 2000 level. It also aims at substantially reducing the area of forests and other ecosystems suffering damage from airborne pollutants. While covering all major air pollutants, the Strategy pays special attention to fine dust, also known as particulates, and ground-level ozone pollution because these pose the greatest danger to human health. Under the Strategy the Commission is proposing to start regulating fine airborne particulates, known as $PM_{2.5}$, which penetrate deep into human lungs. The Commission also proposes to streamline air quality legislation by merging existing legal instruments into a single Ambient Air Quality Directive, a move that will contribute to the Better Regulation initiative.

1.3.1. Costs and Benefits Estimated

The Commission has sought the most cost-effective solution that is consistent with the objective of growth and employment (the Lisbon Strategy) and the EU Sustainable Development Strategy.

The Strategy is expected to reduce the number of premature deaths related to fine particulate matter and ozone from 370,000 a year in 2000 to 230,000 in 2020. Without the Strategy there would still be over 290,000 premature deaths a year in 2020.

It is estimated that the Strategy will deliver health benefits worth at least €42 billion per year through fewer premature deaths, less sickness, fewer hospital admissions, improved labour productivity etc.

This is more than five times higher than the cost of implementing the Strategy, which is estimated at around €7.1 billion per annum, or about 0.05% of EU-25 GDP in 2020. Although there is no agreed way to express damage to ecosystems in monetary terms, the environmental benefits of reduced air pollution are also significant.

1.3.2. Range of Measures Foreseen

Current air quality legislation will be streamlined to help Member States implement it better. A legislative proposal is attached to the Strategy which will combine the existing Framework Directive on air quality, its 'daughter' Directives and a Decision on exchange of information.

The proposed new Ambient Air Quality Directive would cut 50% of the existing legal texts, clarify and simplify it and modernize reporting requirements. For the first time it would require reductions in average $PM_{2.5}$ concentrations throughout each Member State and set a cap on concentrations in the most polluted areas.

At the same time, more flexibility will be given to the Member States. Where they can demonstrate that they have taken all reasonable measures to implement the legislation but are nevertheless unable to comply with air quality standards in certain places, it is proposed to allow them to request an extension to the compliance deadline in the affected zones provided that strict criteria are met and plans are put in place to move towards compliance.

The Commission intends to propose a revision of the National Emissions Ceilings Directive to bring its emissions ceilings into line with the objectives of the Strategy. In addition, a range of other possible measures will be examined, such as the introduction of a new "Euro V" set of car emission standards and other initiatives in the energy, transport and agriculture sectors, the Structural Funds and international cooperation.

2. ITALIAN POLICIES

2.1. General

This chapter reports how European Directives have been acknowledged and implemented in Italy, together with specific national laws, to set the pace to develop coordinated air and traffic management measures in cities.

The situation in Italy is still evolving and the new EU Strategic Lines will be likely followed in the near future. How the Directives and national laws are actuated in cities vary from city to city according to the specific situations, capability of local communities and financial resources available. It is always true, anyway, that restrictions to private car traffic are becoming more and more popular. The boosting of public transport, however, is perceived to be less than enough, despite the many efforts ongoing, such as the construction of tram and light rail connections in several cities, the integration of train and bus transport, the development of underground lines in some cities, and the purchase of new generation buses including natural gas and clean diesel buses. Incidentally, the possibility of adopting BRT (Bus Rapid Transit), the emerging technology for mass transport proposed as an alternative to metro, seems not be considered yet in Italy, despite its cost advantages over the metro itself.

Italy is one of the largest European Countries (population and economy is not too different from France, UK, Germany) having some 65 million inhabitants, and is subdivided into 21 Regions, the largest administrating some 6 million people. Each Region is subdivided into Provinces and Municipalities. The largest municipalities are Rome (some 4.5 million people), Milan (3.5 million), Naples (2.5 million), Turin, Genoa. Some 25 cities have more than 150,000 inhabitants.

The European Directives have been adopted by the Italian Government. As to traffic related provisions, the most important laws and decrees are reported below.

Originally, in May 1991, a Decree established criteria for air quality monitoring, stating characteristics of air quality monitoring stations, which pollutants shall be measured, measurement criteria and criteria for number and position of monitoring stations to be installed according to population and population density.

Again in 1991, another Decree established criteria for the preparation of Regional Plans for recovery and protection of air quality. Such Plans were to be prepared by the Regional Administrations and submitted to the Italian Ministry for Environment and Territory (IMET).

Decree 163, 21 April 1999 defined the criteria for traffic pollution measurement and data evaluation in urban areas, in particular regarding benzene, PM10 and hydrocarbons, as well as the criteria to be adopted for traffic limitation to keep pollution within limits.

The EC Directive 96/62/CE was adopted in Italy by Decree 351, 1999, while Directives 99/30/CE and 2000/69/CE were adopted by Decree 60, 2002. Now the Italian legislation is being revised accordingly and this will improve measurement methods and data quality.

Various pollution reduction programs are ongoing in Italian cities based on such Decrees and following laws.

Directive 2001/81/CE on National Emission Ceilings for SO_x, NO_x, NH_3 etc. was adopted in Italy by Decree 171/2004 and the required National Program for Reduction was sent to the EC in September 2003. According to that Program, Italy expected to exceed the 2010 ceiling for NO_x and NH_3 by some 5%.

2.2. Sustainable Mobility Policies

The Sustainable Mobility Decree was issued on 27 March 1988 by the Ministry of Environment in agreement with the Ministry of Transport and Infrastructures and the Ministry of Health, stating that:

- The Regions shall adopt a Regional Plan for the recovery of air quality parameters.
- The Mayors shall adopt measures to prevent and reduce pollution emissions, if pollutants are expected to exceed allowed limits.
- The Companies over 300 people shall identify a Mobility Manager that defines a plan to reduce private car use for the home-office trips and help reducing congestions through the adaptation of working hours. Mobility plans and achieved results shall be sent to the Municipality yearly.
- The Municipalities shall have a Mobility Manager to coordinate and harmonize company mobility plans and coordinate mobility overall.
- The Municipalities shall promote any measures to reduce private car use, such as promote associations or services for collective use of vehicles, promote multi-property of vehicles and the like, using electrical, hybrid, natural gas or reduced emission vehicles.
- Promote the use of electrical, hybrid, natural gas or reduced emission vehicles by Public Administrations.

Actuating the Sustainable Mobility Decree, the Decree 163, 1999 provides instructions to identify environmental and health criteria for the Mayors to establish traffic limitations and in particular gives power to the Mayors to limit or stop traffic in order to keep air quality within limits. It is valid for:

- All cities above 150,000 inhabitants.
- Cities being at risk for traffic and related pollution peaks.
- Cities where peculiar climate situations may get dangerous for the population.

Here, the Mayors are responsible to:

- Provide an air quality map for the Municipal territory, related to all relevant pollutants.
- Provide an annual report on air quality.
- Define traffic limitation actions, to be applied on an annual basis in those areas where even one of the parameters exceeds limits, and aimed at reducing pollution levels within law limits.
- Update yearly those traffic limitation actions and application areas.

Traffic limitation actions shall be:

- Finalized to the causes of pollution and not be temporary emergency actions.
- Activated for predefined periods (hourly, daily, seasonal intervals etc) and/or predefined types of vehicles based on their pollution level.
  Public transport and social services vehicles are excluded of course.

Among other rules, it is now established that all vehicles shall be subject yearly to an emission test at authorized workshops. A blue tag valid for one year is then attached to the vehicle windscreen. Vehicles without the blue tag are not allowed to circulate in the city central areas and are fined in case they do so.

2.3. Sustainable Mobility Programs promoted by the Ministry of Environment

Several Sustainable Mobility Programs were promoted by the Ministry of Environment after Decree 163, 1999, making some 220 Million Euro grants available to Regions and Municipalities, aimed at the permanent reduction of urban traffic environmental impact and energy consumption.

The basis was to try to gradually reduce the increasing use of private cars in cities, through the management of mobility demand, the improvement of attractiveness of public transport and the use of low impact fuels.

Highlights over the Sustainable Mobility Programs:

- Low impact fuels: to promote the use of methane and LNG, funds have been made available to adapt private cars to methane and LNG and to build gas refuelling stations for public fleets. Moreover, incentives were made available to public and private companies for the purchase of fleets of LNG vehicles for public utility services, taxi services, urban freight distribution, and vehicle rentals with drivers.
- Low impact scooters: an agreement with the National Scooter Association sustained research and production of new motorcycles (electrical and low emissions).

- the so-called Radical Sustainable Mobility Programs: 25 Municipalities have been co-financed for the development of specific programs such as:
    - Collective taxi systems.
    - Automated systems for road and area pricing (ITS applications).
    - Use of gas vehicles for public transport.
    - Fleets of electric bikes, electric scooters or vehicles for public utility applications or for rental in central areas.
    - Demonstration projects regarding electric vehicles and hydrogen vehicles, emission reduction provisions, correlation models between monitored traffic data and air quality data.
- National Car Sharing Program: introduction of car sharing systems has been co-financed for 5 cities. Car Sharing is aimed at serving people needing occasionally a car and allows renting one for some hours or a day at a reduced rate thanks to an efficient ITS booking, pick up and return mechanism. Electric or low emission cars, with proper onboard ITS instrumentation, are used for these applications.
- Mobility Management: funds have been made available to support activities of Mobility Managers set up by the Decree 1988 reported above and to promote a collective use of private cars and alternative transport for work commuters. In additions, discounts on public transport monthly tickets, innovative public transport systems, car pooling initiatives, telework initiatives have been launched, all together to help reducing the use of private cars.
- Financial support to electric/hybrid vehicles: through specific laws, funds have been made available for the purchase of alternative fuel vehicles.

2.4. Other Sustainable Mobility Initiatives and ITS

With funds from the Regional Governments and from the Ministry of Infrastructures and Transport, Municipalities are developing other Sustainable Mobility initiatives, many of which include significant ITS applications. For example:
- Traffic Limited Zones (ZTL) and Pedestrian Areas. A growing number of city centres are defined as Zones with Traffic Limitations (ZTL). Here, only authorized cars are allowed to enter and ITS systems are installed for access control, using either or both RFID (Radio Frequency Identification Devices) and CCTV (Closed Circuit TeleVision) with automatic plate readers. Some central areas are completely closed to traffic and only pedestrian and public transport are allowed there. City centres have become cleaner and more enjoyable after such initiatives.
- Encouragement to Public Transport through ITS improvements. ITS applications such as GPS fleet management and information to passengers are being applied to bus transport systems, in order to provide passengers with a better perception of the service and to encourage using it. In addition, other ITS applications are being developed to promote the use of

Public Transport through the unification of ticketing and payment systems in metropolitan or regional areas. Agreements are signed to this purpose between Municipalities, Bus Transport Companies, Rail Companies, and banks. The ticketing technologies are mostly contactless smart cards, sometimes hybridized with existing magnetic or paper tickets.
- Traffic Management Systems. The largest cities are developing integrated traffic management. In Rome, for example, the Mobility Agency has a large control room centralizing CCTV images, traffic light priority, traffic counting loops, congestion data, and information to drivers via large LED panels installed in focal points in the City. Similar systems are being developed in Turin and Milan. Floating Car Data systems to get actual traffic speed and congestion information are not yet installed in Italy, except in some highways, where FCD data are provided by the same RFID devices used for fee automatic payment.
- Bike routes. Many cities are developing dedicated bike routes, to encourage the use of alternative and ecological personal transport means such as bicycles. Those initiatives are having success especially during weekends and the forecast is good also for weekdays. It also happens that car streets are restricted to allow space for bike routes, hence car drivers feel more discouraged to use the car in the city.
- Use of new public transport vehicles. New and low emission buses, light rail systems and some underground metro are being developed for public transport in urban areas. Bus Rapid Transport applications seem not suited to Italian cities due to urban configuration (historic town planning structure, narrow streets, etc), while light rail looks more attractive as a low pollution and high capacity transport, as in many other European cities.
- New traffic plans: Several cities are revising traffic and mobility plans looking at sustainable mobility paradigms.
- Urban Freight delivery. New strategies for urban freight delivery are under study in Italy for many cities. There is a lot of work on this topic, since a significant amount of traffic is associated to freight delivery. ITS is one of the enabling technologies to improve urban logistics mechanisms. In Venice in 1996, an ITS multi-user GPS fleet management system was developed and tested to allow many delivery service companies plus public transport and public utilities companies to join under the same technology to better manage their services. The initiative was only left to the experimental stage but a new urban logistics plan has been devised afterwards, where one multiservices hub, small multiproduct distribution centres along the city and a standardized pallet system have been foreseen. Another example is Milan, where its urban logistics system has been studied and found to be quite inefficient and polluting, due to how it is left to the unplanned private initiative. Improvements have been devised and possible strategies identified.

3. HOW INTELLIGENT TRANSPORT SYSTEMS MAY HELP SUSTAINABLE MOBILITY

Intelligent Transport Systems applications may help significantly the implementation of policies to reduce air pollution through sustainable mobility actions.

In the following paragraph, some examples will be reported regarding ongoing experiences in Italy and China. Many other good practices have been implemented internationally, showing that ITS best role is in better planning and control of mobility in general and public transport and services in particular.

3.1. Applications of ITS

ITS helps managing sustainable urban growth and, as long as ITS will improve managing traffic and public transport service quality and contribute to reducing vehicles in circulation, it will also produce direct and indirect environmental benefits.

World over it has been recognized that transport infrastructure building has to be supplemented with:

- Intelligent management tools, technologies and systems, to plan and manage transport and traffic. Applications include: centralized, real time traffic monitoring and management; rational planning and control of Public Transport service; advanced traveler information systems; intelligent logistics management, etc.
- Pricing strategies, in order to help restricting car traffic and to internalize external costs according to the principle user/polluter pays (for increased illness costs, road use, pollution abatement, mobility measures, etc.)

The main applications of ITS are the following:

- Transport planning: ITS software applications help very much at transport planning stage, when trade off analyses shall be made to investigate, in simulation, the effects of the various options available. In general, investments to improve planning capabilities are considered among the very best way to spend limited public resources and having the best return: the better the planning activity is, i.e. the better it is done to respond to demands and to analyze alternatives, the fewer are the mistakes and the better is the money spent in actuating well planned measures.
- Public transport systems: Various applications do exist for public transport. Here, reference is made in particular to Bus Public Transport, which will be among the protagonists in any sustainable mobility action and ITS find very good applications in this sector, from bus planning and dispatching, to GPS localization, passenger information, service quality analysis. Planning deserves attention, because the use of advanced optimization techniques and software applications for bus service design may save number of buses and improve the reduction of air pollution, keeping the service level high

and maybe improving it. Bus GPS localization helps real time service control and allows to provide automatic passenger information messages onboard and at the stops, regarding arrival times. Contributing to improve service level and giving such perception in passengers, ITS helps attracting more users to bus transport, hence reducing private car use. Also for this reason, the maximum attention should be paid to improvement of bus transport.
- Traffic management: centralized traffic monitoring and management systems are in place in some pioneering cities and such systems show their importance to facilitate smoothing traffic and help reducing congestions. Traffic light priority, bus priority, floating car data (FCD), parking management and guidance are among the features of these systems. Basically, those systems include traffic counting systems at some sections and intersections, remote CCTV, Decision Support Systems, Variable Message Signs, interfaces to air quality measurement networks, control rooms and communication systems. Traffic management could be improved with TMC services (Traffic Message Channel) which use Floating Car Data to provide travelers via FM radio (Radio Data Systems, RDS) or GPRS with information on congestions and alternative routes.
- Traveler information services: A traveler information centre would be collecting information from the various public transport agencies, traffic management centre, car parks, railway and airlines operations and disseminate to users via radio RDS (Radio Data Systems), GPRS, Variable Message Sign, Internet, SMS and the like.
- Electronic payment systems: road charging systems as well as public transport ticketing are application areas where ITS find their way. Contactless ticketing allows users to use the same smart card to access numerous services, including payment of public transport and parking. Tickets validating machines may be managed by public transport onboard GPS computers to handle zone charging strategies and collect ticketing data.
- Zone access control: implementation of ITS for this purpose include the installation of portals along the streets entering the controlled zone to recognize authorized vehicles and to fine those not authorised to enter. Normally, RFID transponders installed on the car windscreen are used to this purpose as well as CCTV with plate reading software. Traffic counting sensors may also be included in the portals, while a control centre handles the information.
- City logistics management and commercial vehicle operations: Urban freight distribution and transport services represent another important issue, especially in congested cities. It has been shown that freight delivery traffic in urban areas account for some 20-25% of the total traffic, while there is a lot of improvement possible. For example, professional operator delivery fleets can serve some 40-60 destinations per day per vehicle, while a single producer (shop owner, artisan, manufacturer...) delivery vehicle is normally used for much less than 10 destinations per day. Here, too, fuel

and emission saving, as well as congestion reduction, are reachable: it has been observed that professional operator vehicles produce some 3 times less pollution than single operator vehicles, which are normally older, badly maintained and used less efficiently. Urban logistics plans can be developed in order to optimize freight distribution and access to urban infrastructures, while ITS solutions can be adopted to optimize traffic and delivery strategies. Low cost onboard GPS/Galileo localization and communication units and control centres can be used to manage freight delivery strategy and other services such as garbage collection, emergency services, ambulances etc. New services and employment can originate from the new urban logistics management programs.

3.2. An Application Example: The Beijing ITS-TAP Project

This pilot project is supported by the Sino-Italian Cooperation Program, established by IMELS, the Italian Ministry for the Environment Land and Sea and SEPA, the State Environmental Protection Agency of China.

The ITS-TAP project (Intelligent Transport Systems for Traffic Air Pollution) peculiar features are the following:

- It constitutes an innovative approach, where air quality management and traffic management are linked within a coordinated effort.
- It is a best practice example of cooperation between different Institutions, in particular the Beijing Environmental Protection Bureau, the Beijing Traffic Management Authorities and the Beijing Public Transport holdings, which are the main users of the system.
- It is one of the projects within the Beijing "Green Olympics" initiative, to contribute to prepare a better environmental quality for the Olympics.
- It is implementing a zone access control within the city centre, limited by the second ring road, which is dynamic (activated during the peak periods of pollution, based on actual air quality measurements) and selective (limited to the most polluting vehicles), and compensated by boosting public transport when zone access is activated.

The project implementation period is 2005-2008 and the project summary is reported below.

3.2.1. The Environmental Scenario in Beijing
Beijing City has a total area of 16,800 km^2 including 1,040 km^2 of planned urban area and 500 km^2 of built area. Its population is approximately 13,000,000 including 7,000,000 in built area. Beijing's motor vehicles approximate 3,000,000, some 30% of which are old-type ones, tail gas emission far exceeding the European Standard I implemented in recent years.

Due to high density of population, energy source consumption and vehicles in the built area, the air quality problem in the downtown area is relatively serious. According to the monitoring data from Beijing Municipal Environmental

Monitoring Center, the days in the year 2002 during which Beijing's air quality did not meet the national standard were 45%. Beijing city has taken many measures to improve urban air quality, including the use of clean energy sources, prohibition of big coal fired stoves, rebuilding of coal fired boilers, popularization of low-sulfur coal, moving of pollutant producing enterprises in urban area, control of various construction dust emission, use of leadless gasoline, implementation of strict standard for motor vehicle tail gas emission etc. Beijing's air quality has been evidently improved through various efforts. However, it still faces an austere situation.

Presently Beijing has taken some measures to reduce traffic pollution:

- Executing new standards for newly added motor vehicles such as Europe standard I and Europe II emission standard put into practice in 2003.
- Rebuilding old vehicles and strengthening their test and control. Attaching yellow label to the vehicles that have not reached the new standard.
- Constructing and rebuilding roads to expand road network traffic capacity.
- Improving road traffic control system and increasing travel speed etc.

Beijing has established an Air Quality Monitoring System and has provided data support for knowing the situation of air pollution and its development trend in the city. However monitoring networks for traffic environment have not been established and capability of dynamically testing the actual road traffic pollution situation has not been obtained yet. It lacks the control measure and execution capability of improving the condition of traffic pollution through the adjustment of traffic flow and travel speed and the technical monitoring means of executing traffic limit of motor vehicles in specific areas when air pollution is serious.

In order to effectively control the development trend of continuous rise of traffic pollution, create a good environment for Beijing 2008 Olympic Games and provide long term support for the improvement of Beijing's air quality, it was decided that an important integral part of Beijing's urban air quality improvement has to be a plan to establish traffic pollution monitoring network and combined control of motor vehicle travel and public transport improvement.

To this purpose the Beijing Environmental Protection Bureau operates in close cooperation with the Beijing Traffic Management Authorities and with BPT, the Beijing Public Transport holdings.

In the above scenario of active cooperation, the ITS-TAP strategy is to combine private traffic temporary limitation actions with improvement of public transport services, within a framework of maintenance of desired air quality levels.

3.2.2. The Public Transport in Beijing
Beijing Public Transport Holdings, Ltd. (BPT) is a large state-owned enterprise, mainly operating road public transport. By the end of 2004, BPT had over 24,000 operating vehicles of various kinds, among which some 17,000 buses on 750 routes, with annual mileage reaching 1.4 billion km. The over 4.3 billion passenger trips served by bus exceed 80 percent of the total transit modes in Beijing.

By the Olympics year, bus number will be 23,000 on 900 routes, while annual mileage will reach 1.7 billion km. All buses will meet environmental protection requirements. New transport multimodal junctions and 24 centre stations and intelligent operation management are expected to be in place by the Olympics year.

New strategies will be launched through the ITS-TAP project: new service planning technologies will optimize bus shifts, while new generation services such as synthetic voice messages onboard and passenger information onboard and at stops will improve public transport attractiveness reducing the use of private cars: together, those actions will help achieving environmental benefits.

Real time bus localization and service management technology (AVL Automatic Vehicle Localization) will increase operational efficiency, reduce costs and help the Company to manage the new challenges. The results of ITS-TAP will bring around a good example that lays the foundation for future expansion within BPT and China.

3.2.3. ITS-TAP System Strategy and Configuration

The final objective of this project is to help improve air quality of traffic environment in Beijing city and know the pollution situation of Beijing's road traffic, based on the following approach:

- Establishing a monitoring system composed of fixed and movable traffic environment pollution devices.
- Increase travel speed and reduce gas pollution, by improving the command and dispatch capacity of a bus pilot fleet and by a unified planning of the traffic flow in the pilot road sections.
- In the mean time, identify the vehicles entering the highways within the second ring road through a number plate identification system so as to ensure the effective execution of travel limiting measure for yellow-label vehicles.

The project innovative aspect is to help reducing the traffic generated air pollution through a dynamic strategy activated on condition, based on a Decision Support System (DSS). The DSS is a software model operating on actual traffic and environment data such as, when air pollution is predicted to exceed given thresholds, then private cars will be prevented to enter the city central area within the second ring road for a proper period of time. The restriction is not generalized but limited to the most polluting cars, the so-called yellow tag cars. Public transport service will be boosted accordingly.

The Data Center is hosted in the facilities of EPB, the Environmental Protection Bureau of Beijing, while the Bus management system will be operated by BPT, Beijing Public Transport Holdings.

The buses will be those serving the connection between the Olympic Village and the city centre. A brief description of subsystems follows.

Air Quality Monitoring System. It is composed of a network of 6 fixed stations, 3 unconventional mobile stations, 30 saturation stations and 3 sensing instruments for vehicle emissions dynamic measurements. Parameters measured include typical gas and particulate matters, meteorological data, atmospheric stability data.

Traffic Monitoring System. The objectives of this subsystem are to process traffic flow data and to provide facilities to limit access to the inner third ring road to unauthorized vehicles, i.e. the yellow tag vehicles, when air conditions will request so.

According to the tail gas emission test and experimental result of various motor vehicles both at home and abroad, the tail gas emission amount of old vehicles accounting for 10-20% of total motor vehicles can account for 40-50% of total tail gas emission amount of motor vehicles. Therefore, control of the emission and travel area of these vehicles can generate good results. In case of unfavorable meteorological conditions and serious air pollution, it is therefore necessary to prohibit the yellow-label vehicles and the vehicles with high pollutant emission from entering the second ring road.

To reach the objective, the ITS-TAP project will provide section control portals, installed at the 22 entry lanes of the Second Ring Road. Each section is equipped with triple radar sensors (doppler, ultrasonic and passive infrared) for vehicle detection and classification and TV equipment to automatically record the plate number of vehicles entering the third ring road and compare it with the yellow-label vehicle database, to retrieve the vehicles belonging to yellow-label list so as to punish the vehicles violating travel-limiting regulations.

Public Transport Management System (PTMS). Under normal conditions, reasonable dispatch of buses can generate various benefits. If, for example, it is ensured that the need for conveying passengers is met, a reduced number of buses in service, an increase in vehicle use rate, a reduction of staff, lower costs and a reduced energy consumption can generate some environmental benefits. Therefore, it is rational considering both economy and environment.

The system for ITS-TAP involves initially 200 Buses of BPT. An advanced Control Centre runs a service planning and dispatch software and an AVL (Automatic Vehicle Localization) and passenger information software, for real time bus localization and service monitoring. Each bus onboard system includes a state of art automotive computer featuring GPS localization, GPRS communication, touch-screen driver terminal, voice and LED information messages to passengers, passenger counters, external LED displays. Bus arrival time and other messages will be automatically displayed at 20 stops equipped with LED panels.

Data Centre. This subsystem will integrate the central signal processing system of the traffic pollution monitoring network, the traffic flow monitoring and control system and the pilot bus fleet dispatch centre into a GIS platform (Geographic Information System). A Traffic-Environmental Model for the appropriate analysis of pollution and traffic effects and possible forecast is being implemented within a Decision Support System. This model will be fed by the ITS-TAP traffic counting and classification data and by the air quality monitored data.

3.2.4. Training

The ITS-TAP system is made of various specialized subsystems requiring not only the mere technical skills necessary to be operated, but also the implementation of some organizational adjustments in order for the users to take the most from them.

In particular, the Data Centre and the lane traffic monitoring systems interactions with the Beijing Traffic Management organization shall be analyzed and put into practice, while the air monitoring network will be managed by a somewhat well established organization within BEPB. The Public Transport Management System will also have an impact within BPT organization, since the relevant technologies and management structure are new for the Company.

The ITS-Tap scope therefore includes and extensive training program, done both in Italy and in Beijing, dedicated not only to the system technical operators but also to the managers of the involved user organization. To this purpose a careful transfer of experience and practices will be implemented from the Italian partners to the Chinese partners.

3.2.5. Conclusions

ITS-TAP is one of the sustainable transport pilot projects being implemented in Beijing to mitigate negative effects of traffic, that are becoming now the most significant air pollution sources in large cities. Practical actions to reduce social costs of traffic and improve quality of life are becoming a priority in all medium and large cities worldwide.

Among the innovative aspects of ITS-TAP is the attempt to correlate directly the measured air quality and the measured traffic data, in order to proceed with mitigation effects and traffic limitation in a rational way. The strategy is to minimize the lack of mobility through

- the selection of the type of vehicles to forbid from entering the controlled area
- a rational decision about the duration of traffic limitations.

Among the advantages of the approach is the possibility to adapt the strategy to other classes of vehicles and to other requirements that could rise in the future.

The adoption of new technologies for public transport dispatch and real time bus localization and management is aimed not only at saving costs and emissions, but also at attracting more passengers thanks to the perception of a better service provided and therefore, in perspective, at contributing to reduce private traffic percentage, which is a must in city sustainable transport.

Marino Mazzon
Intelligent Transport Systems,
Thetis S.p.A., Venice, Italy

IV. WASTE MANAGEMENT

L. MORSELLI, I. VASSURA AND F. PASSARINI

INTEGRATED WASTE MANAGEMENT. TECHNOLOGIES AND ENVIRONMENTAL CONTROL

Abstract. Waste is generated by activities in all economic sectors involving loss of materials and energy and imposes economic and environmental costs on society for its collection, treatment and disposal. Each material has its own chemical and physical characterization, environmental impact, recycle and re-use options. Waste management can be performed according to an integrated system, which includes reuse, recycling, composting, incineration with energy recovery, and safe disposal, with the objective of the maximum recovery of matter and energy and the minimum impact on the environment.
This paper gives an overview of the trend in waste generation and management in Europe and in Italy, of the principal waste treatment processes and of environmental control supporting management decisions.

1. INTRODUCTION

A widely accepted definition of waste in Europe is:

> any substance or object which the holder discards or intends or is required to discard, to the provisions of national law in force (Directive 75/442/EEC).

Waste is generated by activities in all economic sectors and is generally regarded as an unavoidable by-product of any economic activity. However, the problem of waste management arises also due to the unsustainable consumption systems, typical of developed countries.

A recent estimation of the production of waste in European Countries (Eurostat, 2003) reported that about 560 kg per capita of household waste was produced by the first 15 EU Member Countries.

The generation of waste arises both from production and consumption of goods: from the extraction and exploitation of natural resources, which result in the generation of mine waste and the consumption of energy; afterwards, the processing of raw materials for industrial production generates by-products and other waste streams and emissions, besides consuming energy. The production of energy is a parallel environmental issue, due to its indispensability in all industrial processes, which also causes production of waste and emissions in fuel extraction, refining and combustion (the energy deriving for the most part from the burning of fossil fuels). Finally, the consumption of goods implies a significant waste production, including food scraps, packaging, and all end-life commodities.

Thus, the generation of waste reflects a loss of materials and energy and imposes economic and environmental costs to society for its collection, treatment and disposal.

The final destination of goods has been recognized as a crucial issue, which has been neglected for too much time and could always be considered, in the design of a production process. Its importance is due to the costs, which could be avoided, in terms of economic burden relevant to the treatment and disposal of waste, but also of social acceptability and environmental impacts.

Therefore, it is easy to understand that waste management and policy of sustainable development are closely linked.

In order to take into account all the use phases of the products, not focusing only on some of them, the "life-cycle thinking" has been recently proposed as a strategic approach to be introduced into waste policy.

As it can be seen in Fig. 1, from the formulation and design of the product, to the final recycling and energy recovery, all phases are closely interconnected.

Environmental impacts can be associated to each of these steps; but without an overall view of the cycle it is not possible to optimize the process, identifying the most critical steps, whose improvement could result in a substantial reduction of concerns from an ecological viewpoint.

In particular, as concerns waste management, many processes are indicated by the scheme reported in Fig. 1, which deals with the product after its commercial use: maintenance, repair, reuse, recycling, (energy) recovery, incineration, and final disposal.

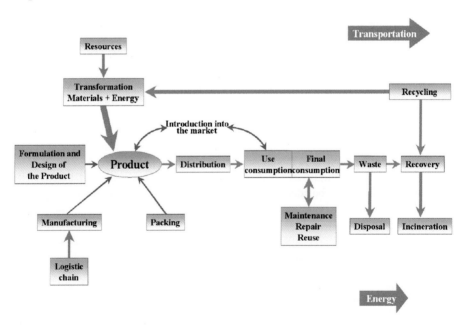

Figure 1. Steps in the production, use, treatment and final disposal of a product.

The production of waste fits directly into the framework of the interaction between environmental and economic systems. In the logic of sustainable development, the intent is to achieve the aim of "more with less", that is to say: more value from goods and services, improved industrial production, economic and social development, coupled with less employment of energy and raw materials, lower amount of solid waste, wastewater, atmospheric emission causing environmental pollution and less heat dispersion.

Since these preliminary observations, waste management can be then regarded as an integrated system of many treatment processes, aiming at the maximum recovery of matter and energy and at the minimum impact on the environment.

Currently, EU waste policy is based on a concept named "the waste hierarchy". It is aimed at maximizing all the activities which can firstly prevent waste production, and then to reuse, recycle, compost it (in general, to recover material with the proper processes, according to their physical-chemical characteristics), and finally to recover energy (through waste-to-energy processes), in order to use landfill as little as possible, and, in any case, in conditions of safety.

The Commission of the European Communities has recently compiled a communication concerning an updated strategy for waste management (Commission of the European Communities, 2005), in which the basic objectives of EU policy are clearly indicated: to prevent waste and promote re-use, recycling and recovery so as to reduce the negative environmental impacts. In order to achieve these goals, the action proposed is to introduce lifecycle analysis in policymaking and to clarify, simplify and streamline EU waste law.

In this approach a further step with respect to the waste hierarchy is made: indeed, a life-cycle assessment could address decision makers to differentiate interventions according to considerations at a territorial level, associated to logistic, industrial and waste characteristics. In fact, even though recycling could be the best solution in principle, the current lack of industrial processes in the governmental district dealing with a particular waste fraction (for instance: plastic material) could lead to different choices (for example: incineration, to exploit at least its heat value).

The same thinking could be applied to different collection systems, which must be located in the particular territorial reality.

1.1. Trends in waste generation and management in Europe and in Italy

At present in the EU, municipal waste is on average managed in the following proportion: recycling and composting 33%, incineration 18%, and finally landfill 49%.

However, the situation is not homogeneous among all the Member States, ranging from those who recycle least (90% landfill, 10% recycling and composting, as in Greece), to those who better meet the requirements of sustainable management (10% landfill, 25% energy recovery, 65% recycling: Denmark and Belgium in particular are the most virtuous).

The major waste producers are Germany (above 55 million tons/year), followed by France (up to 56 million tons/year), Spain (about 35), Italy (about 26) and United Kingdom (up to 25).

It must be added also that for some key waste flows, recycling and recovery targets have been set by European legislation, i.e. packaging, end-of-life vehicles (ELVs) and waste electrical and electronic equipment (WEEE).

As for Italy, the most recent data about municipal waste management are reported in Fig. 2 (APAT & ONR, 2005). It can be observed that the percentage of waste disposed in landfills is still greater than the half, thus classifying Italy above the average of EU Countries. However, a significant improvement in waste management has been performed in the last few years, due to the promulgation of a governmental decree (Ministry of Environment: Decree 22/1997) which implemented the European Community directives 91/156/EEC on waste, 91/689/EEC on hazardous waste and 94/62/EC on packaging and packaging waste and gave the impulse to carry out a structural adjustment of waste management system in Italy.

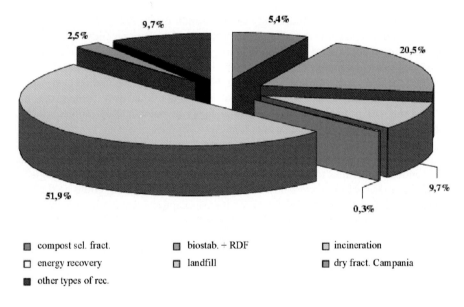

Figure 2. Waste management in Italy (Source: APAT & ONR, 2005).

The management of various packaging waste fractions in Italy is also worthy of mention. Different percentages of each fraction were targeted by the Italian law. In particular, it was:

- 60% for paper and cardboard, and glass;
- 50% for steel and aluminium;
- 22.5% for plastic materials;
- 15% for wood.

According to the last data available (APAT & ONR, 2005), the majority of goals have already been achieved; in particular, for wood, the current technology in Italy allows a much more satisfying result than the set target.

In particular, steel and glass were recycled for the amount of about 55%; the percentage of aluminium recycling reached almost 50%, while it was of 25% for plastic materials (this is the kind of waste which suffers more the lack of suitable and effective industrial processes of recycling); the best performances were those of paper and wood, both around 60%.

To these percentages, the fractions of materials directed to energy recovery must be added; these are around 20% for plastic materials, around 10% for paper and cardboard and aluminium and about 1% for wood.

However, this does not relieve the Italian waste management system from the duty of improving collection and recycling, in particular in those Regions where there is still a late implementation of the law targets on the territory.

1.2. Waste Classification: European Waste Catalogue

According to European directives, waste is classified on the basis of a Catalogue (EWC: European Waste Catalogue), which distinguishes the source (e.g., an industrial process), the specific step of production, and the particular kind of waste. The list is periodically reviewed on the basis of updated knowledge and, in particular, of research results, and if necessary revised. Different kinds of wastes in the list are fully defined by the six-digit code:

$$00\ 00\ 00\ *$$

where:

- the first two digits define the source generating waste (from 01 to 20)
- the second two digits distinguish the specific production unit
- the third two digits characterize a specific waste
- (* is present if the waste is considered hazardous)

However, in order to know physico-chemical and commodity characteristics of wastes, sorting of waste is periodically performed in each Municipality, according to an Italian commonly-used methodology: waste sample is sorted in 18 categories (fines < 20 mm, glass, other incombustibles, metals, aluminium, batteries, drugs, packaging with toxic and/or flammable substances, other hazardous waste, textiles, leather and hide, plastic films, plastic packaging, other plastics, fermentable organics, paper and cellulosic materials, cardboard, wood). Fines are sorted according to the size (10-20 mm, 5-10 mm, 3-5 mm and < 3 mm). Coarse wastes are separated, weighted and sorted in different categories.

This kind of analysis is of the utmost importance to plan the management of waste produced in a Province or in a territorial unit: for example, a separated collection of humid fractions could be improved, if the resulting low heat value falls under the limit suitable for an operating incineration plant. Furthermore, the knowledge of the current status is necessary to assess the treatment and disposal capacity of different plants at a territorial level, improving, if necessary, the dimensioning of the existing system.

2. THERMAL TREATMENTS

Combustion, gasification and pyrolysis are the thermal conversion processes more widely available for the treatment of solid wastes.

They could be distinguished on the base of different working conditions: temperature, pressure, oxygen concentration, reaction rate, etc.

In particular, pyrolysis is a process conducted in an inert atmosphere (which means without oxygen; stechiometric rate = oxygen/fuel = 0), at the atmospheric pressure, at temperatures ranging from 250 to 700°C. It produces gas, condensable vapour and a solid char. The gases produced are: H_2, CO, C_xH_y (different hydrocarbons), N_2, while solid residues are composed by ash and coke.

Gasification is the conversion by partial oxidation at elevated temperature of a carbonaceous feedstock into permanent, non-condensable gases. Thus, it requires an oxygen-poor atmosphere (stechiometric rate < 1), and is managed generally at high pressures (up to 45 bars), at high temperatures (ranging from 800 to 1600°C). Gaseous products are mainly H_2, CO, CH_4, N_2, while the solid one is slag.

Incineration differs from the previous treatments because of the presence of oxygen excess (stechiometric rate > 1) which reacts with the combustible contained in waste. Air is used as comburent, at atmospheric pressure, at quite high temperature (generally ranging from 850 to 1400 °C). It produces gaseous emissions mainly composed by CO_2, H_2O, O_2, and N_2, together with solid bottom and fly ash and heat.

Heat is generally exploited, to recover energy and transform it into electricity or district heating (energy recovery in Europe is now a compulsory requirement). Bottom ash is generally landfilled but in some countries it can be recovered and reused, for example for road construction. Fly ash, instead, can be disposed in landfill only after a process of inertization.

Generally, incineration is applied for the treatment of a very wide variety of wastes, because it needs less restrictive conditions, and can be applied to undifferentiated MSW (Municipal Solid Waste).

Pyrolysis and gasification are less widely applied to wastes, and generally to a narrower range of wastes; they usually need a pre-process for the homogenization and comminution of feedstock. Thus, the overall economic cost is still higher for these two processes, compared to the simple incineration, even though many researches are performed to assess their application to particular waste fractions (for instance, for tyre treatment).

In Fig. 3 the different processes and products of thermal treatment are reported.

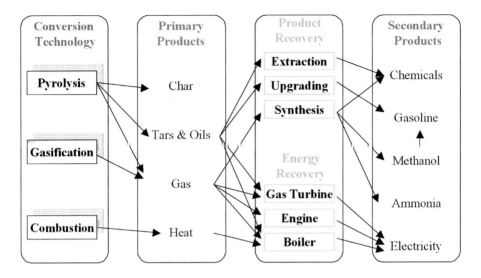

Figure 3. Thermal conversion processes and products (Bridgwater, 1994).

3. COMPOSTING

Organic materials contained in waste, such as food scraps, wood, foliage, etc., can be easily converted by the composting process. These materials undergo an oxidizing digestion process to decompose the organic matter and kill pathogens. The organic material is then recycled as mulch or compost for agricultural or landscaping purposes.

There is a large variety of composting methods and technologies, differing in complexity from a simple window composting of shredded plant material, to automated enclosed-vessel digestion of mixed domestic waste.

The process of composting consists in the following main phases:

- Mesophyte Phase or latency: it may last few hours or some days. During this time, micro-organisms invade the initial matrix.
- Thermophile Phase or Stabilization, which can persist for few days or some weeks. In this phase a severe bio-oxidative activity occurs, with a high oxygen consumption and production of CO_2. Temperature grows up to 50-65°C.
- Cooling Phase or Ageing. It persists for few weeks up to some months, a period during which humification-reactions take place. With the continuous reduction of organic substance, the temperature decreases, reaching the ambient value.

The quality of compost depends on the presence of heavy metals, organic pollutants and non biodegradable substances. With regard to this, a much better performance can be achieved in composting process, if a separate collection of organic fraction is carried out by the consumers, rather than implementing a mechanical separation of humid fraction from undifferentiated municipal solid waste; indeed, in the latter case, there could be the risk of finding pollutant residues in the final product, which could dramatically decrease its employability.

A typical composting process consists of a mix of pruning and OFMSW (organic fraction of municipal solid waste), an accelerated bio-oxidation and maturation; after this, a refining process separates fine fraction from the organic one; finally, by means of a finishing process the green compost is produced.

Among the critical points of the composting process the following ones can be identified for their importance: emission of bad odours, production of leachate, dispersion of particulate and pathogenic microorganisms. Thus, the optimization of process management is very important, both to reduce these impacts and to produce a well stabilized final product. In order to limit air emission, scrubbers or biofilters can be also successfully employed.

4. LANDFILL

The dumping of garbage in landfills is the most traditional method of waste disposal, and it remains a common practice in most countries.

The characteristics of a modern, well-run landfill should include methods to contain leachate, such as clay or plastic liners. Generally, liners are made of several layers: clay, geotextile fibres, mineral layer, a plastic membrane and, over all, gravel.

Leachate is then generally collected and sent to sewage treatment plants.

Disposed waste should be compacted and covered to prevent vermin and wind-blown litter.

Another important parameter to consider, due to its undesired environmental effects especially on air quality (production of odorous species) and climate change, is biogas production. This is due to the degradation processes of organic fraction buried in landfill and can be described according to the following 4 phases:

I. aerobic;
II. anaerobic, non-methanogenic;
III. anaerobic, methanogenic, unsteady;
IV. anaerobic, methanogenic, steady.

During these phases, extensive variations in gas composition occur:

- in the first one, a strong decrease of N_2 (from 80% to around 50%) and O_2 (from 20% to less than 10%) occurs, while CO_2 (from 0 to about 30%) and water (up to 10%) begin to be produced by aerobic reactions;
- in the second phase the same process continues, reducing further the percentage of N_2 (under 30%) and O_2, with a concurring increase of H_2O and CO_2 (which reach the maximum values, up to 10% and nearly 70% respectively);
- in the third phase a rapid drying of the material, and a sharp increase of CH_4 occur, which, at the end of the phase, reaches a value of about 55%;

CO_2 decreases to less then 40%, while the remaining 5% is covered by the residual N_2;
- in the final phase, a stabilization of the percentages achieved by the gases in the previous step take place.

Even if the last European Directive on landfills forbids the disposal of waste containing a significant amount of organic carbon, biogas production continues for many years. Thus, most landfills have been equipped with biogas recovery plants; this device can be used to burn biogas and then to recover energy. This can significantly reduce not only the contribution to greenhouse gases emission converting methane into carbon dioxide, but also the global pollution due to the avoided emission from ordinary power plants, which employ fossil fuels (from a life-cycle point of view).

5. ENVIRONMENTAL CONTROL IN WASTE LIFE-CYCLE

It has been already said that every activity of waste treatment and disposal (recycling, composting, incineration, landfill) causes an environmental impact, due to the fact that waste is generally a complex matrix, which needs a consumption of other matter and energy to be transformed into something useful or to be disposed in safety.

As can be inferred from the previous descriptions of the different treatment activities, environmental pressures are produced on the atmosphere by the emission bearing pollutants, ozone-depletion or greenhouse gases, or other substances inducing photochemical smog; on water, due to the production of sewage sludge, percolate, leachate; on the soil system, due to the deposition of pollutants and to the activity of landfilling.

The determination of possible environmental impacts on the surrounding territory, but also on a regional and global scale, is of utmost importance, in particular for decision makers, who are impelled by public opinion and by the law in force, to manage waste optimizing expenses and resources and producing the lowest environmental damage.

Two important tools for the assessment of this issue, both at a general and at a local scale, are the application of Life Cycle Assessment methodology and an integrated environmental monitoring system (Morselli *et al.*, 2005).

5.1. Life Cycle Assessment

As above stated, a growing importance is assigned to the analysis of the entire life-cycle of a product or a process. Even in the field of waste treatment and disposal, the application of LCA (Life Cycle Assessment) is meeting with growing interest, to gain a better knowledge of the environmental impact associated to the various steps of a process or to compare alternative solutions for the management of the same amount of waste.

LCA methodology is aimed at assessing all the impacts concerned with the entire life cycle of a product or a process. For instance, the environmental impacts of a product is assessed "from cradle to grave", i.e. considering those impacts produced from the extraction of raw materials and the involved consumption of energy, to those due to the final disposal of the product after its use.

The LCA applied to integrated MSW management introduces great potential for development, especially in support of the decisions of planners and companies that manage waste collection, transport and recycling/disposal services.

The scheme of a LCA methodology has been codified by ISO14040 series (International Organization for Standardization, 1997) of standards; a synthetic view is reported in Fig. 4.

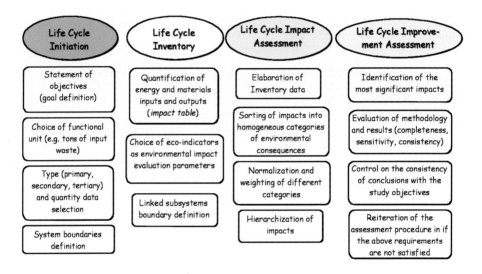

Figure 4. Structure of LCA methodology (ISO14040 series, 1997).

A comparison LCA can be performed to confront two different kinds of treatment or disposal for the same amount of waste. The impact assessment phase allows the researcher to understand which critical steps of the process result in the more significant environmental impacts. These impacts are generally chosen among the main concerns at local, regional and global levels. Generally, the most considered environmental categories for the life cycle impact assessment are: atmospheric acidification; eutrophication; depletion of non renewable resources; global warming; aquatic ecotoxicology; terrestrial ecotoxicology; human toxicology; photochemical oxidant formation; stratospheric ozone depletion.

An important concept introduced by LCA is the "avoided impact": it is associated for instance to energy production meaning that the environmental impact which could be produced by the production of energy according to traditional processes in a country (mainly combustion of fossil fuels) is avoided, just due to a suitable exploitation of the energy content in waste.

The same can be said for material recovery and recycling: it means that this stuff substitutes raw material (avoiding mining and depletion of natural resources);

furthermore, any avoided recourse to landfills means a lowering of environmental impacts, at least in terms of land occupation.

5.2. Integrated Environmental Monitoring System

Nevertheless, since the more remarkable environmental impacts are generally produced in the vicinity of a plant which deals with waste treatment, the investigation of the environmental matrices in the most invested area becomes necessary, in order to complete the conclusions obtained by a LCA application.

An important tool for the evaluation of impacts associated to MSW treatment has been defined Integrated Environmental Monitoring System. This approach can be synthetically described as a sequence of different phases (Morselli *et al.*, 2002).

The first can be defined as the characterization of the contamination source. It means that the industrial plant (or the point source) must be known in terms of ecobalance: this consists in the calculation of mass and energy balance, and if possible, in the normalization of emission rates (for example, for single units of product or waste treated) so as to provide "emission factors", able to compare environmental performance of different plants which work in similar ways.

The second step in the choice of Environmental Indicators. This is a crucial step, because it is necessary to identify quantitative (preferably) parameters which describe the phenomenon which is to be studied. They may be directly ascribable to the contamination source; they have to be easily measurable, from a sampling and analytical point of view; they must be of environmental interest. An example of these environmental indicators, in the case of incinerators, could be heavy metals, or dioxins.

After this, by means of dispersion models, a preliminary investigation of the surrounding areas more or less affected by the emissions from the contamination source is made.

The previous step however does not relieve one from performing an experimental investigation on the environmental receptors (such as plants, soil, atmospheric depositions and particulate); the analysis of the environmental indicators chosen in natural matrices will lead to estimate the real influence of the contamination source in the vicinity.

After all, it is indispensable to perform a data elaboration, taking into account all the data deriving from the analysis of source and of natural sink of pollutant compounds. The final result can lead to the recognition of a cause-effect relationship between a point source of contamination and the pollution found in the surrounding environment. A particular aid can be provided by the knowledge of chemometric techniques, in particular to study the apportionment of pollution to different contamination sources.

6. CONCLUSION

Waste management is a complex subject, directly linked to all the activities concerned with sustainable development. For this reason, the responsibility of politics is fundamental, in sustaining and improving virtuous solutions, involving the population and the productive system in the direction of an advance from economic, social and environmental points of view.

However, it is also an issue in which a real interdisciplinary expertise is needed. The research of innovative technologies, the knowledge of legislation, the ability of applying monitoring and control tools, a management skill: all these capacities have to meet, in order to improve the efficiency of administration of this ever growing environmental issue.

Many possible solutions have been widely pursued, resulting in a decrease of land exploitation, of matter and energy consumption, in an increase of environmental benefits.

Developing Countries could take advantage of the past experience of more industrialized countries, trying to achieve the same results in a shorter time and with less environmental damages.

To this aim, a strong interaction and a wide scientific collaboration is highly advisable, in order to disseminate technical expertise and to cope with environmental problems arising from this matter, from a global point of view, as these days require.

Luciano Morselli, Ivano Vassura and Fabrizio Passarini
Industrial Chemistry Department,
University of Bologna – Rimini Branch, Italy

REFERENCES

APAT (Italian Environmental Protection Agency) & ONR (National Observatory on Waste), 2005. *Rapporto Rifiuti 2005*. Rome, Italy.

Bridgwater A.V., 1994. *Catalysis in thermal biomass conversion. Applied Catalysis* A: General 116, pp. 5-47.

Commission of the European Communities, 2005. *Taking sustainable use of resources forward: A Thematic Strategy on the prevention and recycling of waste*. COM(2005) 666, Brussels, Belgium.

Eurostat, 2003. *Towards environmental pressure indicators for the EU*, second Edition, Eurostat, Luxembourg.

International Organization for Standardisation, 1997. ISO14040. *Environmental management – life cycle assessment – principles and framework*, Switzerland.

Morselli L., Bartoli M., Bertacchini M., Brighetti A., Luzi J., Passarini F. and Masoni P., 2005. Tools for evaluation of impact associated with MSW incineration: LCA and integrated environmental monitoring system. *Waste Management 25*, pp. 191-196.

Morselli L., Bartoli M., Brusori B. and Passarini F., 2002. Application of an integrated environmental monitoring system to an incineration plant. *The Science of the Total Environment 289*, pp. 177-188.

A. MASSARUTTO

ECONOMIC ANALYSIS OF WASTE MANAGEMENT SYSTEMS IN EUROPE

Abstract. The paper sketches the main regulatory economic issues characterizing waste management policies in the EU context. On the background, the evolution of solid waste management (SWM) policies, that has increasingly shifted the attention from end-of-pipe management (waste collection and disposal) to prevention and value-chain management. Economic issues regard in particular the evaluation of waste management options; "market failures" in the WM industry and the dimensions of general interest that require public regulation; the industrial organization of WM services and patterns of private sector involvement; the use of economic instruments and market mechanisms for achieving WM targets. These issues are rapidly presented in their theoretical implications and later discussed on the basis of the experience in EU countries.

1. INTRODUCTION

In the last 30 years, solid waste management (SWM) policies in Europe have been substantially changing their main focus from the mere elimination of waste from urban areas to an integrated management of material flows, aimed at minimizing waste generation and improve as much as possible their valorization potential (Tab. 1).

Table 1. Evolution of SWM regimes in Europe

Regime	Main objective	Key actor	Emphasis on …
Public hygiene (- end 60s)	Removing waste from urban areas	Municipality	Quality of service
			Urban propriety
			"Out of sight, out of mind"
Environmental protection (early 70s)	Minimizing environmental impact of disposal	Legislator	Technology
	Avoid shipments of waste towards low-standard countries		End-of-pipe regulation
Facing the waste mountain (end 70s – mid 80s)	Ensuring adequate disposal capacity face to dramatically increasing quantities and supply shortage	Region	Supply of disposal capacity
			Social consensus
			Economies of scale
Prevention and closed material cycles (90s -)	Minimizing waste flows and increasing the potential for resources recovery	National level	Extended producer responsibility
		Manufacturers of goods	
		Retail sector	

This shift in the policy focus had the consequence of transforming substantially the economic nature of SWM, increasing on the one hand market failures and natural monopoly characteristics, and on the other hand, increasing its industrial complexity, thereby creating more scope for the involvement of the private sector for operating and financing SWM systems.

In this paper, we present the main institutional arrangements characterising the industrial system of SWM in Europe. We start from a brief analysis of the SWM market and the interrelations between different phases; we go then through a comparative analysis of the alternative solutions taking place in different EU countries.

2. AN INTERPRETATIVE MODEL OF THE SWM MARKET

In the value chain of SWM we can distinguish three main activities from which arise three potential markets (Fig. 1). The first one (primary market) regards *collection services*, whose counterparts are waste producers and operators. The second is the market for *waste handling and disposal* (secondary market), whose counterparts are operators of collection services and owners of disposal sites. The third is *separate collection and recovery/recycling* (tertiary market). Each of these markets presents a more or less complex value chain, along which commercial transactions take place, which identify further side markets.

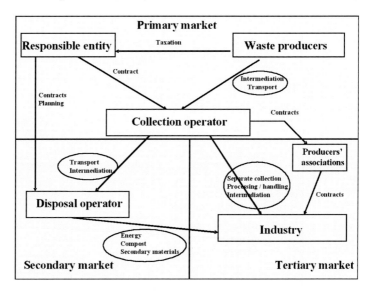

Figure 1. The value chain of SWM.

The interaction between the three markets is analyzed in Fig. 2. Generation of waste (diagram *a*) is here considered as exogenous, basically a function of lifestyle and consumption patterns, eventually influenced by policies aimed at waste prevention. In the figure we have represented two cases, with a greater waste production in the second case.

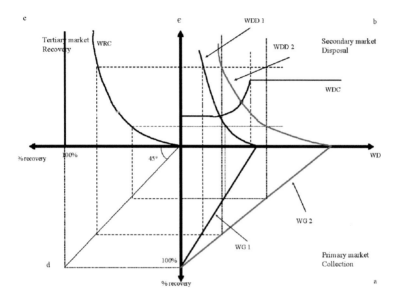

Figure 2. Collection, disposal and recovery market.

Waste can be destined to two alternative flows, separate collection and recovery (sector *c*) and undifferentiated collection and disposal, eventually with recovery of energy or secondary materials (sector *b*). The alternative choices are represented by WG curve, resulting from the difference between total waste and recycled waste: if all waste were addressed to disposal, the concerned quantity could be read at the intercept of WG curve and the *x* axis; the higher the fraction of waste that is diverted to separate collection (until a theoretical maximum of 100%), the lower quantity has to be addressed to disposal. In the figure, two situations are shown, WG1 and WG2, with a higher overall quantity of waste in WG2.

Marginal cost of recycling and disposal are expressed respectively by WRC and WDC curves. In the recovery market, we assume decreasing returns, because of the increasing difficulty in addressing larger fractions of waste with separate collection and/or because the quality of collected materials worsens if the collected fraction is higher. In the disposal market we assume that low cost solutions are available up to a certain point; beyond this point landfills are not sufficient and waste can either be transported elsewhere (with a greater cost due to transport) or handled with high-cost technologies such as incineration or mechanic separation of humid and combustible waste.

For each hypothetical price of final disposal (measured along axis *y*) we can first of all determine the fraction that is convenient to collect separately; reporting it on the WG curve, we find the corresponding quantity of disposal services that is demanded at that price and thus a point of the disposal demand curve, WDD. Repeating the same exercise for all prices, we can draw the whole WDD curve; in the figure we calculated two of them, corresponding to WG_1 and WG_2. The point where WDD and WDC curves meet individuates the market equilibrium and the corresponding optimal level of recovery.

The optimal level of recycling varies in each context, not only because WRC is different (e.g. due to the proximity to markets for recovered materials, social willingness to participate in separate collection, urban density, etc.), but also because of the local structure of WDD and WDC.

Until WG is sufficiently small, local disposal market can easily absorb waste flows; recovery plays only a residual role and is organized in an autonomous way. It is also important to note that elasticity of disposal demand, WDD, depends on the elasticity of recovery supply, WRC. Efficiency of the recycling market influences therefore the equilibrium in the disposal market; it can be interpreted as a side-competitor, whose presence reduces the market power of disposal suppliers but also increases economic risks in the secondary market.

In the past, each activity was developed and managed almost independently.

Collection services (primary market) have been historically the first activities to be organized as public utilities, due to the prevailing public health and urban propriety issues. In most countries, municipalities have the duty to organize collection services that waste producers are obliged to use, and for this purpose they have the right to levy a tax or a charge.

Disposal (secondary market) concerns the elimination of waste, better to say the restitution of used material flows to the environment. In the past, this did not represent – or was not perceived as – a problem, because of the availability of landfill sites in the nearby zones. Nor had it a major impact on costs: until the 70s, the cost of disposal represented a negligible fraction of the total (Ascari *et al.*, 1992).

Waste recovery, in turn, (tertiary market) was carried on spontaneously by private collectors exploiting the very little economic margin allowed by the positive price paid by manufacturing industry for some materials contained in waste (metals, paper, glass, cloth, organic materials, etc.), as still happens today in many developing economies.

The traditional system entails very low disposal costs, waste flows are limited and easily accommodated in the existing facilities (WG_1), recycling is only marginal given its high differential costs with respect to disposal. In the last 30 years, things have been changing substantially.

Without the need to go back over the historical phases that have led to the present situation, we can mention as driving forces the dramatic impact of quantities and harmful potential of waste, the increasing awareness of environmental impact of extensive disposal, the growing difficulties in the development of disposal facilities due to physical shortage and social opposition and, finally, the emerging of an environmental policy aimed at the prevention and the reduction of waste. Moreover, since local disposal capacity becomes insufficient, alternative solutions have to be found, either shipping waste to other areas (thus incurring into transport costs) or recurring to the "backstop technology" of incineration (Buclet and Godard, 2000;

Kraemer and Onida, 1999).

By considering what is reported in Fig. 2, this evolution can be interpreted as follows:

- increase of raw waste generation (WG shifted towards the right);
- WDC shifted upwards because of increasing restrictions and regulations, internalization of costs through performance standards and liabilities, social opposition and need to compensate local communities, etc.;
- WRC initially very inelastic, limited by technical factors as well as from the inadequate development of recycling industry, private separate collectors declining due to the limited economic margins.

The market has initially reacted very rigidly: waste quantities were growing out of control, recycling potential remained limited, and the expansion of disposal capacity was slowed down by social opposition and sometimes by reluctance to invest in facilities characterized by heavy sunk costs, without adequate guarantees. Owners of (scarce) disposal sites already in place started to obtain increasingly high rents: in Italy, for example, landfill prices within a few years grew from 4-5 to 50-100 €/ton (Bertossi et al. 1996).

Two important innovations have subsequently been introduced in the SWM regime. The first one was to put responsibility for disposal planning on regional institutions, thus extending legal monopoly to the secondary market. A vertical separation of collection and disposal is also realized, the former remaining under the competence of municipalities and the latter of regional authorities, although with different degrees of effectiveness and actual commitment of regional authorities in the management of disposal facilities.

In a following phase, a second innovative regime has been introduced, the extended producer responsibility (EPR) (Palmer and Walls, 1999). With some significant variants from one country to another, these policies have obliged manufacturers and distributors of many goods to achieve pre-determined recovery targets of their own products. In this way, part of the recovery cost (WRC) has been transferred from the SWM to the production of goods (and thus into their market price); moreover, industry has been stimulated to search for cost-effective solutions and invest for developing recycling potential. As a result, WRC curve has moved downwards; what was considered in the past a "secondary" and "residual" market is now concerning a fraction between 30 and 50% of total waste flows (RDC and Pira, 2003). On the other hand, the tertiary market, previously competitive, becomes a legal monopoly as the other two.

3. PUBLIC AND PRIVATE SECTOR IN EUROPEAN SWM

3.1. Characterization of European waste management regimes

A regulatory regime can be defined as the set of institutional rules and conventions that govern the organization of a given sector (Buclet, 2002).

Characterizing factors of alternative regimes are in particular:

- property rights on natural resources;
- responsibility for service provision;
- patterns of regulation of entry and exit;
- degrees of freedom left to customers in the choice of the service provider (is it obligatory or not to purchase the service? is it possible for eligible customers to bypass the public utility?);
- property of operating firms and contractual aspects of service provision;
- allocation of financial burden and economic risk in the public budget, operator's shareholders, service consumers and eventually further actors;
- responsibility for setting prices;
- structure of the vertical value chain and governance of transactions among the phases.

Tab. 2 summarizes the main aspects of the current European regime. Of course the situation is not fully homogeneous and many national peculiarities are present. While referring to some national cases as examples, our aim here is to provide a general picture of the most typical arrangements.

3.2. Property rights on the natural resource

The "natural resource" at stake in this case is the assimilative capacity of the environment, particularly considering the disposal of raw or processed waste into the soil (landfilling).

In all EU countries, settlement of waste disposal facilities is subject to public authorization and control, in order to ensure appropriate technical standards. Emission limits, technical requirements, duties of care, aftercare responsibilities, etc. are foreseen in every national legislations, and increasingly harmonized at the EU level.

Environmental legislation is not limited to set emission standards; rather, in the last ten years, it has evolved towards a comprehensive approach outlined in the 5th and later confirmed in the 6th Environmental Programme of the European Commission. According to these documents, and to the subsequent detailed directives 91/156 and 94/62, waste management policy should be targeted to the whole material flow and not to the mere generation of waste at the end of the production-consumption cycle.

The basic principles of European SWM policy are set down in Dir. 91/156 and transposed, though with some relevant differences, into national legislation. These are in particular:

- the *polluter-pays principle*: waste producers should be charged for the total service cost, including external costs;
- the *self-sufficiency* and *proximity principle*: each territorial unit should take care of its own waste; units should be designed in such a way as to allow self-sufficiency while achieving economies of scale;
- the minimization of negative impacts arising from final disposal of waste;
- the *ladder principle*: waste management should prioritize reduction of

waste and valorization (through direct reuse, recycling or recovery of secondary materials or energy); only waste that is not suitable for valorization ("ultimate waste") can be landfilled.

While considering recycling as a priority, European policy is also committed to avoid illegal disposal practices, which very often could be disguised as recycling, particularly when recycling is indirect and scrap materials are used as inputs in different production cycles. For this purpose, three categories of materials are individuated: those belonging to the green list can be freely traded as any other material, while the orange list requires an administrative burden for giving the proof of effective recycling, and the red list implies that only specifically authorized facilities can be allowed to treat them.

Some national legislation requires the public property of disposal sites, while in other cases private property is also admitted. Whether publicly or privately owned, the capacity of disposal sites is not freely available on the market. Public regulations (arising from regional plans) often force facilities to provide their services to specific customers (e.g. neighbour municipalities). This occurs in particular for incinerators and landfills for untreated waste. In turn, facilities treating industrial waste enjoy higher freedom and their capacity is normally sold in the market, with the public sector only supervising the whole thing.

3.3. Public good content and responsibility for service provision

3.3.1. Primary market: Collection

Traditionally, the public service is represented by the collection of waste from the streets, where households can deposit it following apposite regulations.

Industrial and commercial waste is normally not included in the public service, although this sometimes occurs for small industry and laboratories. Industry and commercial activities have the faculty to use the public service, and pay the related tax; yet in most cases they provide directly by addressing a specialized firm in the waste disposal sector or an authorized broker. Disposal of industrial waste is in general supplied on the market; waste transfer, at least within Europe, is normally possible although being regulated in order to avoid illegal practices.

The household waste collection service is normally linked with street cleaning and other minor activities aimed at urban propriety. Domestic and sometimes other customers are compelled to use the service under specified conditions, and normally pay a tax regardless they actually produce waste or not (Massarutto, 2001).

The responsibility for waste collection and successive elimination lays everywhere on municipalities, which are also universally the administrative level that is financially responsible for expenditure and in most countries raise dedicated taxes; in many countries, municipalities have the faculty or even the obligation to join up for the collective provision of the service.

3.3.2. Secondary market: Disposal

As we have seen in section 2, disposal is usually under the control of planning authorities located at the regional level. Planning is especially aimed at guaranteeing sufficient disposal capacity in any management unit and, at the same time, creating the conditions that enable operators to invest.

Table 2. Prevailing institutional regimes in the European SWM markets

	Collection (Primary market)	**Disposal (Secondary market)**	**Recycling (Tertiary market)**
Environmental and quality regulation	Municipal regulations concern the use of public spaces for abandoning waste	All disposal sites require authorization. Emission standards – Technical prescriptions. Self-sufficiency principle – Polluter-pays principle	Recycling targets imposed by national legislation and enforced through economic instruments. Shipments of waste regulated according to pollution potential
Responsibility	Municipalities (*sometimes associated*) are responsible to provide collection and usually raise a dedicated tax	- collection operator (France) - municipality (UK) - regional plan (legal monopoly)	Manufacturers and retail distributors, mainly through responsible bodies created through EPR (e.g. packaging waste)
Entry and exit	Legal monopoly (household waste; could be extended to commercial/small industry). Free entry (commercial and industrial waste); authorization required, based on the provision of adequate guarantees	Free entry (subject to authorization and in the frame of planning usually individuating min and max capacity to be put in place)	Free (but market is organized by a responsible entity having strong market power)
Eligible customers	Commercial and industrial waste	Responsible entities are normally free to choose the preferred supplier (following the prescription of general rules and regional planning). Regional plans sometimes only supervise, sometimes provide precise regulations and associate waste flows to specific treatment facilities	All customers are free, but the responsible entities are the only legal mediator
Contractual arrangements	Allowed management forms include direct labour organizations, municipally-owned companies, mixed PPP and delegation through competitive tendering	- Spot market (very rare, only for peak demand and managing of emergencies) - Long-term contracts for purchasing disposal capacity - Regulated contracts framed by regional planning - Vertical integration between collectors and owners of disposal sites, both private and public	Market transactions (often framed and regulated by responsible bodies)
Value chain structure	Outsourcing of operational activities, especially by publicly owned companies and	Landfills often integrated with quarrying. Incineration often integrated with equipment manufacture and/or	

	direct labour orgs	energy sector	
Key operators	Municipally-owned companies Private concessionaires (usually SME, often owned or participated by multinational companies) Equipment producers	Publicly owned companies Private companies usually under PPP agreements or DBFO Manufacturers of technology and equipment – construction industry – engineering services	Responsible bodies Recycling industry Separate collection – processing – handling Innovative SME – RTD
Responsibility for setting price	Municipalities usually enabled to recover all costs through dedicated taxes (exc. UK) Contracts with operators define the maximum revenue (usually cost-based for publicly owned orgs, tender-based for delegation)	Varying: - Free negotiation - Regulated by regional plans and/or administrative norms - internal transactions in case of vertical integration	Framed by national agreements between responsible bodies and main actors (municipalities, collectors, handlers and recycling industry)

Again, national experiences differ substantially. In Germany, for example, disposal is normally supplied by public facilities owned and sometimes even directly operated by regional authorities, with a strong emphasis on the proximity and self-sufficiency principle. Municipalities are then obliged to confer their waste to these facilities and pay the price set by the regional authority.

In France and in the UK, planning is less pervasive: Departments and Counties limit their role to the supervision and licensing of private sector activities, and in the regulation of the behaviour of waste handlers (municipal collection systems, in our case). In Italy, waste disposal planning has been based on very detailed prescriptions concerning location and typologies of technical facilities; these were later realized directly by municipalities, often within compulsory associative agreements.

Migration of waste between different areas should be limited, according to the proximity principle; and in fact it represented an exceptional case, motivated by shortage of capacity and failure to develop new sites (due to planning inefficiencies, social opposition, etc.).

Another reason originating waste migration was asymmetric legislation. Before the European directives on landfills, incinerators and waste trade were approved, technical standards were very different. Although transport costs are significant and reduce the economic benefits of arbitrage, differentials of treatment costs were in some cases high enough to justify the sending of waste to other countries. Something similar occurred due to the unprecedented increase of separate collection in some countries such as Germany: especially in the first phase, this was not corresponded by sufficient investment in recycling capacity; as a result, increasing quantities of secondary materials (paper, plastics, glass) invaded the nearby countries with destabilizing effects on their scrap markets. In fact self-sufficiency was introduced in order to eliminate these spillovers and avoid the consequent "race to the bottom".

In recent times this trend has inverted. While rules were increasingly harmonized, the main determinant behind waste migration has been the excess of capacity in countries such as Germany or the Netherlands, motivated by the unexpected boom of separate collection and recycling. As a result, nearby areas that are still in the transition between landfill and incineration are often sending their waste to these countries. Analogous movements occur within national countries, between regional areas.

3.3.3. Tertiary market: Recycling

A peculiar regime has been created in the tertiary market (recycling). As argued in section 2, this sector was marginal to the SWM industry until the 90s, and was characterized by free economic initiative of business operators. Entities responsible for collecting waste were also free to decide whether to organize separate collection or not, although in some countries the national legislation and the regional plans have introduced obligations to set up separate collection and also introduced mandatory targets.

In the 90s, this regime has changed deeply with the introduction of extended producer responsibility (EPR). Industrial and commercial subjects – located along the vertical value chain of a given material or product – are forced or encouraged to accept responsibility over the achievement of a publicly-determined recycling target, and to finance this effort through their own expenditure, later on reversed on product prices.

These new entities, artificially created by environmental legislation, can be interpreted from the economic point of view as providers of a club good – namely, the achievement of the recycling target and the avoidance of a collective penalty – with appropriate measures that are typically aimed at avoiding free-riding behaviour. Adhesion to industrial consortia is obligatory in some countries (such as Italy), while in other cases consortia are legitimated to use their monopoly power against those who do not participate. Organizations set up by producers in most cases enjoy a monopoly face to local authorities; normally an agreement is found, with the mediation or supervision of national bodies, defining prices that producers' organizations will pay for separate collected materials according to their quality, pre-treatment and commercial characteristics. They will also take care of organizing processing activities aimed at transforming them into marketable products and finally promote downstream investment for absorbing these materials again in the production chain.

3.4. Ownership of service infrastructure

Ownership of assets is not a major problem in the primary market, since collection is basically a labour-intensive activity, while assets are simple and can be easily transferred elsewhere (no sunk costs). Collection infrastructure – bins, lorries, temporary storage and processing facilities – can be owned by municipalities, operators or even by customers, under a wide spectrum of alternative combinations.

More complex is the situation in the secondary market. Disposal facilities used to be private until the 70s. During the 80s and 90s, landfills were often owned by local authorities, while their operation was managed directly or contracted out; where this did not occur, despite the attempt to introduce some price regulations, private companies were often able to confiscate significant scarcity rents.

In more recent times, this trend has been somewhat inverting. Landfill of raw waste has a decreasing role, even because it is fiercely opposed by legislation as we have seen above; modern landfills more often receive the leftovers of previous processing phases. The reduction of quantities and of harmful potential of this pre-treated waste allows easier location of new facilities and thus increases supply elasticity. Waste arising from processing activities is normally considered equivalent to industrial waste, and responsible operators can therefore dispose of materials on the open market.

In turn, processing facilities – representing now the most important component of the final cost – are in most cases owned and operated by local authorities, even if many plants still belong to privates or are operating under various forms of concession or project finance arrangements. Publicly-owned companies and mixed venture capital companies are diffused in all European countries. As we noted above, entities responsible for disposal planning are normally located at the regional level; these sometimes provide the service with their own facilities (Germany, Netherlands), sometimes indicate facilities and authorize local public operators to realize them (e.g. Italy), and in some other cases supervise and encourage private sector initiative (e.g. France, UK). Even where the private sector is involved, however, the economic risk is reduced by the provision contained in regional plans, obliging municipalities to confer their waste to authorized facilities at a conventionally regulated price.

In the tertiary market, collection facilities are sometimes owned and operated on behalf of municipalities; in Germany, in turn, separate collection is the responsibility of producers' associations, that also own and run the facilities (although they often delegate this task to private firms or municipal operators). The same applies to handling and processing facilities. Recycling, in turn, is provided by the market, although producers' associations promote in many ways asset development.

3.5. Property of operators and contractual arrangements

In the entire EU, municipalities (individually or associated) are responsible for providing household waste collection and can discharge their responsibility with a variety of management solutions ranging from direct labour to full delegation.

France is undoubtedly the country with the highest level of delegation to the private sector: private companies operate as concessionaires, even if they normally do not charge consumers directly. The peculiarity of the French model is the fact that the concessionaire for collection is also delegated the responsibility of finding disposal solutions, that often take place in plants that are controlled by the same operators. Industry is quite concentrated and vertically integrated; arrangements in the secondary market often foresee mixed venture companies and joint ventures between private companies.

In the rest of Europe, despite a certain diffusion of delegation, normally disposal sites are individuated by the municipality and/or by the regional plan. Municipally-owned enterprises are massively diffused, particularly in larger cities, especially in Italy and Germany. Direct labour organizations are also diffused in all countries except in the UK.

An emerging trend in most publicly controlled management systems is the outsourcing of many operational activities to private subcontractors; on the other hand, initiatives with higher degrees of complexity and economic risk (e.g. new treatment plants) are often realized in collaboration with other public or private firms and financed with the recourse to the private capital market.

The UK has adopted a compulsory competitive tendering model: municipalities are obliged to entrust the service after a competitive bid, to which former public employees can participate with the creation of their own cooperative firm. After 10 years of application, the British model has now reached its maturity, with 100% of the service units delegated to the private sector; the majority of cooperatives did not survive in the long term and have been absorbed by private companies. The peculiarity of this system with respect to the French one, however, is that only occasionally it ends up with the full delegation of responsibility over the waste management cycle to a single private firm; rather, the preferred option is that of delegating the "blue-collar" activities, while holding in public hands the strategic management of the service. Bids concern simple activities, very often fractioned in a great number of different contracts. The main difference with the French system is that in the UK, municipalities enter into separate contracts with owners of disposal facilities. Collection tenders do therefore include already the destination of waste and the price: the decisive variable for awarding contracts is the cost of collection alone. As a result, industry concentration is lower than in France.

Contractual relations between primary and secondary market are quite varied. Wherever regional planning is in place, it normally singles out one or several facilities, to which municipalities can (or are obliged to) confer collected waste at pre-determined conditions. For this reason, normally collection contracts already specify the destination of waste; collection operators do not need to be vertically integrated with disposal, since the control of disposal sites does not allow particular market advantage. Nonetheless, sometimes disposal operators are owned through complex structures involving different layers of government and other public or semi-public institutions.

This is not occurring everywhere. In France, as we have seen, regional plans have a more modest role; disposal facilities are organized by the private industry, and concessionaires have also the responsibility of finding disposal solutions. This has favoured vertical integration, since ownership of disposal sites could internalize the scarcity rent and be used as a barrier against potential competitors.

In the UK, municipalities enter into separate contracts with owners of disposal facilities; again planning has mostly a supervisory function. Ownership of disposal facilities might well belong to the same companies that compete for collection contracts; yet this is not a necessary condition.

In Italy, although legislation attributes this responsibility to the regional plan, implementation of the latter has not always been complete or timely; collectors were therefore encouraged or forced to develop their own solution, what has been realized especially by municipally-owned companies. A vertically-integrated market, though in public hands, has therefore originated.

In the tertiary market, in the past, municipalities or collection operators went directly into contracts with the recycling industry, sometimes delegating third parties as collectors, handlers or brokers. After the introduction of EPR, producers' associations have taken over the market risk and the responsibility of organizing the vertical value chain. In Germany, the competent body, Duales System Deutschland (DSD), organizes separate collection through direct agreements with specialized operators. It then establishes framework contracts with recycling industries. In the other countries, responsible bodies do not organize separate collection themselves but rather rely on municipalities; their main function is to negotiate the price at which municipalities – better to say, collection operators delegated by municipalities – have the right to deliver separate collected materials.

3.6. Allocation of financial burdens, economic risks, price setting and price regulation

Municipalities usually finance this expenditure through the own budget; in most countries, a dedicated municipal collection tax is foreseen, through which municipalities are in the condition of recovering the costs. The correspondence of tax revenues and service costs, however, is not necessarily total: sometimes municipalities prefer to finance at least a part of the cost out of the general budget.

The two extremes are the UK – where no dedicated waste collection tax is foreseen, and the service cost is entirely financed through the municipal budget like many other urban services – and Germany, where the recovery is total.

Sometimes, more complicated financial structures are foreseen, with the provision of a charge raised directly by the operator (typically, on a fee-for-service base).

In the secondary market, the situation is quite different. Since the cost of the landfill is much lower than that of incinerators and other disposal solutions, the equilibrium disposal price tends to allow an economic rent to the landfill owners; in turn, technological plants are at permanent risk of underuse if the municipalities are left free to choose the preferred disposal solution. Transport costs limit the possibility of using the same plant for treating waste coming from big distances. This means that investment in incinerators entails a significant sunk cost; a huge economic risk is thus associated with it. In fact, in many countries, the realization of disposal facilities is accompanied by more or less strict compulsory provisions for municipalities to confer their waste to them (Massarutto, 2006).

Regulation of disposal prices is normally associated with planning; its main purpose, however, is to reduce the scarcity rent and not to increase operational efficiency. Its degree of success is varying. In most cases, rents are indeed present and accepted in order to encourage investors in developing facilities. In Italy, for example, disposal prices have rapidly moved from a range of 20-40 to 100-200 €/ton or more.

In the tertiary market, two delicate points have to be raised.

The first one concerns the price that producers' associations award to municipalities for separately collected material. Since the objective of the consortium is to persuade municipalities to engage in separate collection, this price should compensate the cost; however, the relevant cost here is not the industrial cost of the collection system, but rather the differential cost with respect to ordinary disposal. Therefore, in congested areas (where disposal costs are higher due to scarcity rents), this price can be much lower than in less congested areas. If the consortium is left free to negotiate with municipalities – what generally happens in all European countries – it ends up with appropriating a part of the economic rent of disposal facilities.

The second issue concerns the relations between industrial sectors themselves, particularly in case a single consortium is set up for different materials (what occurs for example in the packaging sector). The risk here is that the consortium uses its monopoly power in order to discriminate against certain actors (e.g. producers of a particular material, such as PVC, or small retail distribution premises) or adopts technical standards or other measures that raise barriers to competition on the goods market, or artificially alter the prices by introducing hidden cross-subsidies between materials.

None of these problems has found a comprehensive solution in European countries, although industrial consortia are regulated by very complex and evolving statutory norms (Massarutto, 2006; Buclet and Godard, 2000).

4. CHALLENGES AND PRESSURES FOR CHANGE

The above outlined structure of European SWM is on the one hand open to substantial private sector participation, on the other heavily regulated both in terms of quality standards (emissions, "best practices", targets of waste valorization, self-sufficiency, minimization of waste transfers, etc.) and patterns of industrial organization. While a substantial harmonization has been achieved in the former aspect, many important differences characterize the latter; as we have argued, the most important ones regard the extension of legal monopoly, that in some cases regards only the primary market (collection), in other cases is extended to disposal and sometimes recycling.

As we have argued above, the first case, while apparently more open to competition, in fact encourages SWM operators to achieve vertical integration in order to internalize the monopoly rent of disposal facilities; the second case, in turn, is more open to competition for the market in each of the three market segments, but requires that public sector planning is good enough in order to achieve economies of integration between the three markets. Therefore, two alternative sources of inefficiency can be predicted, and are well documented by empirical examples. In the former structure, incumbent operators acquire relevant market power, and justify the case for the introduction of regulatory systems aimed at reducing this power. In the latter case, public planning might incur into errors (e.g. in the setting of recycling targets, in the sizing of disposal facilities), while competitive tendering is more suitable for some phases (namely, collection) than for others (e.g. incineration).

Both risks are rather enhanced in the EU system by the provision for self-sufficient regional markets stated by Dir. 91/126. While this principle can be justified by the need to avoid that asymmetric regulations or insufficient control capacity make illegal exports easier, in a well-harmonized internal market this is not necessarily the case. Once evidence of excess supply of treatment facilities is present in many EU areas such as in Germany and in the Netherlands, pressures to reduce somewhat the rigidity of the self-supply principle are now emerging.

A second important trend is determined by the polluter-pays principle. While in most countries SWM is a financially self-sufficient service, innovative trends can be detected both in the way SWM revenues are gathered and in the increasing role paid by goods manufacturers. In the first case, tariff structures are being implemented with the aim of encouraging a proactive role of waste producers, e.g. limiting undifferentiated waste and adopting as far as possible separate collection systems. In the second case, that is particularly relevant for separate collection and recycling, manufacturers are required to finance collection systems and processing, thus transferring this cost on the price of final goods and alleviating the burden of local taxpayers.

Finally, outsourcing of activities is particularly important wherever public ownership and management dominate. As a result, the market share of private operators is growing almost everywhere, even if with rather different outcomes and market structure. While in one case SWM companies are large, integrated and operate on an EU-wide market, in the other, local SMEs specialized in given processing activities are the most typical partner of public companies. In both cases, a lively market along the value chain (e.g. specialized equipment, engineering services, processing technologies) has been developing.

Considered together, these trends allow predicting a future SWM market that will be increasingly open to private sector participation, although requiring innovative regulatory mechanisms for overcoming market failures.

Antonio Massarutto
University of Udine, Italy

IEFE
Bocconi University, Milan, Italy

REFERENCES

Antonioli B., Fazioli R. and Filippini M., 2000a. "Il servizio di igiene urbana tra concorrenza e monopolio", *Working paper series, Quaderno 00/07*, University of Lugano, http://www.bul.unisi.ch/cerca/bul/pubblicazioni/eco/pdf/wp0007.pdf

Antonioli B., Fazioli R. and Filippini M., 2000b. "Analisi dei rendimenti di scala per il servizio di igiene urbana in Italia", *Economia delle fonti di energia e dell'ambiente*, No. 2.

Ascari S., Di Marzio T. and Massarutto A., 1992. *L'igiene urbana*, FrancoAngeli, Milano.

Bertossi P., Kaulard A., LolliA. and Massarutto A., 1996. "Per una nuova politica industriale nel settore dell'igiene urbana in Italia", *Economia delle fonti di energia e dell'ambiente*, No. 3.

Biagi F. and Massarutto A., 2002. "Efficienza e regolamentazione nei servizi pubblici locali: il caso dell'igiene urbana", *Economia Pubblica*, No. 2, pp. 79-115.
Brusco S., Pertossi P. and Cottica A., 1995. "Mercato, cattura del regolatore e cattura del controllo", *Economia e Politica Industriale*.
Buclet N. (ed.), 2002. *Municipal waste management in Europe: European policy between harmonisation and subsidiarity*, Kluwer, Amsterdam.
Buclet N. and Godard O., (eds.), 2000. *Municipal waste management in Europe: a comparative study in building regimes*, Kluwer, Amsterdam.
Davies S., 2001. *Mergers and acquisitions in the European waste management industry 2000-2001*, Public services international research unit, University of Greenwich (www.psiru.org)
Eunomia-Ecotec, 2003. *Costs of municipal waste management in the EU*, Report to the EC-Dg Env, http://europa.eu.int/comm/environment/waste/studies/eucostwaste_management.htm
Kaulard A. and Massarutto A., 1997. *La gestione integrata dei rifiuti urbani: analisi dei costi industriali*, FrancoAngeli, Milano.
Kinnarman T.C. and Fullerton D., 1999. *The economics of residential solid waste management*, NBER working paper 7326, National Bureau of Economic Research, Cambridge Ma.
Kraemer L. and Onida M., (eds.), 1999, *I rifiuti nel XXI secolo: il caso Italia tra Europa e Mediterraneo*, Edizioni Ambiente, Roma.
Legambiente and CCTA, 2005. *Rifiuti Spa. Radiografia dei traffici illeciti*, http://www.legambiente.com/documenti/2005/0125_rifiutiSpa/rifiuti_spa.pdf
Massarutto A., 2001. Dalla tassa alla tariffa: cosa cambia veramente per il settore dei rifiuti, *Economia delle fonti di energia e dell'ambiente*, 3, pp. 33-76.
Massarutto A., 2005. A policy roadmap for the evaluation of liberalization opportunities and outcomes of regulatory reforms, working paper series in *Economics, 02-05*, Dse, Università di Udine, http://web.uniud.it/dse/working_papers/working_papers_eco.htm
Massarutto A., 2006. "Waste management as a service of general economic interest: options for competition in an environmentally-regulated industry", *Utilities Policy*, forthcoming.
Palmer K., Sigman H. and Walls M., 1996. The cost of reducing municipal solid waste, *Discussion paper 96/35, Resources for the future*, Washington DC.
Palmer K. and Walls M., 1999. Extended product responsibility: an economic assessment of alternative policies, *Resources for the Future, Discussion Paper 99-12*, Washington DC (http://www.rff.org/disc_papers/PDF_files/9912.pdf)
Poli C., 2002. *Le strategie di risposta delle imprese operanti nella filiera idrica e nell'igiene urbana*, in Vaccà S., a cura di, "Problemi e prospettive dei servizi locali di pubblica utilità in Italia", FrancoAngeli, Milano.
Szymanski S., 1996. "The impact of compulsory competitive tendering on refuse collection services", *Fiscal Studies*, vol. 17 No. 3, pp. 1-19.
Walls M., 2003. How local government structure contracts with private firms: economic theory and evidence from solid waste collection and recycling, RFF *discussion paper 03-62, Resources for the Future*, Washington DC (www.rff.org/disc_papers/PDF_files/0362.pdf)
Waste Management Council (AOO), 2003. *The waste market: the Netherlands and neighbouring countries*, http://www.aoo.nl/images1/aoo_nl/bestanden/AOO2003-12.pdf
RDC-Environment and Pira International, 2003. "Evaluation of costs and benefits for the achievement of reuse and recycling targets for the different packaging materials in the frame of the packaging and packaging waste directive 94/62/EC" – *Final consolidated report, European Commission*, DG Environment, www.europa.eu.int/comm/environment

I. PAVAN,
E. HERRERO HERNÁNDEZ AND E. PIRA

HOSPITAL WASTE MANAGEMENT

An Italian Experience from a Medium/Large State Hospital

Abstract. This paper focuses on the practical aspects regarding hospital waste management in a public hospital in Northern Italy. Our institution is mainly a Trauma Centre offering a wide range of medical and surgical highly specialized services, as well as hosting laboratories and university training facilities. This organizational complexity generates various kinds of waste. The classification of the different types of waste according to the Italian and European legislation, as well as the planning, storage, handling and disposal strategies are discussed. Finally, educational aspects such as staff training program for waste handling and economic issues regarding preventive strategies for waste reduction and cost/efficiency analysis are considered.

1. BACKGROUND

The Italian legislation accomplishes European directives and regulations to ensure that hospital waste is managed in an environmentally sound and health-protecting way. According to the hierarchy of waste management, the competent authorities and healthcare institutions have to promote waste prevention as a priority, and secondarily ensure reuse/recycling and energy recovery. Hospital waste collection, transport and disposal must always be optimized. These strategies together can save money and protect the environment. To establish an effective waste management policy, a complete check-list of purchased items has to be filled. The types and quantities of purchased materials, as well as their packages must be carefully selected to respect, whenever possible, the "reduction at source" principle. It is also important to assess how these materials are used and how much waste is thus generated. Once this information is clear, the ways to eliminate, substitute, reduce and recycle have to be carefully analyzed.

2. THE SPECIFIC CASE OF HOSPITALS

Hospitals constantly generate great amounts of waste, posing a considerable economic and environmental problem. Hospital waste composition is complex; however, contrarily to what is generally assumed, hospitals generate mostly non hazardous waste. Paper, plastics and food are some of the materials, analogous to ordinary urban waste, contributing to the majority of waste originated by healthcare institutions. This simple observation has to influence hospital waste policies. For instance, the simple decision of buying recycled paper can be very rewarding for such kind of institutions. Many hospitals actively promote ideas and projects for waste reduction among their staff and impressive savings can be obtained.

Many kinds of waste material are produced by hospital activities:
- non hazardous hospital waste;
- hazardous hospital waste with risk of infection: it is characterized by potentiality to cause infections and requires treatment and decontamination;
- hazardous hospital waste with no risk of infection: materials posing a significant danger to health including chemical risk derived from drugs, cytotoxic agents and mercury from dental amalgams;
- hospital waste that requires particular disposal treatment (radioactive substances and materials);
- various waste material: includes every kind of generic material such as paper, plastics and foodstuff.

The first step towards a correct hospital waste management policy is the precise *identification and classification* of the different kinds of waste. This allows correct segregation, treatment and disposal. Erroneous mixing of different waste can have serious consequences on health, environment and hospital budget. The most frequent error is mixing municipal-type waste with hazardous one, as hazardous waste needs particular treatment and normally has a higher economic impact. It has been estimated that large hospitals can achieve remarkable savings simply by ameliorating waste separation.

In the European Union, each type of waste has an identification code (C.E.R. code of European waste) made up of six numbers in groups of two. The *first two* numbers identify the *origin* of the waste, the *second two* identify a particular *characteristic* of waste and the *last two* numbers identify the waste *precisely*. An asterisk after the waste code identifies hazardous waste (Tab. 1).

Table 1. Example of hospital waste identification according to C.E.R. code

First two numbers	Second two numbers	Last two numbers	Final Code and examples
Origin of the waste	Particular characteristic	Precise identification	
Waste produced by hospitals, veterinary clinics and research laboratories	Waste from illness diagnostics	Varies according to waste kind	
18	01	01	180101= scalpels and other cutting instruments
18	01	02	180102= anatomic parts and body organs, blood/plasma sacks

18	01	03	180103*= any infected material
18	01	04	180104= waste not requiring particular caution when handled (paper waste bins, clothing, sheets)
18	01	06	180106*= hazardous chemical substances
18	01	07	180107= chemical substances other than those listed in 180106*
18	01	08	180108*=Cytostatic agents
18	01	09	180109= Drugs other than those listed in 180108*
18	01	10	180110*= waste produced by dental treatment (mercury and metals)

3. OUR EXPERIENCE IN HOSPITAL WASTE MANAGEMENT

Hospital C.T.O. (Traumatology and Orthopaedics Centre) is a medium/large public hospital in Turin, Piedmont, in Northern Italy. Founded 40 years ago, it houses also highly specialized university departments and facilities (teaching rooms, a library). The activities vary greatly, and include services (outpatients, wards, operating rooms, clinical and toxicology laboratories) dedicated to diagnostics and treatment in the fields of traumatology and orthopaedics, occupational health, surgery (general, plastic, and neurosurgery), nephrology, cardiology, rehabilitation, as well as special units for burns and spinal lesions. Such a structural complexity makes waste management a challenging and complex task.

In 2004, our hospital waste production had the characteristics listed in Tab. 2:

Table 2. Quantity, identification codes and description of our yearly hospital waste

Quantity (tons)	C.E.R. Code	Waste Description
265	180103*	waste not requiring any particular attention to be disposed of (directly to the incinerator)
33	180107	"other chemical substances not classified as hazardous"
2.5	180106*	waste classified as "hazardous chemical substances"
1.1	180109	waste deriving from various non-cytotoxic drugs
1.0	090107	waste deriving from paper and photographic film
0.9	090101	deriving from photographic developing solutions
0.7	090104	waste deriving from photographic fixing solutions
0.3	200121	waste deriving from neon lamps
0.3	080318	waste deriving from toner-photocopiers and printers

The type of container to be used for the different kinds of waste has also to be planned. The Container code depends on the waste code, and Tab. 3 shows our yearly use of each kind of containers.

3.1. Waste Collection and Management Procedure

The handling of waste material is performed by third party companies temporarily employed (usually for a renewable three year period), which offer turnkey projects including the supply of the right kind and quantity of containers (based on the requirements of the previous year). The hospital structure dedicates divided space for the storage of empty and full containers.

Full containers are then transferred to their final destination which is determined by law (e.g. infective materials 18 01 03* to controlled incinerator). The external company carries out the following operations to handle the waste:

1. They supply the empty containers to the different hospital departments;
2. They collect the full containers;
3. They move them to the dedicated space;
4. They transport them and dispose of them in the correct manner as established by their codes.

Waste is collected according to the following timetable: on Mondays and Tuesdays before 10.00 a.m., on Wednesdays between 11.30 a.m. and 01.00 p.m., on Thursdays and Fridays after 04.00 p.m. On Mondays the weekend waste has to be collected early and on Fridays it is important to collect the week waste later, so that it will not accumulate during the weekend.

3.2. Container Coding

The operators mark each container with the date, department of origin and waste code. Containers are marked and chosen according to the waste they contain and their composition and capacity has to be adequate. The characteristics and coding of containers are listed in Tab. 3.

Table 3. Characteristics of containers used for the different waste in our hospital

Container Coding	Waste Description	Container Composition and Capacity
180101	Cutting instruments	high density polyethylene 5-7 litres vol.
180102	Anatomic parts and body organs (also blood and plasma sacks)	high density polyethylene 25 litres vol.
180103*	Any infected organic material	high density polyethylene 40 litres vol. (600 containers); 25 litres vol. (380 containers)
180104	Waste not requiring particular caution when handled (paper waste bins, clothing)	high density polyethylene 25 litres vol.

180106*	Hazardous chemical substances	high density polyethylene 25 litres vol.
180107	Chemical substances other than those listed in 180106*	high density polyethylene 20 litres vol.
180109	Boxes for unused (deteriorated) drugs	40 litres vol. (10 containers)

This marking system enables the waste to be located in the correct site inside the hospital, and to be moved according to codes for final destination.

The high density polyethylene containers must have a classified chemical composition, stating that they do not contain any substances that may damage the environment, as most are incinerated. Finally, the containers with hazardous hospital waste are destroyed by incinerators, whilst the others are washed with a special solution in dedicated machines and re-used.

4. TRAINING FOR PERSONNEL INVOLVED IN WASTE HANDLING

European legislation covers global occupational risk. In Italy, this main law is known by its number 626/1994.

The responsibility for its application falls on the managing director who is qualified to do so by a training course.

The managing director is legally obliged not only to enforce the law but also to inform and train the personnel about the risks involved in their tasks.

4.1. Training Programme

The training programme is held by an expert in Occupational Safety and Health. In the specific case of hospital waste management, the first step is to teach which code number/s is/are to be applied to which substance/s and to explain the kind of specific risk for each substance or material (i.e. biological risk, chemical risk, radioactive risk – only for specific substances from the analysis laboratory or nuclear medicine applications).

The second step is to teach which kind of personal protective device (PPD) is to be adopted in the handling of specific hospital waste (e.g. gloves, masks, shoes or boots, clothing etc).

The personnel are obliged to use these PPDs correctly when carrying out their tasks. The third step involves good practice of personal hygiene, such as changing clothes on arrival at the workplace, storing them in an unexposed environment, putting on the PPDs and overalls, removing them before eating, showering at the end of the work shift and changing back into personal clothing.

5. CONCLUSIONS

Hospital waste management is a dynamic process, therefore each step of the whole procedure must be analyzed for cost/efficiency and evaluation of waste production has to be accurately monitored. To do so, once a year the Hospital Waste Commission meets to verify the progress and quality of the programme set up. The costs are carefully evaluated and cost/efficiency ratios discussed. It should be kept in mind that waste prevention should always be the main goal, as waste production affects the environment and has a remarkable economic impact. Hospital waste management and disposal costs vary between 0.30 Euro/kg (for non hazardous waste) and 1 Euro/kg for hazardous waste. In a medium/large public hospital this implies a yearly cost estimate of 300,000-400,000 Euros (approximately 3-4 million Yuan), remarkable amounts deserving careful management.

Ivo Pavan, Elena Herrero Hernandez and Enrico Pira
Department of Traumatology, Orthopaedics and Occupational Health
University of Turin, Italy

REFERENCES

Decreto Legislativo 5/02/1997 n° 22, Suppl. Ord. N.33 G.U.R.I. 15/02/1997, No. 38.
Decreto Legislativo 19/09/1994, n. 626, *Gazzetta Ufficiale* n. 265 del 12/11/94 – Supplemento ordinario No. 141.
European Directives 91/156 CEE; 91/689/CEE; 94/62/CEE, available online at EUR-Lex, The portal to European Union law website: http://europa.eu.int/eur-lex/en/index.html
Indagine sulla produzione dei rifiuti sanitari in Regione Piemonte (2003). Available online: http://extranet.regione.piemonte.it/ambiente/rifiuti/dwd/approfondim/sanitari/indagine_03.pdf
Legge 15 dicembre 2004, *Gazzetta Ufficiale* n.32, 27/12/2004.
Reusable totes, blue wrap recycling and composting. (2002) *Environmental Best Practices for Health Care Facilities*. EPA publication, available online at:
http://www.ciwmb.ca.gov/WPIE/HealthCare/EPATote.pdf
Waste reduction activities for hospitals. California Integrated Waste Management Board. http://www.ciwmb.ca.gov/BIZWASTE/factsheets/hospital.htm

V. ENERGY EFFICIENCY AND RENEWABLES

1. RENEWABLE RESOURCES

R. BARILE

SOLAR ENERGY

Principles, Applications and a Case Study

Abstract. This paper is subdivided into three main paragraphs: basic principles of solar radiation, main applications, and a case study of a rural electrification in China. The first paragraph will introduce the basic principles of solar energy, highlighting the advantages and disadvantages of this kind of renewable energy in comparison with other types of energy supply. The second paragraph will introduce two main applications of energy production with photovoltaic systems: grid connected systems and stand alone systems. Both cases will be broadly described and followed by an explanation of the principles of sizing, with basic formulas for the systems' yearly energy yield computing. An example of three different Italian cities will conclude this part. In the last paragraph, a case study of a decentralized rural electrification in Inner Mongolia will be described. This case study is part of a joint project on Energy and Environment between Italy and China, called "Solar Village in China".

1. SOLAR ENERGY PRINCIPLES

Solar energy, derived from fusion processes of hydrogen contained in the sun, is usually utilized in a double form: the first is the passive one in which heat produced by solar rays is used for different applications, e.g. thermal collectors for the production of sanitary warm water, solar ovens and the capture of light and heat in buildings. The second form is the active one that turns solar radiations into electric energy through photovoltaic (PV) devices, which were named after the physical effect discovered in 1839. In this contribution we will exclusively deal with solar photovoltaic energy.

1.1. Advantages and Disadvantages

The main characteristic that marks the production of energy from a photovoltaic solar source is its availability in a distributed form all over the world. Solar radiation is available, obviously with different intensities, over the entire surface of the earth. Fig. 1 shows an isohyetal map of the earth, in which the omnipresence of solar radiation in different intensities can be observed: in the equatorial regions the intensity is distinctly higher than that of the areas around the polar regions.

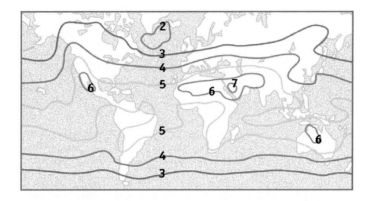

Figure 1. World Availability of Solar Radiation.

However, the availability of solar radiation is limited to daylight hours, therefore, it is necessary to store the electricity produced during the day in order to be able to use it during the night as well. Moreover, particularly intense and prolonged cloudiness in certain regions of the world causes a reduction of energy production. A great advantage of on-site local energy production is that it makes it unnecessary to transport energy over long distances, or to transport fuel to produce it. For example if we consider energy production with a diesel generator in remote, hard-to-reach areas, we must add in the economic assessment the considerable cost of fuel transport on-site. Furthermore, this production of energy also entails an emission of CO_2 into the atmosphere. Another great advantage of energy production from a photovoltaic solar source is the complete absence, during production, of harmful and polluting substances into the atmosphere. Being understood that the industrial process of photovoltaic panel production surely entails the emission of a certain quantity of polluting substances, if electricity produced from fossil sources (such as oil, gas, etc.) is used in the factory, the use of the finished product for producing energy is completely emission-free for its entire life cycle (about 25-30 years).

As already mentioned, the production of electricity from a photovoltaic solar source is limited to daytime, which is why it is necessary to store it during the day for the night use. Energy is stored using chemical batteries, which introduces a loss of energy into their charging and discharging process. Furthermore, their average life, as for the best technology on the market, is around 12 years, less than half the duration of photovoltaic modules. Therefore it will be necessary to replace the storage system once or twice. This disadvantage is not present in grid-connected photovoltaic systems, where the power grid provides an endless storage buffer.

The production of energy from a solar source is, by its very nature, in flux. In fact, the presence of clouds causes a fluctuation in electricity production and, in the case of a total absence of the sun, an almost total reduction. In isolated photovoltaic systems the storage subsystem is therefore necessary not only for the night time powering of electrical loads, but also to provide energy in the event of a total or partial absence of the sun.

1.2. Components of Solar Radiation

Out of the terrestrial atmosphere the solar radiation is propagated in a symmetrical way and its value in terms of density of power, called constant solar, is equal to 1350W/m^2. Once the solar radiation enters the atmosphere, the molecules of the present substances cause various absorptions and consequent emissions, reducing its intensity and creating a component of diffused solar radiation. The concept that expresses the absorption of solar radiation in the atmosphere is the Air Mass. The Air Mass (AM) represents the atmosphere's thickness that direct solar rays meet on their path to the sea level. The mass of unitary air AM1 points out the thickness of standard atmosphere crossed by perpendicular solar rays to the terrestrial surface, measured at the sea level in a day of clear sky. At European latitudes the mass of air will be superior to the unity (AM 1.5; AM2) considering the increase of the path of direct solar rays.

The standard conditions commonly used for both the measurement of photovoltaic modules and for plants design are: AM 1.5, irradiance 1000W/m^2 and solar cells temperature 25°C.

Once the solar radiation reaches the terrestrial surface, a part of it called albedo reflects itself. Commonly it is assumed that global solar radiation is constituted of the above mentioned components, i.e. direct, diffused and reflected.

Usually, the value of global solar radiation is used for PV plants sizing, since it is also the most available to retrieve among the existing data. Solar radiation is very variable; therefore, in order to carry out a plant sizing, the site's statistic data must be as near as possible to the plant.

Another remarkable factor to calculate the quantity of solar energy that will be picked up and therefore produced in form of electric energy by a photovoltaic module is the module disposition in comparison to the terrestrial surface. The modules have two angles of inclination: the first one, referred to the horizontal angle is called tilt, while the other one, referred to the geographical south is called azimuth. In order to maximize the collection of solar radiation, the plant of PV modules must be directed to the south if situated in the northern hemisphere, and to the north if situated in the southern hemisphere. This orientation allows the maximum exposure of the modules' surface to the sun during the day. As for the tilt angle, the situation is different according to the typology of plant. As an initial data, the angle of tilt can be considered equal to the latitude of the site which must be optimized. In the case of a plant connected to the national electric network (grid-connected), the best inclination will be the one that maximizes the collection of energy during the year. In the case of isolated plants (stand alone), the best inclination will be the one that maximizes the collection of energy in the month with less sunlight. In technical literature, the use of specific algorithms allows to infer the value of solar radiation falling on the tilted plan of a generic β angle, and turned towards a generic θ angle from the value of solar radiation falling on a generic horizontal plan. As a first approximation for grid connected systems, the optimal inclination is obtained with a decrease of up to ten degrees in proportion with the angle of latitude, while for stand alone systems, the optimal inclination is reached with an increase of up to twenty degrees in proportion with the angle of latitude.

If the angle of inclination is modified, and supposing that the orientation is perfectly towards the south, the energy profiles picked up by the photovoltaic modules vary during the year. For example, the solar global radiation profiles collected during the year by tilted, southward surfaces in the village of Tang Jia Xiang in Tibet (latitude 29°N 53') China, are reported below (Fig. 2).

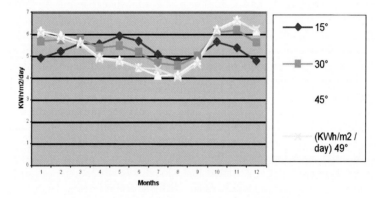

Figure 2. Tang Jia Xiang Insolation Profiles.

In stand alone plants, we also need to compare the solar radiation profile with the annual profile of electric absorption of the load.

2. SOLAR ENERGY APPLICATIONS

A possible classification of different types of photovoltaic systems is reported below:

(1) Stand alone small direct power systems.
(2) Stand alone small direct and alternating power systems.
(3) Stand alone medium power systems.
(4) Stand alone hybrid medium power systems (PV-Diesel).
(5) Grid-connected systems.

2.1. Stand Alone Systems

The description of an isolated photovoltaic system for the powering of a rural dwelling is reported below. The system is made up of the following components:

- a photovoltaic generator made up of 10 EniTecnologie PN16 modules;
- a module support structure;
- wiring between modules and the PV field boxes, between the PV field boxes and the charge regulator, and finally between the charge regulator and the battery;
- assembly and wiring accessories;
- two PV field boxes for the parallel connection of the modules;
- a regulation and control box;
- a battery bank.

2.1.1. Photovoltaic Generator
The photovoltaic generator must have a guaranteed minimum power of 1,500 Wp and consists of 10 modules of 150 Wp each. The whole generator is subdivided into subfields, so as to obtain a power division as fair as possible and a rated voltage suitable for recharging the 24 volts batteries. Therefore, the generator is subdivided into two subfields, being the connection in parallel to five photovoltaic modules.

The two subfields that make up the photovoltaic generator are connected to the regulation and control box, to regulate and control the battery charge, according to preset parameters identified as voltage thresholds.

The electrical circuits, wiring, and signalling and measurement equipment must fit the type of installation and be manufactured in compliance with the standards and regulations in effect.

A temperature probe installed in the battery compartment is supplied for the correction of the battery end-of-charge voltage on the basis of the temperature.

2.1.2. Support Structure
The photovoltaic modules supporting structure is made of hot-galvanized steel. The optimum tilt of the modules' plane, with respect to the horizon envisages an amplitude angle of 60°, so as to guarantee the maximum delivery of energy during the least sunny period.

2.1.3. Regulation and Control Box
The purpose of the regulation and control box is to contain all the equipment used for regulating the battery charge for the photovoltaic generator; it also contains all the signalling and measurement equipment required by the specifications.

The regulation of the battery charge takes place by means of regulators with solid-state power devices, controlled by highly reliable control logic. All the rated characteristics of the electronic devices used are redundant so as to ensure the functioning required, limiting power losses to a minimum even in the worst environmental conditions.

For the correction of the battery end-of-charge temperature, a special temperature probe installed near the battery is used.

The box contains the over voltage arresters on each line coming from the sub fields. All equipment is contained in a metal case. The regulator disconnects the load in the event of low battery voltage.

2.1.4. Field Boxes

The PV field boxes, necessary for the five modules parallel connection, are made of die-cast aluminium with protection level IP65. Inside, bipolar disconnecting switches with metal cartridges are installed to hold the disconnecting strings, the support bar, the terminals for outgoing wire connection, and the blocking diode, all with appropriate dimensions; the wires entry points are protected by adequate wire clamps.

2.1.5. Wiring and Accessories

The system is supplied with the following connecting wires:

- modules-PV field box connection, with FG7(O)R type wire;
- PV field box-charge regulator connection, with FG7(O)R type wire;
- regulator box-battery connection, with FG7(O)R type wire.

2.1.6. Battery Bank

The elements of the battery bank are sealed, valve-regulated type in compliance with CEI 21-63 standards; they have a long life expectancy and are widely used as an energy source for telecommunication systems, industrial automation, naval uses, conventional electrical, solar and wind power plants.

The electrolyte is a gel, the tubular positive plates have antimony-free lead rheophores, the negative plates are made of pasted lead, and the separators are plastic.

These elements do not need any type of maintenance or electrolyte toping up through the end of their lifetime. They are supplied with all connection accessories.

2.1.7. Battery Container

In accordance to the specifications, a 316 AISI stainless steel container is supplied: it has adequate dimensions to host the battery elements and the temperature probe, and to facilitate connections and cleaning.

The pipe for emission conveyance and the battery blocking system is also provided. The wires entry points are equipped with suitable AISI 316 wire clamps.

2.1.8. Examples

The "Terraegna" Shelter in the Abruzzo National Park power plant is constituted of:

- 14 PV modules of 65 Wp each for a total amount of 910 Wp of installed power;
- an assembly structure;
- a battery with 540 Ah/C 10 at 24 V for a two-day autonomy without sun;

- a battery container;
- a charge regulator board: 40 A, 24 Vdc;
- an inverter type MASTERVOLT Mod. Mass Sine 24V/500W;
- a user switch board;
- 8 electronic lamps of 15 W;
- one refrigerator of 150 litres;
- a parallel string board;
- cables, protection pipes;
- grounding.

2.2. Grid Connected Systems

The system is made up of modules connected in series/parallel, so as to optimize the functioning voltage of the inverter, which transforms the direct current electricity produced by the modules into alternating current, within the voltage range accepted by the distribution grid.

The photovoltaic modules are fastened to the metal support structures, which are in turn installed on the solar surface of a building, suitably tilted with respect to the horizontal plane and oriented southward.

The direct current electricity produced by the photovoltaic modules, converted into alternating current electricity at 230V-50 Hz, is thus fed into the electrical company's distribution grid as single-phase current for systems with power up to 5kWp, and three-phase 380V-50 Hz current for systems with higher power (Italian standard).

This energy is measured by means of a specific dedicated meter, installed by the grid operator, and computed according to the provisions of the Energy Authority.

The system is designed and constructed in accordance with the standards and regulations in effect at Italian and European levels, with particular reference to the directives issued by the following institutions: CEI/IEC; ENEL; UNI/ISO.

The productivity of a photovoltaic system, net of losses within various components of the system – wires, converter, fouling, etc. – estimated at around 20-25% of the installed power, is calculated by the irradiance values (Wh/m^2) of the installation site. These, compiled in Italian standard UNI 10349, must be optimized, with respect to a collection area with a certain inclination depending on the latitude of the installation side, and to conversion efficiency (12-15%) of the photovoltaic modules used.

2.2.1. Energy Production

The following tables (Tab. 1, Tab. 2) summarize the data referring to the annual productivity of a 10kWp photovoltaic system installed in the cities of Milan, Rome, and Catania, expressed in kWh.

Table 1. Annual Energy Collection

Site and Latitude	Milan, 45°27'N	Rome, 41°59'N	Catania, 37°30'N
Optimal tilt of modules, southern exposure (azimuth 0°)	35°	30°	30
Occupied Area, m^2	100	100	100
Annual Total Energy Collected in kWh	14,810	18,110	20,380

On the basis of the ground-level irradiance values and the optimized plan of the modules indicated in the table, and assuming an average system efficiency of 75% at various functioning levels, the systems' energy productivity, meant as electricity fed into the local distribution grid, is reported in Tab. 2.

Table 2. Annual Energy Productivity

Site	Milan	Rome	Catania
Net Energy fed into grid in kWh	11,107.5	13,582.5	15,285

2.2.2. *Examples*

The Bologna Airport power plant (Fig. 3) is constituted of:

- 43 strings of PV module EUROSOLARE PL 800, each of them made up of 27 connected module series for a total number of 1161 modules;
- an assembly structure realized in hot-dip galvanized iron;
- ten field boards inside a protection box with surge diverters, isolators and input/output terminals;
- two parallel boards located inside the building;
- two inverters 40kVA each, with digital control, and modem for remote data transmission.

Figure 3. Bologna Airport 80KWp.

3. CASE STUDY: SOLAR VILLAGE IN CHINA

3.1. Introduction

The Chinese Ministry of Science and Technology and the Italian Ministry for Environment and Territory signed an agreement for the implementation of joint projects focused on energy and environmental protection. EniTecnologie and the Inner Mongolia Natural Energy Institute were chosen to be the executive partners for the implementation of this project. One of the project's goals was to demonstrate a way to solve the power supply problem in some rural areas of the Inner Mongolia province, using solar energy applications, meeting end users' requests for a quality of life improvement and economic development. Husbandry is the only economic activity in this rural area, and it implies continuous grassland degradation, with an inevitable increase of land desertification. Solar power water pumping systems can be a resource to develop other economic activities such as agriculture. A key issue covered by the project is the follow-up strategy (Barile *et al.*, 2005) which will enable training of local personnel for the plant maintenance. In other words, the dissemination of know-how is directly supported by the project's financial resources. This unique feature intends to provide a show case for possible further projects.

3.2. Description of the Project

To reach its objectives, the project uses the following approach:

- Minigrid electrification of a rural village in the north of Inner Mongolia, through the installation of a 20kWp Hybrid Photovoltaic Diesel generator, supplying energy for street lamps and house lighting, as well as for two

electro pumps for water extraction from a well, and for other possible applications.
- Electrification of single scattered households through the installation of 151 Solar Home Systems.
- Three water pumping systems with PV generator, without storage subsystem, for the irrigation of surrounding fields. The first system has a 8kWp PV generator, and it is installed close to an artificial swimming pool for water collection, while, the other two systems have a 4kWp PV generator, and are installed close to existing wells.
- Monitoring of the above mentioned systems for a period of at least one year. Purpose of the monitoring is to evaluate the end users' energy consumption, and to establish the best system size for future installations.
- Dissemination of the project results in a final workshop, after the evaluation of the data collected during the monitoring period. The overall photovoltaic power installed is equal to 110kWp, with a number of six electro pumps, for a total of 185 families served by electrification.

3.3. Project Organization

The Italian side is the General Contractor with the responsibility for the whole project. This role includes the project coordination and scheduling, and the construction of the Solar Power Plants. Once the commissioning and start up of the PV systems will be completed, the systems will be handed over to the Chinese side. The Chinese partner is the Project Supporter; therefore it will be responsible for the civil work, the power plant installation, and the systems' monitoring for one year, with the support of the General Contractor. Furthermore, the Chinese partner will ensure the execution by local companies of the solar power systems maintenance, funded by local fee service. This provides a unique feature of this project, in that it poses the basis for further development and dissemination, trying as much as possible to avoid stop and go situations, which are known to be a barrier to the diffusion of PV technology.

Inner Mongolia is located in the northern and north-eastern part of China. The longitude of the region goes from 97°12' to 126°04' and the latitude from 37°24' to 53°23'. The average height of the region is about 1000 m above the sea level. The region measures 1.18 million km^2, and within this area, the plateau occupies 53.4%. The solar radiation in Inner Mongolia is very good. The sunshine hours are long, and the air is clean, which gives the light a better penetrating rate, resulting in good solar energy on the ground. Finally, the northern and western parts of the region do not have electrification.

3.4. Objectives

Mongolians live on grassland and their traditional economic activity is husbandry. Without power supply, we can say that their life style has not changed for the last century. However, on the other hand, new policies give herdsmen land use right, giving them freedom to decide for their production. People start to be more active, and to improve their quality of life with an increasing level of farming. Ecological conditions for grass are already poor in the region and when an increasing number of animals started to eat grass, the grassland began to suffer from a severe degradation process, becoming a desert in some cases. The reduced grassland area in turn leads to more concentrated grass consumption by animals, thereby producing a vicious circle.

The target group in this project is the rural population of Inner Mongolia. The reason to choose such a target group is mainly because they live in a region – the western part of China – for which the Chinese government decided to adopt a strategic development policy. Furthermore, being poverty alleviation another governmental policy, this poor population represents a proper target group. Finally, Mongolians are also one of the largest among 56 minority ethnic groups in China. Dealing with different ethnic groups and stabilizing social development are also among the government's objectives.

The population lives in a scattered way. Being their main economic activity animal husbandry, they must live separately to leave enough grassland for their animals. It is quite normal to see a stand-alone household in the rural area of Inner Mongolia.

Single households were equipped with Solar Home System (Fig. 4), a 520Wp system with the following characteristics:

- four EniTecnologie PV modules PN16/140 with 133Wp each;
- one field board;
- one battery and control board including a Charge Regulator EniTecnologie 24V 30A, an Inverter TC15/24V 1200VA continuous power, a Battery 24V with 12 elements type 5 OPzS 385/70, one plug and energy meter;
- three 15W electronic lamps, and electrical distribution.

Another aim of the project is to develop agriculture through the existing water resource. The solar water pumping system has the following characteristics: 8kWp PV generator with EniTecnologie modules PN16/150 with 150Wp each, two field boards for the parallel connection of the strings, two submergible electro pumps Caprari of 2.2kW each; one raft and water pipes floats, two water counter litres and two inverters ES Solar Drive 4000.

The minigrid electrification of one village was carried out using a hybrid PV-diesel generator of 20kWp power.

The generator supplies the following loads:

- 40 street lamps;
- the electrification of 34 houses with three 15W electronic lamps and one plug;
- two submergible electro pumps, one for animal beverage and one for irrigation purposes;
- two submergible electro pumps Caprari of 2.2KW each;
- one raft and water pipes floats;
- two water counter litres;
- two inverters ES Solar Drive 4000.

3.5. Conclusions

The experience has been positive, both for the co-operation between the Chinese and Italian groups, and for the positive outcome for a number of families now having access to electricity. A close cooperation between local companies and the Chinese project management is one of the main issues to guarantee the success of the project.

Figure 4. Solar powered house in winter.

Roberto Barile
Photovoltaic Business,
EniTecnologie S.p.A., Rome, Italy

REFERENCES

Markvart T., 1994. *Solar Electricity,* John Wiley & Sons Ltd.
Green M. A., 1982. *Solar Cells,* Prentice-Hall.
Barile R., Valentini C., Mingyi X., Zhizhang L., 2005. Solar Village in Inner Mongolia, China. *20th European Photovoltaic Solar Energy Conference and Exhibition,* Barcelona, Spain 2005.

R. BERTANI

GEOTHERMAL ENERGY

Abstract. Despite the limitations of Planet Earth's conventional energy resources, the demand for energy is continuously rising as a result of increasing population and industrialization. The utilization of fossil energy resources, which led to great technological and social developments in the past, now is causing increasingly disastrous effects on the global environment. In this situation there is urgent need to deploy sustainable and environmentally clean energy sources. An important contribution could be made by rapidly expanding the use of renewable energy sources such as geothermal energy. A number of countries stand out as having made utilization of geothermal resources a national priority. This paper will illustrate the major aspects of geothermal energy, its generation, the relevant power plant technologies for electricity generation, the utilization of geothermal heat for direct uses, with focus on some showcases of best practice. The impact of geothermal energy in China is the topic of the second half of the present document.

1. INTRODUCTION: WHAT IS GEOTHERMAL ENERGY AND HOW DOES IT WORK?

Geothermal energy is the heat from the Earth. The Earth's interior supplies heat thus providing us with warmth and power that do not pollute the environment. This heat originates from Earth's formation, over 4 billion years ago, when dust and gas consolidated. Estimates of the Earth's core temperatures – at 6,000 km depth – exceed 4,000°C. Another powerful source of heat is the radioactive decay of rocks long-lived isotopes of Uranium, Thorium and Potassium.

The heat outflows from the Earth's core to the surrounding rocks, the mantle. When temperatures and pressures grow to be high enough, mantle rock melts, becoming magma. Then, because of its density, lower than the rocks', the magma upraises toward the Earth's crust, carrying the heat from below. When it reaches the surface it may flow as lava. Most often it remains below the Earth's crust, heating the surrounding rocks and water, reaching up to 370°C. Some of this hot geothermal water travels back up through faults and cracks, reaching the surface as hot springs or geysers, but most of it stays underground, trapped in cracks and porous rock. This natural stock of hot water is called a geothermal reservoir.

Heat is a form of energy and geothermal energy is literally the heat contained within the Earth that generates geological phenomena on a planetary scale. "Geothermal energy" is often used to indicate that part of the Earth's heat that can be recovered and exploited by man. Earth's total heat content is of the order of 12.6×10^{24} MJ, and that of the crust of the order of 5.4×10^{21} MJ. As a comparison, the world's total electricity need is about 6×10^{13} MJ per year, i.e. 100 million times less! The thermal energy of the Earth is therefore immense, but only a fraction can be utilized by man. So far, our utilization of this type of energy has been limited to areas in which geological conditions permit a carrier (water in the liquid phase or steam) to "transfer" the heat from deep hot zones to or near the surface, thus giving rise to geothermal resources.

1.1. Geothermal systems

The presence of volcanoes, hot springs, and other thermal phenomena must have led our ancestors to surmise that parts of the interior of the Earth were hot. However, it was not until a period between the sixteenth and seventeenth century, when the first mines were excavated to a few hundred meters below ground level that man deduced, from simple physical sensations that the Earth's temperature increases with depth. The normal temperature gradient is about 2-3°C/100m, which implies an average temperature of about 60°C at 2,000 m depth. But in "geothermal areas" the gradient can be up to ten times higher. The total heat flux from earth to surface is about 80 mW_{th}/m^2.

Geothermal systems can therefore be found in regions with a normal or slightly above normal geothermal gradient, and especially in regions around plate margins where the geothermal gradients may be significantly higher than the average value. In the first case the systems will be characterized by low temperatures, usually no higher than 100°C at economic depths; in the second case the temperatures could cover a wide range from low to very high, and even above 400°C.

What is a geothermal system and what happens in such a system? It can be described schematically as a system conveying water in the upper crust of the Earth, which, in a confined space, transfers heat from a heat source to a heat sink, usually the free surface. A geothermal system is made up of three main elements: a *heat source*, a *reservoir* and a *fluid*, which is the carrier that transfers the heat. The heat source can be either a very high temperature (> 600°C) magmatic intrusion that has reached relatively shallow depths (5-10 km) or, as in certain low-temperature systems, the Earth's normal temperature, which, as we explained earlier, increases with depth. The reservoir is a volume of hot permeable rocks from which the circulating fluids extract heat. The reservoir is generally overlain by a cover of impermeable rocks and connected to a superficial recharge area through which the meteoric waters can replace or partly replace the fluids that escape from the reservoir through springs or are extracted by boreholes. The geothermal fluid is water, in the majority of cases meteoric water, in the liquid or vapour phase, depending on its temperature and pressure. This water often carries with it chemicals and gases such as CO_2, H_2S, etc. Fig. 1 is a greatly simplified representation of an ideal geothermal system.

Fluid convection is the most important mechanism involved in the heat transfer for a geothermal system. Convection occurs because of the heating and consequent thermal expansion of fluids in a gravity field; heat, which is supplied at the base of the circulation system, is the energy that drives the system. Heated fluid of lower density tends to rise and to be replaced by colder fluid of high density, coming from the margins of the system. Convection, by its nature, tends to increase temperatures in the upper part of a system as temperatures in the lower part decrease.

Drilling wells into the geothermal reservoirs to bring hot water to the surface allows the use of this clean and sustainable resource. Geothermal experts and engineers work to locate underground areas that contain geothermal water, in order to carry out drilling operations for geothermal production wells. Once the hot water and/or steam travel up the wells to the surface, they can be used to generate electricity in geothermal power plants or for non-electrical (direct uses) purposes.

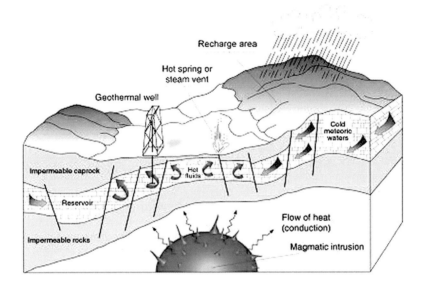

Figure 1. Schematic representation of an ideal geothermal system.

Many power plants need steam to generate electricity. The steam rotates a turbine that activates a generator, thus producing electricity. Conventional power plants use fossil fuels to boil water for steam. Geothermal power plants, instead, use steam produced from geothermal reservoirs sited from a few hundred to a few thousand metres below the Earth's surface. This steam is produced without any artificial or natural combustion, i.e. without man-induced generation of CO_2 gas. There are three types of geothermal power plants: *dry steam, flash steam,* and *binary cycle*.

1.1.1. Dry steam power plants
Dry steam power plants draw from underground resources of steam. The steam is piped directly from underground wells to the power plant, where it is directed into a turbine/generator unit. The first geothermal field ever exploited, and still in operation at present, Larderello in Italy, is one of the very few known underground resources of steam in the world.

Conventional steam turbines require fluids at temperatures of at least 150°C and available with either atmospheric (backpressure) or condensing exhausts. Atmospheric exhaust turbines are simpler and cheaper. The steam, direct from dry steam wells or, after separation, from wet wells, is passed through a turbine and exhausted to the atmosphere. With this type of unit, the produced steam consumption (from the same inlet pressure) per kilowatt-hour is almost twice the one of a condensing unit. However, the atmospheric exhaust turbines are extremely useful as pilot plants, stand-by plants, in the case of small supplies from isolated wells, and for generating electricity from test wells during field development. They are also used when the steam has a high non-condensable gas content (> 12% in weight). Condensing units, which have more auxiliary equipment, are more complex than the atmospheric exhaust units and the bigger sizes take up to twice as long to construct and install. The specific steam consumption of the condensing units is, however, about half that of the atmospheric exhaust units. Condensing plants of 55-60 MW_e capacity are very common, but recently plants of 110 MW_e have also been constructed and installed.

1.1.2. Flash steam power plants
Flash steam power plants, the most common, use geothermal reservoirs of water with temperatures higher than 180°C. This very hot water flows up through wells in the ground under its own pressure. As it flows upward, the pressure decreases and some of the hot water boils into steam. The steam is then separated from the water and used to power a turbine/generator. Any leftover water and condensed steam are injected back into the reservoir, making it a sustainable resource.

1.1.3. Binary cycle power plants
Binary cycle power plants operate on water at lower temperatures of about 100°-180°C. These plants use the heat from hot water to boil a working fluid, usually an organic compound with a low boiling point. The working fluid is vaporized in a heat exchanger and used to turn a turbine. The water is then injected back into the ground to be reheated. The water and the working fluid are kept separated during the whole process, so there are little or no air emissions. By selecting suitable secondary fluids, binary systems can be designed to utilize geothermal fluids in the temperature range 85-175°C. The upper limit depends on the thermal stability of the organic binary fluid, and the lower limit on technical-economic factors: below this temperature the size of the heat exchangers required would render the project uneconomical. Apart from low-to-medium temperature geothermal fluids and waste fluids, binary systems can also be utilized where flashing of the geothermal fluids should preferably be avoided (for example, to prevent well sealing). In this case, downhole pumps can be used to keep the fluids in a pressurized liquid state, and the energy can be extracted from the circulating fluid by means of binary units.

Small-scale geothermal power plants (below 5 MW_e) have the potential for a widespread application in rural areas, possibly even as distributed energy resources.

The largest installed capacities correspond to dry steam and single flash units, covering 2/3 of the total. Binary units, despite their low position in this ranking because of their smaller capacity ratings, are becoming increasingly more common.

There were a total of 490 geothermal units operating in 2005 (see Tab. 1).

Table 1. Power plant distribution by plant type (early 2005 data)

Plant type	Installed capacity (MW_e)	Percent	Installed capacity (number of units)	Percent
Dry steam	2545	28%	58	12%
Single flash	3294	37%	128	26%
Double flash	2293	26%	67	14%
Binary/combined cycle/hybrid	682	8%	208	42%
Back-pressure	119	1%	29	6%
Total	**8933**	**100**	**490**	**100**

2. ELECTRICITY GENERATION

The installed capacity has increased by approximately 960 MW_e since year 2000; that is, only about 190 MW_e per year were added during the 2000-2005 period. The world net electricity generation for 2003 was 15.8 million GWh/y, while the geothermal generation was only 0.057 million GWh/y. (Bertani, 2005; Huttrer, 1995 and 2001).

Fig. 2 represents a world map showing the countries that generate electricity using geothermal resources, and their installed capacity in early 2005.

3. DIRECT USES

Distributed energy resources refer to a variety of small, modular power-generating technologies that can be combined to improve the operation of the electricity delivery system. Cascade utilization of residual heat from discarded geothermal fluid is an excellent way of increasing the efficiency of the overall cycle.

Hot water near the surface can be also used directly for heating. Direct-use applications exploit the geothermal resources having a lower temperature, usually referred to as medium-low enthalpy resource, which are however the most widespread all over the world. Any process that needs heat can take advantage of a local geothermal source and in this sense the only limit of application is our creativeness: heating/conditioning of buildings, district heating, de-icing of roads, greenhouses for flowers and vegetables, recreational and therapeutic pools, industrial washing of wool, drying crops, drying timber, fish and crustacean farming, pasteurising milk, dyeing of fabric, sterilization of soil without chemicals, open air cultivation of crops and so on.

In 2005 (Lund, 2004 and Lund *et al.*, 2005), some 72 countries were making a direct use of geothermal energy for about 28,268 MW_{th}, that is 273,372 TJ/year.

The energy saving for direct uses of geothermal energy is equivalent to:

- 270 million barrels
- 41 million tonnes of oil/year
- 3.5 days or 1% of world's production
- 37 million tonnes of carbon/year
- 118 million tonnes of CO_2 /year
- 0.8 million tonnes of SO_x /year
- 22 thousand tonnes of NO_x /year

The rate is increasing rapidly, with 11% per year in capacity since year 1975, with a doubled rate in the last 5 years.

Cascade utilization is the most efficient way of using geothermal energy (Fig. 3).

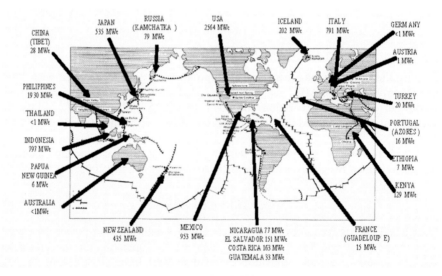

Figure 2. Geo-thermoelectric installed capacity worldwide in early 2005.

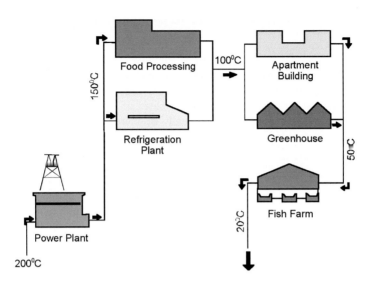

Figure 3. Cascade use of heat and electricity.

3.1. Illustrated case #1

Milgro-Newcastle, USA: its greenhouse is the largest potted plant grower in the US. The facility is located in Southern Utah, at an altitude of about 1,500 m, and heating is required, especially in winter when the outdoor temperature falls below zero. The area is rich in geothermal resources, and there are several wells. The produced water, at a temperature exceeding 90°C, is associated to a fault system located southeast and flows to northwest through the permeable sedimentary intercalations of the Escalante Valley crossed by the wells. The Utah facility, about one million square feet, is supplied with geothermal water through two production wells at shallow depth (600 m), and two re-injection wells made for disposal at the end of the circuit.

Operating costs of this facility are not available, but the total maintenance budget of $16,000 per month, (including maintenance of structures, vehicles, electrical systems, plant growing equipment and geothermal system), is lower than that required by the Milgro's conventionally heated greenhouses in the Los Angeles area.

3.2. Illustrated case #2

Toronto, Canada: a three-storey Secondary School just north of Toronto, in Ontario, is 16,800 m². Built in 1992, the school has 38 classrooms, 19 laboratories and workshops, a library, administrative offices, a chapel, a greenhouse, a cafeteria, three gymnasiums and a child-care centre. Some 2,400 staff and students frequent it.

A ground-source system has been installed mostly on the basis of two economic reasons:

- The electrical utility offered a very attractive ground-source incentive.
- The use of a ground-source system, reducing the required equipment room, was leaving more space for classrooms, which, in turn, allows government granting to school boards. On the basis of the utility bills, the ground-source system saves $9,420 annually in energy compared to a system that uses a central chiller and gas boiler.

3.3. Illustrated case #3

Maguarichic, Mexico: it was an isolated village in the State of Chihuahua where the power was supplied by a 90 kWe diesel generator that ran 3 hours/day; homes had no refrigerators, thus the villagers rarely had meat, cheese, milk, etc. They were not aware of national events since there was no TV.

In 1997, a 300 kWe binary plant, using fluid at 150°C with a flow rate of 55 t/h, at the cost of $3,000/kW was built. Now villagers have street lights, refrigerators, electric sewing, and tortilleria machines and ice cream for the kids.

3.4. Illustrated case #4

Nangong village, Beijing: it is a normal small village in the south west suburb of Beijing. In recent years, comprehensive geothermal development and usage have made great achievement, and Nangong became a sparkling star in this area. In year 2000 the drilling of a 3,000 m depth productive well has been completed, producing 2,700 tons/day of water at 75°C. This water is used in a two stage process: first the heat, second the mineral contents for medical and gymnastic facilities. Six major projects are on-line: geothermal greenhouses, fish farming and fishing centre, entertainment centre, district heating system and geothermal exhibition/education centre. The greenhouse occupies 1.5 km^2 with 12 buildings, for flowers, vegetables and season fruits. The fish breeding and fishing area is the biggest indoor centre for fishing entertainment in China, with 220,000 m^2 of covered area. The hot spring water entertainment (18,000 m^2) serves 1,500 customers per day, with pool water temperature at 20°C all year round and other heated gymnastic facilities.

The villagers of Nangong have already benefited from geothermal development, with their houses geothermally heated and high quality hot spring water available at home. A new greenhouse system for 20,000 m^2 and a new hot spring water gymnastic centre of 30,000 m^2 are planned.

4. GEOTHERMAL ENERGY IN CHINA

4.1. Introduction

China is a global leader in the use of geothermal resources for direct applications and has rich geothermal resources with a long history of utilization, with a thermal power contribution of more than 2,000 MW$_{th}$ (Shibin at al. 2005; Tian et al., 2005; Yi et al., 2005 and Zheng, 2005).

Hot springs have been used for space heating and for treatment of diseases since the Ming dynasty. Today there are more than 1,500 sites throughout China where geothermal energy is in direct use. In total, they have produced energy for an equivalent of 5 million tons of standard coal. Geothermal resources are spread widely across China. Evidence of abundant geothermal resources is widely spread around the country, including 2,500 thermal springs and 270 geothermal fields. Most of the hot springs are located in the Provinces of Fujian, Yunnan, Sichuan, and the Xizang (Tibet) Autonomous Region. The following figures illustrate the major utilization of geothermal water in China (Fig. 4):

- 8,000,000 m^2 of geothermal space heating
- 700,000 m^2 of greenhouses
- 3,000,000 aquaculture sites
- 1,600 spas

High temperature geothermal resources are concentrated along the Himalayan Belt, which is an extension of the Mediterranean Geothermal Belt, passing through southern Tibet, western Sichuan and Yunnan, turning south through Thailand.

One of the best geothermal resources was found in one of the sites in the Yangbajing Field in Tibet, with a geothermal fluid temperature of 330°C at a depth of 2,000 m. This indicates the excellent mid- to high-temperature geothermal resources in the region.

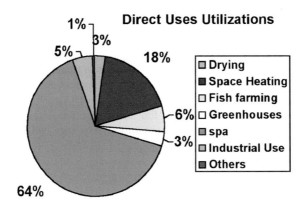

Figure 4. Direct uses in China.

4.2. Tibet

The total energy consumption in Tibet is estimated at 1.4 million tons/year in terms of standard coal, subdivided into:

> 1.05 million tons/year of traditional fuel. The consumption of traditional fuel such as firewood and animal waste has increased in recent years due to the shortage of low cost alternatives, accelerating soil degradation in this arid region.
> 0.35 million tons/year of high quality energy.

At present the total installed power capacity in the Lhasa grid is made up of 24 MW_e geothermal, 19 MW_e hydroelectric, 16 MW_e thermal.

Electricity is in serious shortage in Tibet, especially in the Lhasa area which has a very limited network output capacity especially during winter. Moreover, the natural environment is progressively being damaged by the increasing forest cutting and the use of animal waste. Except for a few major cities (Lhasa, Xigaze, Nagqu), the two million inhabitants of Tibet live in small rural centers, scattered over a surface area of 1.5 million km^2, at an average altitude of more than 4,000 masl. Due to such severe geographic and climatic conditions, communication is extremely difficult and electricity distribution in Tibet is still limited.

As a matter of fact, at present, only 30% of all populated centers in Tibet have been served by electricity, the capital city Lhasa with its 100,000 inhabitants being the main centre. Although fossil fuels are costly to import into the region, their consumption is growing and will continue to grow in the future by necessity, causing future stress to the overall economy and to the very fragile environment of the Tibetan Plateau. Further development of the hydroelectric potential might not result a cost-effective option with the current technology, due to Tibet's environment and social-economic conditions. Wood and animal dung are to be preserved for ecological reasons. It is believed that geothermal energy could account for the bulk of Tibet's energy needs.

Moreover, the cascade uses of geothermal energy, exploiting the heat from discharged brine in greenhouses, district heating or other applications, can contribute significantly to fulfil the energy needs of such remote areas of China, where the importation of non-locally available fuels is very expensive and pollution problems are particularly relevant in the fragile ecosystem of the Tibetan Plateau.

The Tibetan Plateau was created by the subduction of the Indian and Eurasian constructive plates, uplifting since the end of the Cretaceous, and was generated by collisional events and characterized by seismic and geothermal activities.

Upon these geological conditions, 672 geothermal active areas (spots) appeared on the Tibetan Plateau, and the total natural heat discharge has been estimated at about 40,000 MW_{th}. The exploration of geothermal resources in Tibet has begun in Yangbajing geothermal area at the beginning of the '70s, with geological studies, geophysical surveys and drilling exploratory wells. Geothermal electricity production in Tibet is so far limited to the Yangbajing, Nagqu and Langjiu fields, with 27 MW_e installed capacity.

Since 1975, the Yangbajing geothermal field has had an initial exploration and development utilization. Until 1996, 80 wells were drilled with a total depth of about 20,000 meters. The maximum downhole temperature obtained was 172°C in shallow reservoir (150-200 m), which is the main production zone for Yangbajing power plant. A deep geothermal reservoir has been recently tapped by two deep wells in the northern zone of Yangbajing, with a maximum downhole temperature of 330°C. Until 1993, the total installed capacity of Yangbajing geothermal power plant was 24 MW$_e$ (9 units, 8 operating).

Until 1997, the total electric power generation attained 1,060 GWh, with an annual power generation from Yangbajing power plant of 110 GWh/yr, which accounts for 41% of the total in Lhasa Grid and up to 52% during the winter time. However, the power station equipment has become obsolete through its 26 years of running. The problem of corrosion and scaling in wells and pipes has not been solved yet. Yangbajing geothermal power plant operates about 4,300 hours annually on average. Geothermal electricity occupies about 40% of the electric network of Lhasa.

The total area of the field is 14.6 km^2, of which 7.4 km^2 represent the area in the southern part, while the northern part is large 7.3 km^2. The field potential has been estimated at about 35 MW$_e$. Until now, only the shallow reservoir has been utilized for electric generation. In over 20 years of exploitation, side effects, such as land subsidence, dry up, and disappearance of hot springs, boiling springs and geysers, pressure drawdown in the production wells, etc. have made their appearance. The shallow reservoir shows visible signs of decline of its performances (flow rate of wells and fluid enthalpy). The output of production wells cannot reach the designed capacities. Many phenomena show that the production from the shallow reservoir is much greater than the natural recharge. There are about 60 production wells in both the northern and southern areas, and now only 14 wells are used for power generation. This shows that the exploitation area of geothermal field has been reduced. The maximum temperature drawdown in production wells is 21°C in both the northern and southern area during the exploitation and utilization.

The geothermal resource is in depletion, and the environmental pollution is quite serious. The waste brine of the geothermal power station includes abundant deleterious elements. About 40% of this brine is reinjected into the southern area of the field, at the same depth of the reservoir production zone, and the relevant brine is discharged into the river, with serious environmental consequences.

The reinjection system has been built through three construction periods from 1990 to 1998. Eight reinjection wells have been drilled, and six of them are currently in use.

A high-temperature deep geothermal reservoir was discovered in the northern area of Yangbajing Geothermal Field: the well encountered high-temperature fluids (257°C) at the depth of 1,459 m, with a steam-water discharge of 300 t/h that illustrates that the single well electricity-generation potential reached 12 MW$_e$. A second one was identified with a temperature of 330°C at 1,850 m.

4.3. 2008 Olympic games

From October 29th to October 31st 2002, Beijing hosted the 2002 Beijing International Geothermal Symposium (BIGS). The main hosts were the Beijing Municipal Administration of Land Resources and Housing, and the Beijing Bureau of Exploration and Development of Geology and Mineral Resources (Jiurong et al., 2005).

The purpose of the symposium was to receive inputs from international geothermal experts and to discuss the use of geothermal energy for the 2008 Summer Olympics to be held in Beijing.

Beijing will host the Summer Olympic Games in 2008, and the Olympic Green, where the most important facilities for the great event are to be built, will be sited in the northern part of the city. A few geothermal wells have been drilled around the Olympic Green, indicating that there is a rather good geothermal potential in the area. The geothermal energy will be used for space heating and thermal baths facilities of the Sports Village, which will be more than 29 hectares, while the floor area of the department buildings is planned to be 360,000 m^2. According to the planning, 10 geothermal wells, including 6 production wells and 4 reinjection wells, will be drilled for the project.

Beijing is rich with low-temperature geothermal systems, and the area identified with geothermal potential is over 2,300 km^2. The geothermal water in the area of Beijing, that has a temperature of 40 to 88°C, is used for various direct purposes, including space heating, thermal baths facilities, greenhouses, fish farming, swimming pools and recreation etc. With the rapid development of the city, the geothermal development is rather fast and recently 20 to 40 geothermal wells have been drilled every year. At present, there are over 200 geothermal wells producing around 10 million m^3/year of geothermal water.

Coal is the dominant energy source of Beijing, causing serious air pollution. In the past years, measures to avoid air pollution, including the promotion of geothermal utilization, have been taken. The Olympic Games in Beijing will be featured as the Green Olympics, and geothermal energy is planned to be used for space heating and domestic hot water supply for the Sports Village.

The area of the Olympic Green has a positive geothermal potential according to the geological exploration of the area, and successful geothermal wells have been drilled nearby. The target geothermal reservoirs are limestone and dolomite aquifers buried at more than 2,000 m deep. The geothermal water around the area has a temperature of 55-75°C, and its components ($HSiO_2$, F, Sr etc.) are good for human health. A geothermal resources assessment was carried out in 2003. The result shows that the geothermal reserve in the area is rather big, but the allowable production, which is closely related to the water recharge of the geothermal system, is quite small.

It is estimated by volumetric method that the geothermal reserve in a 49 km^2 area around the Olympic Green is 7.34×10^{18} J, and the water storage in the geothermal reservoir is 2.5×10^8 m^3. It is important to estimate the allowable production from the reservoir on a sustainable basis. It has been calculated that the annual allowable production for the Urban Geothermal Field in Beijing accounts for 3.13% of the total water reserve in the geothermal reservoir.

If this is used in the area of the Olympic Green, the annual allowable production is 80×10^4 m^3/year. The reservoir pressure will not rapidly decline over a long period of time. It is planned that geothermal energy will be used for space heating (supplemented by heat pump system) and bathing facilities of the Sports Village, which will have a floor area of 36×10^4 m^2. If the temperature of geothermal water is assumed to be 65°C, the annual hot water need is 134×10^4 m^3, of which 78×10^4 m^3 for space heating in about 150 days per year, and 56×10^4 m^3 for bathing. It is clear that the allowable production of geothermal water for the area (80×10^4 m^3/year) is smaller than the total amount of hot water need (134×10^4 m^3).

Therefore, it is essential to reinject the tail water from the heating system. The tail water temperature is planned to be 22°C. Ten geothermal wells at around 3,000 m deep, subdivided into two groups, will be drilled. Each group will be drilled in a small surface area, composed of 3 production wells and two reinjection wells. Most of them will be directional wells. The production capacity of each well is estimated at 60 m^3/h (1440 m^3/day). Considering that there is abundant heat in place in the geothermal reservoir, while the water recharge is very limited, it is essential to reinject the used geothermal water from the space heating system, so as to support the reservoir pressure and use the geothermal in a sustainable manner.

Of course, reinjection is a kind of complicated technique related to the management of the geothermal reservoir, and careful experiments including tracer tests have to be carried out, in order to avoid premature thermal breakthrough. The space for reinjection and production wells is more than 1 km for the area of the Olympic Green.

Ruggero Bertani
International Geothermal Association,
Enel GEM AdB Renewable Energies – Geothermal Division, Italy

REFERENCES

Bertani R., 2005. World geothermal power generation in the period 2001-2005. *Geothermics 34*, pp. 651-690.

Huttrer G.W., 1995. *The status of world geothermal power production 1990-1994*. Proc. WGC 1995.

Huttrer G.W., 2001. The status of world geothermal power generation 1995-2000. *Geothermics 30*, pp. 1-27.

Jiurong L. and Jianping C., 2005. *The Geothermal Resources and Development Plan in the Olympic Green*, Beijing, China. Proc. WGC2005.

Lund J.W., 2004. 100 years of geothermal power production. *Geo-Heat Center Bulletin 25* n. 3, Oregon, USA, pp. 11-19.

Lund J.W., Freeston D.E. and Boyd T.I., 2005. Direct application of geothermal energy. *Geothermics 34*, pp. 691-727.

Shibin L. and Zhu H., 2005. *The Status and Trend Analysis of Geothermal Development and Utilization in China*. Proc. WGC2005.

Tian T. and Huang S., 2005. *Sustainable Development of Geothermal Resource in China and Future Projects*, Proc. WGC2005.

Yi Z., Yong J. and Miyazaki, S., 2005. *The Present Status of Utilization of Geothermal Energy and Resources Research by the Aid of Japanese Government in Yangbajing Geothermal Field*, Tibet, China, Proc. WGC2005.

Zheng K., 2005. *50 Years of Geothermal Development in Beijing, China*, Proc. WGC2005.

M. CHIADÒ RANA AND R. ROBERTO

RENEWABLE ENERGY FROM BIOMASS: SOLID BIOFUELS AND BIOENERGY TECHNOLOGIES

Abstract. Biomass is a renewable source of energy that can give a special contribution in reducing the use of conventional fuels from oil and in lowering emissions of high carbon dioxide from large coal use, particularly in a wide country such as China. Moreover, the use of biomass as a fuel can help to meet the increasing energy demand of a country with a high rate of economic growth.
In this chapter, the discussion is limited to solid bio-fuels as wood, woody and agricultural crops residues, and about sludge and wastewater treatment processes for the production of biogas.
The main technologies used to obtain energy from biomass fuels are well known, concerning the fields of direct combustion, and Italy offers a wide number of case-studies that could result in very interesting applications in Chinese rural areas. The solid-biomass gasification and pyrolysis processes are quite new technologies now under investigation, with documented experimental test facilities and study cases, but they are not yet on-the-market technologies.

1. INTRODUCTION

Biomass is the oldest fuel known to man. It is a renewable energy source and can be found in nature or derived as a residue from human activities. Biomass can be converted into solid, liquid or gaseous fuels and used to produce heat, electricity or in a combined power plant.

The use of biomass for energetic purposes is gaining greater and greater importance nowadays. In addition to its environmental benefits, it can also contribute to a substantial decrease in the use of fossil fuels, which may result in a reduced energetic and economic dependence on other countries.

China is a country in rapid economic growth, with an increasing energy demand. In 2003, the total primary energy supply was around $1.4*10e6$ tons of equivalent oil, excluding electricity trade, and around 15.5% of that amount came from biomass resources, i.e. from the combustion of renewables and waste (IEA data, from http://www.iea.org).

China has a large population, of more than 1.2 billion inhabitants, 64% of which live in rural areas (National Bureau of Statistics People's Republic of China, 2001). In 1998, it was estimated that more than 60 million people lived without electricity, 70 million used to experience shortages of cooking fuel and the threat of desertification was affecting the life conditions of 120 million people (Jingjing et al., 1998). This economic and social context makes it necessary to support the use of the so-called bio-energy, as a benefit for both natural and social environment. It is also required to promote the use of efficient and modern technologies, in both domestic appliances and industrial power plants, and to enhance environmental consciousness, since inappropriate and non-controlled utilization of energy from biomass has caused ecological deterioration and environmental impact.

In the present chapter, we will focus on how to obtain energy from biomass, in particular from solid bio-fuels, like wood, woody and solid crops residues, with a mention of other bio-fuel productions, e.g. biogas.

2. WHAT IS BIOMASS

The term biomass refers to a wide range of organic materials: wood, energy crops, organic residues from animal breeding, organic fraction content in municipal solid waste, and organic sludge from municipal waste and food industries. Sometimes what differentiates biomass from waste is just a matter of legislation and national directives. For instance, in Italy solid residues from olive oil production passed from being considered waste to being considered biomass fuel in a few years time, while wine grape residues from alcohol distillery are still considered waste and can be used as fuel only in large plants, such as incinerators. Apart from these aspects related to the various national legislative systems, biomass is a versatile fuel that can be used, after proper treatment, to produce heat and electricity in a clean and efficient way.

Biomass is mainly solid raw material, and it has to be treated before being converted into energy, in order to obtain what is known as bio-fuels. Bio-fuels are then classified on the basis of the aggregation state (solid, liquid or gaseous) and their state determines also the choice of suitable power plant technologies.

Biomass can be divided into many categories as follows: energy crops, by-products, biological by-products and waste, described below.

2.1. Energy crops

This category covers fast growing crops, specific for sustainable energy production. There are two main methods used to produce energy crops: energy plantations, in which an area is devoted exclusively to the production of such crops, and the simultaneous production of both energy and non-energy crops.

Co-production of energy and non-energy crops has the benefit of providing farmers with revenue between harvests of energy crops, since they typically require several years of growth before getting the first harvest. Another important aspect to keep in mind is that the use of land specifically for bio-energy causes competition with other important land uses, particularly food production. This is the case of developing countries, where sometimes the land under crop or the available manpower and technologies are just enough to ensure food production. In some cases, the co-production approach can solve this problem and helps to meet other environmental and socio-economic criteria for land use.

Typical energy crops are short rotation forests such as trees of poplar, robinia pseudoacacia, eucalyptus, etc. or herbaceous and yearly crops as whole plant mischantus, harundo donax, etc.

2.2. By-products

By-products from agriculture or forestry residues, e.g. by-products from cereal harvesting (straw from rice, cotton, corn, etc.) and tree pruning (bark branches, saw dust, etc. not used in timber industry), can be used to obtain energy. This practice helps to reduce the costs of agriculture production and to improve the exploitation of natural resources. By-products are usually ready to be harvested in fields after few treatment procedures, as drying (by sun or by oven), and by employing the same technologies and machinery used in standard agricultural activities. These materials are harvested during the short period of the crops collecting, and problems arise in stocking and keeping them in dry conditions for the whole year-long.

2.3. Biological by-products and waste

By-products from animal breeding or industries can also be used to obtain bio-fuels and energy. This includes the use of stable, liquid dejections and residues from food industries, and vegetable fibres to produce bio-fuels to feed combustion or biogas plants. Biological waste includes the humid fraction of municipal solid waste, sludge and wastewater treatment, and residues from food industries. Organic waste is subject to the national law concerning waste, and it needs proper treatment in the respect of specific sanitary rules. Organic waste may be used to produce electric and/or thermal energy, if burnt in an incinerator, or to produce biogas.

3. WHY TO CHOOSE BIOMASS AS A FUEL

Energy from solar radiation on Earth has been estimated to be about 8.000 times greater than the energy consumed for human activities today (data from *LIOR CD Rom Collection*). Thanks to the photosynthesis process, biomass can be considered as a stored form of solar energy, with carbon locked within it.

Fossil fuels and biomass differ in the length of time over which carbon dioxide is released. All fossil fuels, when burnt and converted into energy, release almost instantaneously great amounts of carbon dioxide in the atmosphere, which had taken millions of years to be fixed as carbon. On the contrary, the amount of carbon dioxide released while burning biomass is equal to that used, within a relatively short period of time (namely years), by the vegetable to grow up. In particular, as far as carbon dioxide is concerned, there is no difference in leaving a tree decomposing on the ground or converting it into energy. This is why biomass is a renewable and carbon neutral form of energy, not contributing to the greenhouse effect. As known, carbon dioxide is one of the greenhouse gases, together with methane, water vapour, ozone and nitrogen oxide. It is estimated that more than 18 billion tons of carbon dioxide per year (*LIOR*) are released in the atmosphere due to the use of fossil fuels and it is acknowledged by the experts that the increasing concentration of carbon dioxide results in global warming and climate change.

Providing the balance between plant growth and biomass consumption is maintained, the system is sustainable and helps to fight climate change.

Moreover, the land use for agricultural and forestry scopes favours the protection of the soil's hydrological and geological equilibrium, and the use of local renewable resources brings employment and finance to rural areas.

Another strong incentive for the use of bio-fuels is the necessity to grant the security of domestic non-finite energy supplies, since most of the countries do not have enough domestic fossil fuels resources.

The main problem in using biomass as a source of energy is due to the costs and environmental impact of transportation, since biomass usually has low density and low specific energy content. Except biogas, bio-fuels need to be transported by road, by rail or by ship, which means higher environmental impact and costs.

4. BIOMASS CHARACTERISTICS AND ENERGETIC CONTENT

Biomass is a complex mixture of substances, consisting mainly of carbon, oxygen and hydrogen and smaller amounts of nitrogen, potassium, phosphorus and sulphur plus other trace elements.

Biomass characteristics are quite varied, since they depend on the nature of the raw material they come from. From an energetic point of view, the main biomass properties are moisture content, heat value, physical structure, weight, volume, density and ash content.

The moisture content can range from around 10% for straw to 90% for sewage sludge material. It strongly influences the energy content of the material and the choice of the energy conversion technology. Biomass fuels with low moisture content are in fact more suitable for thermal conversion and electric energy production, while biomass fuels with high moisture content are more suitable for biochemical processes, such as fermentation conversion. In the case of wood, the moisture content depends mainly on the harvesting length and method and on climatic conditions. The length and the method of storage have also great importance, since the final woody bio-fuels (i.e. chopped wood, pellets, briquettes, etc.) can absorb water from the air.

As for fresh-cut ligneous-cellulosic biomass, the water content usually varies from 40% to 60%, while for green trees it can be even 80%. For open air dried woody biomass, the moisture content is around 12-18%, depending on the season and on the air humidity. Oven dried solid bio-fuel (i.e. pellets and briquettes) has a moisture content of about 10%.

The heat value is strongly affected by the water content of the biomass material, with a trend that is almost inversely proportional. The heat value can be referred to the unit mass, expressed as J/kg, or to the unit volume, expressed as J/m^3. For solid bio-fuels, the heat value referred to the mass does not differ much, but can differ even by a factor of ten if referred to the volume, depending on the material and on the methods of growing, harvesting and storage.

Tab. 1 lists the main characteristics of some fuels (samples of solid bio-fuels from Chinese production; see Cuiping *et al.*, 2004a for more details), i.e. moisture content, ash content, proportions of volatile matters and fixed carbon, together with calorific value. In the last line, the bituminous coal is listed for comparison.

Table 1. Characteristics of some solid bio-fuels – proximate analysis (Cuiping et al., 2004a)

Group	moisture [wt%]	ash [%]	volatile matter [wt% db]	fixed carbon [%]	calorific value [MJ/kg]
Rice straw	8.11	15.25	61.10	15.54	14.66
Wheat straw	8.63	12.45	63.96	14.96	16.56
Corn straw	9.31	13.12	62.74	14.83	16.64
Soybean	9.35	6.08	68.95	15.62	16.96
Corn cob	6.41	7.55	70.24	15.80	16.38
Cotton stalk	7.66	6.11	67.36	18.57	17.91
Cotton shuck	10.23	6.88	62.16	20.73	17.88
Peanut shuck	9.36	12.15	61.64	16.85	18.62
Peanut stalk	8.56	9.12	66.67	15.65	15.75
Sesame stalk	7.66	6.11	68.93	17.30	15.92
Broad bean stalk	7.62	5.04	68.44	18.90	16.31
Rape stalk	6.15	3.60	72.99	17.26	16.65
Foliole eucalyptus	6.51	5.55	67.75	20.19	19.33
Rubber plant	8.88	9.90	62.92	18.30	18.14
Willow tree	9.08	6.17	69.20	15.55	18.79
Poplar	7.91	2.63	74.04	15.42	18.57
Pine tree	8.61	0.89	76.05	14.45	19.38
Spruce	9.21	5.36	71.04	14.39	18.93
Phoenix tree	7.75	5.28	68.68	18.29	17.96
Birch tree	9.06	2.35	74.91	13.68	19.34
Metasequoia	7.38	2.21	74.30	16.11	19.62
Bituminous coal	2.83	20.08	28.01	49.08	34.00

The higher the density of the biomass, the lower the costs due to transportation and storage. This allows competitiveness between bio-fuels and traditional fuels, particularly in the case of large-scale power plants.

The biomass solid and inert part on combustion is ash. It can vary from about 0% to more than 20% in weight on dry basis. For practical use, ashes influence chemical processes, combustion and flue gas emissions due to element contents (metals, alkalis, and solid oxides) and physical properties, such as the melting point and temperature behaviour of the solid particles.

A simple way to characterize a biomass for energy use is to apply some standardized procedures on bio-fuels and to quantify the lower heating value, the density, the water content and the ash content. In Europe, the CEN (Committee for European Standardization) is studying different standards for solid bio-fuels (SBF) and for solid recovered fuels (SRF) derived from wastes, in order to regulate the market for this kind of fuels.

Liquid and gaseous bio-fuels can be derived from the chemical treatment of solid biomass. The most famous liquid bio-fuels are biodiesel and bioethanol, which are equivalent to conventional diesel oil and to petrol/gasoline fuel respectively.

Biodiesel is derived from fat material and oil seeds, while bioethanol is derived from sugar matter as sugar cane and beets.

Gaseous bio-fuels are obtained in two different ways: by thermal treatment, such as pyrolysis and gasification processes (please refer to the following paragraphs and to the study case at paragraph 8.4), and by anaerobic fermentation. The major constituents of biogas and landfill gas are methane, carbon dioxide, water vapour and some minor constituents. The gas composition depends on the biomass from which it is generated, for instance biogas has usually higher methane content than landfill gas. A reference biogas containing 60% of methane is characterized by a density of around 1.11 kg/m^3 at 293 K and by a minimal calorific value of 21,500 kJ/m^3; one m^3 of such biogas thus corresponds to about 0.73 kg of coal as energy content.

5. SOLID BIO-FUELS

Solid fuels from biomass range from woody products (wood itself or production residues) and herbaceous by-products (straw, rice husk, etc.). The main solid bio-fuels available on the market are logs, chips, briquettes and pellets made of wood, while by-products have a limited market and varied aggregation conditions (e.g. straw in bales and rice husks in loose heap). In order to produce each different type of solid bio-fuel, the following main production activities are required: biomass production, harvesting, collection, compacting, transportation, processing and storage. Some of these will be better described in the following paragraphs.

5.1. Growing and production

The biomass obtained from forest residues is near 0.5-0.8 t per hectare, rising to 1.5 t/ha from the maintenance of forests and urban parks. From one hectare of cereals cultivation, the straw produced is about 5 t for wheat and about 15 t for corn. Energy crops guarantee yield of more than 15 t/ha per year, with cut-off every 3-4 years.

Short rotation forestry has the objective to produce the maximum quantity of woody biomass in the form of coppice. From that, the term SRC (short rotation coppice) defines a cultivation of specified wood plants, as willow, robinia, poplar, etc. Energy crops cover also grass and herbaceous (annual) productions of straw, hay and stalks. Today in Europe, agricultural lands for non-food production are interesting also to support farmers with further incomes from energy crops. In China, where food production has to feed the most populous Asian country, things are different and only by-products from agriculture can be used for energetic purposes (Cuiping et al., 2004b).

5.2. Harvesting

Harvesting operations have a significant effect on the energy ratio balance and on the cost of the overall biomass system; hence the use of appropriate technologies and systems is essential for a sustainable introduction of biomass to the energy market or in rural areas.

There are many different technologies and methods for harvesting, depending mainly on the kind of growing, the nature of the site and of the infrastructure, and on the social and economical context. Some of the techniques used for woody fuels are the two-pass harvesting system (consisting in the extraction of wood residues piled at the site after adapted logging), the one-pass, fully integrated, system (in which all products are harvested in one operation) and the harvesting in conjunction with thinning operations. The last method requires labour-intensive operations, resulting in higher woody fuel costs.

5.3. Transport

Since woody fuels are relatively low-value commodities with low energy content, transport methods and distances to the conversion plant have a strong impact on the final cost of the biomass fuel and on the overall low environmental impact of bio-fuels. High transport costs are one of the limiting factors for the construction of large-scale biomass power plants, which have higher efficiencies and lower emissions than small plants.

Looking at woody biomass, the problem is that forestry resources are usually distributed over a wide area while conversion into commodities occurs at specific points. Another important issue is the quality and structure of the forest road network.

Liquid bio-fuels, such as biodiesel and bioethanol, can be stored in tanks and pumped from place to place through pipelines as for traditional oil products. Road, rail and sea transport are also possible using tankers mounted on truck decks, on trailers pulled by tractors or in vessels. However, these last choices lead to a greater environmental impact.

To ensure cost-efficient transportation without compromising quality, methods for compacting and chipping unprocessed fuel materials have been designed and tested, though they are time-consuming and expensive.

5.4. Processing

Biomass usually needs to be treated before being converted into energy, in order to obtain what is known as bio-fuel. In particular, solid bio-fuels pass through a long preparation process (cut, transport, drying, size adjusting, pressing etc.) before the energy recovery process can start. These operations can result in a considerable reduction of fuel storage, transport and handling costs as well as in the reduction of the plant's investment, maintenance and personnel costs.

The use of an automatic fuel-feeding combustion system provides high convenience for the user; however, it requires a relatively homogeneous fuel, in both characteristics and particle size.

Breaking up of woody biomass to usable fuel size in the forms of wood chips can be achieved with a chipper, a hammer mill or similar other devices. It is important to note that reduction of the size of woody fuel components should be delayed as long as possible to shorten chip storage time and thus minimize the risk of biomass loss and deterioration. Wood chipping can be achieved by specialized, purposely designed machines (chippers) but also by large agricultural tractors equipped with a chipper and high tipping trailer. A hammer mill breaks up forest residues by the crushing and tearing action of fixed or pivoting hammers mounted on a rotor.

Agricultural residues, like bales of straw, can be shredded directly at the combustion plant since the material is relatively soft and does not require the use of specialized machines. The combustion of whole bales of straw is not recommended since they tend to burn not uniformly and it is more difficult to maintain a constant heat output.

Regardless of the type and location of the breaking up device, the quality of the material produced is of vital importance to the operation of the energy plant. Biomass fuels of poor quality or consisting of over-sized or under-sized particles, can reduce the efficiency of handling and combustion equipment at the energy plant.

6. BIOGAS PRODUCTION

Biogas is produced from the fermentation of biomass (from solid and liquid organic waste and both animal and vegetable by-products and residues) in the absence of air as a result of the activity of groups of micro-organisms, which are able to transform organic matter into methane and carbon dioxide. This technique is usually adopted in farms and in landfills, to feed biogas engines for electric energy and heat production. In larger biogas plants, the biogas is distributed by pipeline to local users, as natural gas.

The widespread biogas technology is based on simplified plants with a well-stirred reactor as a tank with control of the biomass temperature and a moving roof or a plastic cover to content the biogas production. At the end of the production cycle, the residue is a liquid odourless fertilizer, which can be used on the ground for the new production of biomass and cultivations.

In landfill biogas production, waste naturally produces biogas. For safety reasons, after extraction with special wells for collecting the biogas in pipes, it has to be burnt directly or used as bio-fuel.

7. ENERGY FROM SOLID BIO-FUEL: AVAILABLE TECHNOLOGIES

In the following paragraphs, the main ways of converting solid biomass into energy are described.

7.1. Direct combustion

Direct combustion is the most common way of converting biomass into energy, both heat and electricity, and worldwide it already provides over 90% of the energy generated from biomass. It is well understood, relatively straightforward, commercially available, and can be regarded as a proven technology. It is the simplest and most established thermo-chemical primary conversion technology, and biomass combustion systems can easily be integrated within existing infrastructures.

The main combustion technologies that can be pointed out are fixed-bed combustion, fluidized bed combustion and dust combustion. The choice of the most suitable technology depends mainly on the characteristics of the bio-fuel (moisture and ash content, particle sizes) and on the nominal capacity of the system.

7.2. Gasification

Like direct combustion, gasification is a high-temperature thermo-chemical conversion process, but the desired result in this case is the production of a combustible gas, instead of heat. This can be achieved through the partial combustion of the biomass material in a restricted supply of air or oxygen, usually in a high-temperature environment of around 1,200-1,400°C. The product of gasification, after appropriate treatment, can be burned directly for cooking or heating, or it can be used in secondary conversion technologies, such as gas turbines and engines for producing electricity or mechanical work. Producing electricity from gasification is a more complex process then using the direct combustion/steam cycle process, thus biomass gasification power systems nowadays have higher investment costs. The main types of gasifier designs are updraft (or counter current), downdraft (or co-current) and fluidized bed (bubbling, circulating or pressurized). The gas thus produced needs to be cleaned (since it contains tars and oil) or to be cooled, depending on the specific technology adopted. The cleaning phase is the most critical and now it limits this technology to experimental and demonstrative applications, or to direct combustion of the syn-gas, avoiding that treatment process (see paragraph 9.4 for further details).

7.3. Pyrolysis

Pyrolysis is the name for thermal decomposition occurring in absence of oxygen, and also, it is the first step in combustion and gasification processes, where it is followed by total or partial oxidation of the primary products. The goal of pyrolysis is to produce a liquid fuel, named bio-oil or pyrolysis oil, which can be used as a fuel for heating or power generation. The produced oil is a very complex mixture of oxygenated hydrocarbons and can be used in refining to produce a range of chemicals, fuels and fertilizers. The main benefit of the pyrolysis process is that it generates a liquid fuel, which is easier to transport then either solid or gaseous fuels. The pyrolysis plant can be located near the biomass resource supply, thus resulting in a consistent reduction of fuel transportation costs and emissions.

Pyrolysis technology, in comparison with combustion and gasification, is in an early stage of development and thus the development costs are still very high and not well established, but this also means that cost reductions can be expected.

Current trends in the research and development of pyrolysis involve the so-called fast pyrolysis, a high temperature process in which biomass is rapidly heated in the absence of oxygen and as a result decomposes to generate mostly vapour, aerosols and some charcoal. As a result, after cooling and condensation, we have a liquid fuel, which has a heating value about half the one of conventional oil.

8. ENVIRONMENTAL IMPACT AND EMISSIONS

As every kind of energy production and usage process, also the so-called bio-energy has an impact on the environment, from the moment of the actual production of the bio-fuel ending with its usage in a power plant (flue gas emissions). Apart from the carbon dioxide emissions, which are considered to be neutral in the case of biomass (short carbon cycle), some gases emitted through the chimneys of biomass plants can be dangerous and pollutant.

In Tab. 2, some of the parameters that should always be monitored during the combustion process are listed. In particular, these parameters are subject to national directive prescriptions and limitations.

Table 2. *Emission limits for power plants burning biomass and solid bio-fuels in Italy (extract from D.lgs. 152,06 norme in materia Ambientale)*

Nominal Heat Output (MW)	> 0.15 ÷ ≤ 3.0	> 3.0 ÷ ≤ 6.0	> 6.0 ÷ ≤ 20.0	> 20.0
	[mg/Nm^3]	[mg/Nm^3]	[mg/Nm^3]	[mg/Nm^3]
Total particulate	100	30	30	30 - 10 *
Total organic carbon (TOC)	-	-	30	20 - 10 *
Carbon monoxide (CO)	350	300	250 - 150 *	200 - 100*
Nitrogen oxides (as NO_2)	500	500	400 - 300 *	400 - 200*
Sulphur oxides (as SO_2)	200	200	200	200

* peak value and day average
limits referred to dry flue gases and nominal reference conditions (0°C and 0.1013 MPa)

9. CASE STUDIES

9.1. Biopellets

For biomass large-scale plants, transportation costs of the bio-fuel can be very high and prohibitive, especially in the case of a raw material with low density (e.g. rice husk with a density of 150 kg/m^3). The possible solutions consist in drying the raw material to increase the heating value and subsequently pressing it to produce pellets. Doing so, density increases from 150 kg/m^3 to 600 kg/m^3. The energy transformation cost from raw material to pellets is limited to 10% of the total energy content of the pellets themselves. This technique is particularly recommended for medium and long-distance transport by road and by boat, since the volume is reduced by a factor 4.

From an economic point of view, for the production of wood pellets it is better to use the residues from wood industry than the whole tree, since the dry wood dust needs no more treatments before pellet compression, thus being the best raw material. Other biomass residues are also ready for the compression phase if starting from dry conditions, as straw or rice husk.

Wood pellets can be used as solid bio-fuels also for small burning appliances in residential heating systems. The fuel quality, in terms of ash and water content and of heating value, is controlled and the solid aggregation is suitable for pneumatic or mechanical moving systems for feeding the burner. The pellets used in large-scale burners usually have a diameter of 10÷20 mm, while the pellets used in small-scale appliances are smaller in diameter, around 6÷8 mm (Fig. 1).

Efficient domestic wood burning appliances are now available on the market New wood pellet stoves can reach an efficiency of more than 80% (as the ratio between output and input energy), which is 5-10% more than what the traditional stoves fed by wood logs can do.

Quality problems associated with wood pellets can occur as a consequence of incorrect storage in high humidity places or with rain exposure, or of low compression during the production (sometimes low compression is used to save energy), resulting in a high dust content and thus in a lack of aggregation.

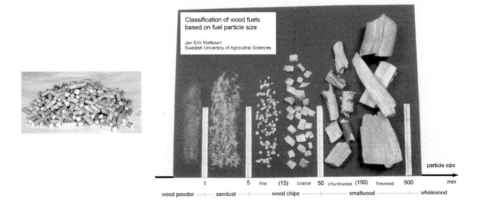

Figure 1. *6 mm wood pellets and classification of wood fuels based on particle size (ENEA and European Standard CENTS15588).*

9.2. Woody-fuel burning appliances: efficiency and emissions of residential systems

Small residential heating systems using solid bio-fuel are small boilers, stoves and fireplaces. In accordance to the new laws and regulations that will be soon introduced in Italy and in the European Union, these heating appliances need to be designed to reach higher efficiency in the combustion and heating exchange processes, and lower levels of pollutant flue gas emissions (Fig. 2).

The total efficiency of small appliances for heating purposes, as stoves and fireplaces, is rarely higher than 65%. The same is true for cooking stoves, where the heat produced by the combustion is used to cook and bake, or for water heating systems or hot tap water.

In the Italian and European Union market, these products are subject to the technical standards of CE mark, thus they have to meet various requirements on industrial production, safety use, mechanical resistance, installation prescriptions, and soon also to respect the limits about efficiency and flue gas emission.

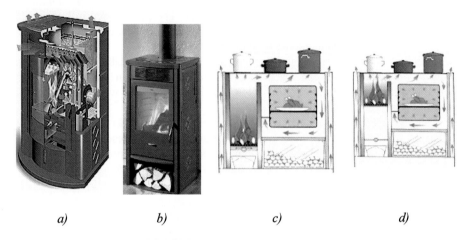

a)   *b)*   *c)*   *d)*

Figure 2. New models of a wood pellets room-heater (a) and of a wood log stove (b), sections of a cooker/heater stove (c and d).

9.3. Electric energy production from biomass power plants

The standard power plants for electric energy production fed by biomass rely on direct combustion technology on a moving grate and on the Rankine steam cycle, consisting of a boiler, a turbine and other components for the thermodynamic steam cycle. In particular, the technology of Organic Rankine Cycle (ORC) is very interesting for electric energy production processes in the case of small plants (0.2÷2.0 MWe). Compared to the standard steam cycle, the ORC is a closed cycle that does not require specific control systems. Generally, the heat carrier in the boiler is thermal oil, which does not require pressurized equipments as in the case of steam production, since the thermal oil works at high temperature but at normal pressure. The whole process is convenient also from the economical point of view, since the costs of production and specialized personnel are reduced.

With regard to the environmental impact, the emission limits for large-scale combustion plants in the European Union are very tight. In particular, in Italy the emission limits are the same for these plants and for waste incineration processes. For this reason large-scale power plants need a complete flue-gas cleaning system, usually based on a dust remover in two or more stages, a NO_x control system with urea or ammonia injection in the combustion chamber, a combustion control system for unburned materials and carbon monoxide controls. This kind of plants can also be used in the production of steam and heat for industry purposes or heating system net, as a Combined Heat and Power plant (CHP).

The choice of the plant size depends on both the energetic and the legislative requirements. An increment of power plant size may lead to higher efficiency but also to higher construction and maintenance costs.

As an example, two real cases are briefly described below:

a) rice husk power plant in Crova near Vercelli (Italy): 35 MW of input energy, 6 MW at the steam turbine, moving grate, air inlet control in two stages (primary and secondary air), flue gas cleaning systems consisting of cyclone and bag filters, no flue gas re-burning and ammonia injection;

b) wood power plant in Airasca near Turin (Italy): 48 MW of input energy, 14 MW of electric energy production, moving grate, air inlet control in two stages (primary and secondary air), flue gas cleaning systems consisting of cyclone and bag filters, possibility of flue gas re-burning and ammonia injection (Fig. 3).

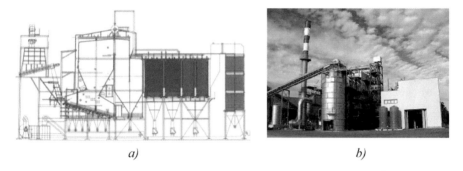

a) *b)*

Figure 3. Section and general vision of wood chips power plant of 14 MWe.

9.4. Electric energy from rice husk power plant: gasification by fluidized bed reactor

The product of the biomass gasification process is a mixture of fuel components (i.e. CO and H_2) and inert gases (e.g. N_2). A typical composition is: CO_2 10÷12%vol, H_2 8÷15%vol, CO 15÷20%vol, CH_4 ~1%vol, O_2 ~2%vol, and N_2 45÷60%vol. This mixed-composition gas can be transported through the local gas distribution network and burned as it is for cooking or heating purposes in small villages. The gas obtained through biomass gasification can also be burned in gas engines to produce electric energy, with the possibility of recovering heat from the cooling system of the engine and at the stack gas.

Since the gas produced from gasification contains tars and oil, to preserve the environment and public health it needs to be cleaned and chemically treated before being used. At the present state of the art, gasification plants are constructed mainly for experimental and demonstrative purposes. The gasifier technology (i.e. the reactor) is known and proved, while the gas cleaning system for tars and oils removing, required to satisfy the prescribed limits in engine or residential applications, is complicated and very expensive. The size of the gasification plant can present technological limits, due to the higher complexity of the chemical plant for gas cleaning and to a more efficient gas production.

Reactors without any moving system, as a down-draught reactor, are simple small gasifiers, involving a simple batch operation or stage feeding process, obtained by adding a tight rotating valve. The gas produced adopting such systems has a very poor quality, not adequate for an engine application, and needs a complete cleaning treatment in order to completely remove all traces of tar and to reduce the content of corrosive acids.

The quality of the gas produced in large-scale plants, by adopting more complex reactors, such as moving bed systems (e.g. fluid bed gasifiers), is quite better. The syn-gas produced is a chemical reactor that can be used as a part of the cleaning system, in addition to the bed chemicals (calcium) for acid separation in solid phase.

As said before, the biomass gasification process is still at a research and experimental stage and is not yet an on-the-market technology, due to the high content in tars and the control of the quality of the produced gas.

As an example, two real cases are listed below (Fig. 4):

a) down-draught gasifier: 25 kWe (ENEA test facility in Trisaia, Italy). Input: 150-450 kWt with untreated and dry (< 20%) biomass; output gas with LHV = $3.8 \div 4.5$ MJ/Nm^3

b) bubbling fluidized bed (BFB) gasifier 160 kWe ("*Cina*" project at the ENEA test facility in Trisaia, Italy (*Marzetti et al.*, 2004)).
Input: 1.00 MWt with untreated and dry (< 20%) biomass; output gas with LHV = $4.5 \div 6.0$ MJ/Nm^3

a) *b)*

Figure 4. Biomass gasification plants: a) test facility with fixed bed gasifier (150-450 kWt) and b) bubbling fluidized bed gasifier (1 MWt – 0.16 MWe).

10. CONCLUSIONS

Renewable energy from biomass is an actual alternative to the use of fossil fuels and it represents an opportunity especially for small rural communities, as it can help their economy and the maintenance of the territory.

As far as solid biomass is concerned, there are two critical aspects to consider: its low energy density, resulting in transportation and usage problems, and its limited production (e.g. crops or residues) from agricultural and forestry territory exploitation.

While syngas/pyrolysis processes are under investigation by research activities, advanced direct combustion technologies for the conversion into thermal or mechanical energy are nowadays available. If combined with flue gas cleaning systems, they represent one of the most environment-friendly techniques of energy production. Therefore, it is desirable to expand the use of these technologies, to achieve higher efficiency in the energy conversion and heating exchange processes and to reduce pollutant emissions.

Mario Chiadò Rana and Roberta Roberto
ENEA Research Centre of Saluggia, Italy

REFERENCES

Biogas from waste and wastewater treatment (2000). LIOR CD-Rom Collection, Brusselss (Belgium), LIOR International.

Biomass combustion (2000) LIOR CD-Rom Collection, Brussels (Belgium), LIOR International.

Biomass gasification (2000) LIOR CD-Rom Collection, Brussels (Belgium), LIOR International.

Cuiping L., Yanyongjie, Chuangzhi W. and Haitao H., 2004a. Chemical elements characteristic of biomass fuels in China. *Biomass and bioenergy*, 27, pp. 119-139.

Cuiping L., Yanyongjie, Chuangzhi W. and Haitao H., 2004b. Study on the distribution and quality of biomass residues in China. *Biomass and bioenergy*, 27, pp. 111-117.

http://www.aboutbioenergy.info/

http://www.iea.org

Jingjing L., Jinming B. and Overend R., 1998. *Assessment of Biomass Resource Availability in China*. Beijing, China Environmental Science Press.

Jungfeng L. and Runqing H., 2003. Sustainable production of energy in China. *Biomass and bioenergy*, 25, pp. 483-499.

Leung D. Y. C., Yin X. L. and Wu C. Z., 2004. A review of development and commercialization of biomass gasification technologies in China. *Renewable & Sustainable Energy Review*, 8, pp. 565-580.

Lin D., 1998. The development and perspective of bioenergy technology in China. *Biomass and bioenergy*, 15, pp. 181-186.

Marzetti P., Braccio G. and Zhi R. Y., 2004. Il progetto "CINA". Un esempio di collaborazione internazionale nel settore dell'utilizzazione energetica delle biomasse mediante gassificazione. *Gestione Energia*, 3, pp. 17-23.

National Bureau of Statistics People's Republic of China, 2001. From http://www.cpirc.org.cn/en/e5cendata1.htm

Qiu D., Gu S., Catania P. and Huang K., 1996. Diffusion of improved biomass stoves in China. *Energy policy*, 24, pp. 463-469.

Xiaohua W. and Zhenmin F., 2002. Sustainable development in rural energy and its appraising system in China. *Renewable & Sustainable Energy Review*, 6, pp. 395-404.

Zhao M. and Zhou G. S., 2004. Estimation of biomass and net primary productivity of major planted forests in China based on forest inventory data. (2005). *Forest Ecology and Management*, 207, pp. 295-313.

L. PIRAZZI

WIND ENERGY

The Wind Resource – Technology – Industry – Economics – Market

Abstract. Wind energy is becoming more and more important in the energy sector in several countries. This is mainly due to the evolution of the wind technology, which has strongly reduced the cost of the electricity produced, and the simultaneous increase of the cost of conventional energy sources. The principal aspects of wind energy, and in particular siting, technology and market are considered. The wind resource of a site is the first step to develop and evaluate in a wind project, in order to understand the best way to exploit the area through the right choice of the wind turbines, their power, size and number.

1. THE WIND RESOURCE

Wind energy is a form of solar energy present everywhere in all countries worldwide, but in different ways in terms of time, speed, frequency and direction. In particular, when people are thinking of exploiting the wind resource, the wind speed is the most important parameter to be considered. In fact, with a doubling of average wind speed, the power in the wind increases by a factor of 8, so even small changes in wind speed can generate large variations in the economic output of a wind project.

Initial assessment of the wind resource at a given site involves the analysis and evaluation of anemological data from nearby meteorological stations in order to have an approximate idea about the wind characteristics of the interested area. Then, a decision will be taken, on the basis of the value of the average wind speed estimated, whether or not to perform more detailed measurements.

1.1. Siting

First of all in the choice of a wind site, the following principles must be always taken into account:

- The power in the wind is proportional to the cube of the wind speed
- The power in the wind is proportional to the area swept by the rotor
- The power in the wind is proportional to the density of the air
- A wind turbine can extract a maximum of only 59.3% of the total power in the wind

Wind speed changes continuously in a very short time, therefore in order to avoid big mistakes in the evaluation of the energy output we usually refer to the average wind speed. Areas with annual wind speed averages higher than 6 m/s at 10 m above surface terrain are considered good and can guarantee interesting economic revenue.

The rotor area determines how much energy a wind turbine is able to harvest from the wind. A typical 1,000 kW wind turbine has a rotor diameter of 54 metres, i.e. a rotor area of some 2,300 square metres.

The density of the air is a function of air temperature, humidity and elevation above sea level. At 1,200 m a.s.l., air density is about 10% less than at sea level, and consequently is wind power. At normal atmospheric pressure and at 15° Celsius, air weighs some 1.225 kilograms per cubic metre.

1.2. Wind atlases

By the end of 2005, the wind atlas methodology had been employed in more than 100 countries and territories around the world. National wind atlases exist for about 30 countries; many of these atlases contain wind data and wind statistics on disk.

In addition, in the same year a global wind map has been carried out by two researchers of Stanford University in the US. This global wind map has identified the wind power potential through the analysis of more than 8,000 wind speed measurements.

CESI S.p.A in 2002 performed the Wind Atlas of Italy, which is a full general atlas of Italy's wind resources, which can provide a framework for singling out the most promising areas for purposes of regional energy planning and wind farm siting. The work started with the simulation of wind flow all over Italy in co-operation with the University of Genoa. An up-to-date wind flow model was used to obtain preliminary wind maps at various heights above ground from geostrophic wind data provided by meteorological institutes (ECMWF of Reading, U.K.). Subsequently, these maps were adjusted by comparison with data recorded by over 240 wind measuring masts throughout Italy.

The Atlas was published in both paper and electronic form. In addition to synthesis maps, it comprises three series of 27 maps each showing wind speeds at heights above ground of 25 m, 50 m and 70 m, respectively. A further series of 27 maps shows the theoretical specific annual electrical energy yield at 50 m above ground (namely the production in MWh/MW of a notional wind turbine of 50 m hub height (Fig. 1)). In 2003 CESI also further developed its methodology for assessing the feasibility of wind farms in mountain areas above 1000 m altitude and applied it to the whole country to get a picture of Italy's usable potential up to elevations of 2000 m. Unlike former investigations concerning a few sample areas, this last phase could rely upon the newly-made Wind Atlas. In addition to windiness, account was taken of factors (terrain slope and roughness, soil use, distance from the electrical grid and roads, environmental and other constraints etc.) which can heavily affect the feasibility and production costs of a wind farm. It was found that Italy does, in principle, have an economically exploitable wind potential in the 1000-2000 m altitude range. However, the estimated wind capacity that could be installed there turned out to vary considerably, albeit always in the order of hundreds of megawatts, depending on the assumptions and criteria that were taken into account.

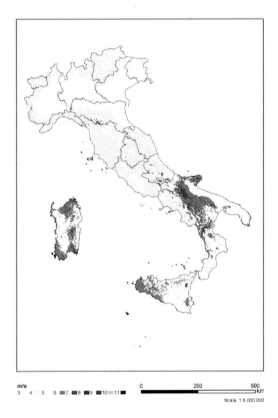

Figure 1. Map of annual mean wind speeds at 50 m above ground taken from the Wind Atlas of Italy.

1.3. Roughness and wind shear

In the lower layers of the atmosphere wind speeds are affected by the friction against the surface of the earth. In general, the more pronounced the roughness of the earth's surface, the more the wind will be slowed down.

Forests and large cities obviously slow the wind down considerably, while concrete runways in airports will only slow the wind down a little. Water surfaces are even smoother than concrete runways, and will have even less influence on the wind, while long grass and shrubs and bushes will slow the wind down considerably.

In the wind industry, people usually refer to roughness classes or roughness lengths, when they evaluate wind conditions in a landscape. A high roughness class of 3 to 4 refers to landscapes with many trees and buildings, while a sea surface is in roughness class 0. Concrete runways in airports are in roughness class 0.5.

Wind shear is referred to the variation of wind speed with different heights. The wind speed generally increases with height as shown in Fig. 2.

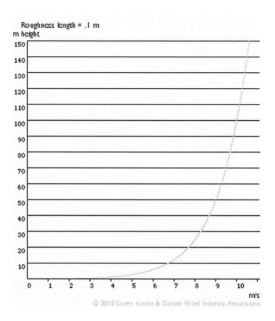

Figure 2. Wind shear profile.

In this way a roughly estimation of the wind speed can be done, but in order to obtain more accurate wind speed measurements, it is clear that it will be important to measure the wind speed as near to the hub height of the chosen aerogenerator as possible.

1.4. Turbulence

Turbulence is a serious constraint reducing wind energy output and causing undue stress on the wind turbines. Wind quickly changes its speed and direction. Turbulence is common in areas with a very uneven surface terrain characterized by obstacles such as buildings, trees, etc. Towers for wind turbines are usually made tall enough to overcome the effect of such nearby obstacles.

1.5. Wind energy

Wind turbines use the kinetic energy of the wind flow. Their rotors reduce the wind velocity from the undisturbed wind speed far in front of the rotor to a reduced air stream velocity behind the rotor. The difference in wind velocity is a measure for the extracted kinetic energy which turns the rotor and, at the opposite end of the drive train, the connected electrical generator. The power theoretically extracted by the wind turbines can be described by the equation:

$$P = \rho/2 \, c_p \, \eta \, A \, v^3 \qquad (1)$$

With the air density ρ (kg/m^3), the power coefficient c_p, the mechanical/electrical efficiency η, the rotor disk area A and the undisturbed wind velocity v, in front of the rotor. In ideal conditions the theoretical maximum of c_p is $16/27 = 0.593$ or in other words 59.3% of the energy content of the air flow can be extracted by a wind turbine. Under real conditions the power coefficient reaches not more than $c_p = 0.5$ because it includes all aerodynamical losses of the aerogenerator.

2. TECHNOLOGY

The nature of technology research and development has changed as the industry has matured and become more commercial. Wind energy is now in a state of rapid development and implementation also through the involvement of the large utilities and oil and gas companies.

Research and development has been an essential activity in achieving the cost and performance improvement that have brought this technology to be considered one of the most attractive in the energy field.

2.1. Wind turbine technology

Basic R&D continues to be of high importance and will contribute to further cost reductions and improved effectiveness in all technology development lines. For the time being the large sized turbines are used both onshore and offshore, but this technology will probably diverge in the near future.

The average new turbine size is continuously increasing. Currently in Germany, there are 9 running aerogenerators in the power range 4.5-6 MW, while the average size in the country in the year 2004 was 1.7 MW.

Testing of several multimegawatt turbines has produced good results. A 3 MW turbine has been developed in both a geared and gearless configuration. The directly driven version (no gearbox) was installed in Norway in 2003.

In the United States, GE Wind Energy completed a design conceptualization for a 3 MW to 5 MW prototype turbine that includes advanced controls and diagnostic systems and blade load alleviations. It also built a 2.5 MW prototype for testing at ECN's site.

The Italian company Leitner is progressing well with its first turbine installed at the end of 2003. The three bladed, gearless turbine has been certified for class 1 sites and a new machine with larger 77 m rotor for wind class 2 sites was erected at the same site at the beginning of 2006.

2.2. Wind turbines classification and applications

On the basis of the power and dimension of the wind turbines the following classification is generally used:

- Small wind turbines
  < 100 kW rotor diameter < 20 m
- Medium wind turbines
  100 – 1000 kW rotor diameter 20 m – 55 m
- Large wind turbines
  > 1000 kW rotor diameter > 55 m

Small wind turbines and hybrid systems (wind–diesel or wind-photovoltaic-biomass-hydro-diesel in different combinations) have several types of applications:

> Water pumping.
> Generation of electricity for remote areas.
> Stand-alone systems or connected to small networks.
> Cathodic protection, navigational aid, telecommunications, weather stations, seismic monitoring.

Medium and large wind turbines are used alone or more frequently in small and large numbers (wind farms) to produce electricity in offshore (medium sized until 2000) and onshore applications.

2.3. Grid integration

From the technical point of view, the integration of a large amount of wind energy into electricity networks is possible with existing technology. Denmark has already achieved a contribution of almost 20% from wind energy at national level, without particular problems. However, in most of the countries with high expectations from wind energy, large-scale integration into the electricity distribution system is perceived as a constraint. Transmission constraints, operational policies, and a lack of understanding of the impacts of wind energy on utility grids are a few aspects that should be afforded shortly in order to facilitate the future deployment.

2.4. Operational experience

Nowadays wind turbines are very reliable and perform with few operational difficulties. Average availability is generally over 98%. Capacity factor, which is an index mainly dependent on the site characteristics, is ranging from 0.20 to 0.35, with higher values in Ireland and New Zealand, where capacity factors over 0.35, and in some areas exceeding 0.40, are quite frequent. Capacity factor is the ratio of the net electricity generated, for the time considered, to the energy that could have been generated at continuous full-power operation during the same period.

Wind turbines are designed with a life of 20 years or more. Danish studies have found that consumables such as gearbox oil and brake pads are often replaced at intervals of one to three years. Parts of the yaw system may be replaced every five years, and vital components exposed to fatigue loading, such as main bearings might be replaced once in the design light.

3. ITALIAN INDUSTRY

3.1. Medium size wind turbines manufacturers

In the manufacturing of medium-sized wind turbines Vestas Italia is the only company operating in Italy. Vestas Italia is a 100% owned Vestas subsidiary. The company was founded in 1998 with the objective of launching Vestas' experience and technology into the Italian Market. Vestas Italia is composed of two units: a Sales & Service Business Unit for Italy and the East Mediterranean area (South Balkans, Turkey & Cyprus, Switzerland, North Africa and Middle East) and a Production Unit making V52 850 kW turbines for the whole world.

Vestas Italia is located in the industrial area of Taranto. By the end of 2005 Vestas Italia had about 500 employees. The main activities of the Vestas Italia Sales Business Unit are sales, marketing and maintenance of wind installations. Vestas Italia can supply all kinds of wind power installations: from single ones to complete turnkey wind power plants.

The Vestas Italia Production Unit has a factory with three production lines dedicated to the V52 850 kW and V47 660 kW medium sized turbines. The Vestas Italia factory can produce over 400 WTGs per year. It produces blades, nose cones and nacelle covers and performs the assembly of all the WTG components. In 2005 the Italian factory in Taranto manufactured 140 turbines for the Italian market and another 225 for other markets (Greece, United Kingdom, Sweden, Ireland, Netherlands, France, China, USA and other countries).

By the end of year 2005 Vestas had installed more than 1,000 MW in Italy and had reached a market share of 62% of total accumulated capacity.

Vestas Italia is QSE (Quality, Safety and Environment) management systems certified and has obtained the following certifications: ISO9001 for Quality management systems, ISO14001 certification for Environmental Management Systems, OHSAS 18001 for Occupational Health and Safety management systems. Germanisher Lloyd issued all the certificates.

3.2. Small size wind turbine manufacturers

In the small machine sector, Jonica Impianti is currently manufacturing wind turbines of 20 kW, with a total capacity of 60 units per year. This sector is expected to extend its market when Law No. 387, which was issued on December 29th 2003 for implementing EU Directive 2001/77/EC on the promotion of renewable energy sources, becomes fully operational and also thanks to the reduction from 100 MWh to 50 MWh of the threshold energy amount requested for obtaining a green certificate.

3.3. Component suppliers

The component suppliers, in the chain of wind technology, are playing a fundamental role through their activities devoted to produce a large number of components both for aerogenerators and wind farms.

The main Italian wind component suppliers are:

- ABB for engines and generators
- Bonfiglioli, Coman for reduction gears
- Ring Mill for forging
- Elettromeccanica di Marnate (EDM)-SEA for transformers
- Pirelli for cables
- Monsud, Leucci, Pugliese for towers
- Moreover, the DAVI Wind Tower Division is very active in producing towers for most wind manufacturers.

4. ECONOMICS

Cost of energy is the correct economic performance measure for wind energy system. The cost of wind-generated electricity continues to fall steadily. This is driven by technological development and increased production levels, together with the use of larger machines. For the recent multimegawatt machines, the installed costs per unit capacity might not be lower, but the overall economics continue to improve. This is because the turbines are on taller towers, which give rise to higher wind speeds and improved energy yields.

4.1. Investment costs

The main investment cost of wind energy projects is the cost of wind turbines. Because of the commercial nature of wind farms, there is very little firm cost data available, but a more or less correct estimation can be done in most countries. In Italy the total average investment cost is around 1 million Euro/MW, the cost of electrical and civil work corresponds to 30% of total investment cost for smaller turbines (capacity < 1MW) and 20% of total investment cost for larger turbines (capacity > 1MW).

The grid connection, including transformer, delivery station, grid reinforcement etc., and the foundation are the main auxiliary costs, followed by consultancy, land, financial cost and road construction.

4.2. Operation and maintenance costs

The operation and maintenance costs are increasingly attracting the attention of manufacturers due to their influence on the overall cost, particularly on offshore applications. These costs are related to a limited number of cost components: insurance, regular maintenance, repair, spare parts, and administration.

The maintenance and operational costs are in the interval between 10 and 15 Euro/MWh of produced wind power seen over the total lifetime.

4.3. Costs of wind energy

Energy production is the most important parameter affecting the cost per kWh generated by wind (Fig. 3). Turbines located at good wind sites with high capacity factor, around 0.3, and investment cost per kW of 900 to 1,100 Euro/kW, produce electricity at a cost of 4-5 cEuro/kWh. In this case the following additional assumptions have been considered:

- Operation and maintenance costs are assumed to be 1.2 cEuro/kWh as an average over the lifetime of the aerogenerator
- The lifetime of the aerogenerator is 20 years
- The discount rate is assumed to range within an interval of 5% to 10% a year.

In the case of sites with a capacity factor around 0.25 the cost ranges approximately to 6-8 cEuro/kWh. In Europe, high wind areas in coastal positions are mostly to be found in the UK, Ireland, France, Denmark and Norway. Medium wind areas are generally found at inland terrain in central and southern Europe in Germany, France, Spain, Holland, Italy, but also in inland sites in Sweden, Finland and Denmark. In many cases, local conditions significantly influence the average wind speed at the site. Therefore, strong fluctuations in the wind regime are to be expected, even for neighbouring areas.

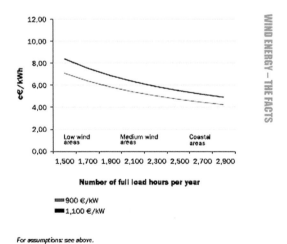

Figure 3. Calculated Costs per kWh Wind Power as a Function of a Wind Regime at a Chosen Site (Number of Full Load Hours).

5. MARKET

The positive trend in terms of wind exploitation, which began two decades ago, is still continuing and, in 2005 a new record was achieved with 11,769 MW of new installations and an expectation that further important goals will be reached in the near future. The total installed wind power capacity at the end of 2005 was 59,322 MW.

5.1. Market development

With this new record of wind farms connected to the grid in 2005 and an annual growth rate corresponding to 25%, 5% more than in the previous year, many countries in all continents showed a great interest towards wind energy acknowledging its great and strategic value.

Large wind turbines, ranging from 1.5 to 2 MW, are rapidly and steadily increasing their presence even in complex terrain strongly contributing to enlarge wind farm capacity.

Germany (18,428 MW), Spain (10,027), the USA (9,149), India (4,430) and Denmark (3,122) are the countries with the highest wind power capacity. Great results have been achieved by India and China, two of the most promising markets (Fig. 4).

Europe is still leading the market and has already achieved the 2010 target set by the European Commission of 40,000 MW, five years in advance, providing around 3% of EU's electricity consumptions in an average wind year.

Asia also had in 2005 a strong growth, almost 50% of new wind capacity, bringing the total to 7,135 MW. The largest market is represented by India, followed by China and Japan. The Chinese market has boosted in anticipation of the country's new Renewable Energy Law, just entered into force on January 1st 2006. China installed some 500 MW of new capacity in 2005, bringing its total wind capacity up to 1,260 MW, and doubled it by the end of 2006, exceeding 2500 MW of installed wind capacity. China's goal, according to the National Development and Reform Commission (NDRC) is 5,000 MW by 2010 and 30,000 MW by 2020. New Energy Finance estimates that China will beat these targets, with 9,700 MW by 2010 and 54,000 MW by 2020.

The Australian market nearly doubled in 2005 with 328 MW of new installed capacity, bringing the total up to 708 MW.

The African market also increased its wind capacity in 2005, mainly thanks to Egypt and Morocco.

5.2. Incentives

Demand for wind has been created through a spectrum of capital grants, tax incentives, and production credit mechanism. Investments in wind sector are stimulated by assuring a higher value for the electricity generated. This enables wind to compete, in some favourable situations, with conventional energy sources. The level and form to support wind energy varies from country to country, mainly depending on the base cost of electricity.

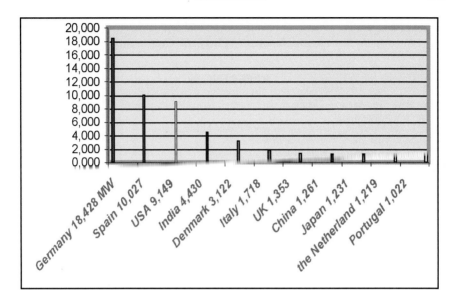

Figure 4. Cumulative wind installed capacity at the end of 2005.

The support mechanism nowadays running in Italy is mainly based on the Renewable Energy Sources (RES) quota obligation imposed on operators who, in the reference year, have produced or imported electricity from non-renewable sources exceeding 100 GWh/year (electricity from Combined Heat and Power (CHP) plants, auxiliary service consumption and exports of energy are excluded from this computation). These operators must feed into the Italian grid, within the end of the subsequent year, an amount of RES equalling at least 2% of this non-renewable electricity. For non-renewable energy referring to the period 2004 to 2006, this mandatory percentage has then been raised by 0.35% a year by the Legislative Decree 387 of 29th December 2003 transposing RES Directive 2001/77/EC (for instance, in 2005 the RES quota has become 2.35% and so on).

5.3. Constraints

The main constraints at global level can be divided into two categories: technical and non technical. Grid connection is the main technical problem in many countries, due to the lack of electricity network in some windy rural areas or the presence of weak grids that need to be reinforced before accepting important amounts of electricity generated by wind. Difficulties to obtain building consent and lack of support mechanisms represent a serious drawback to wind deployment. Sometimes, more frequently where people have not yet experienced wind farm installations, a strong opposition can arise from minor environmentalist groups, which are very active especially at the local level.

Luciano Pirazzi
ENEA Renewable Sources and Innovative Energetic Cycles, Italy

REFERENCES

EWEA, 2004. *Wind Energy – The Facts*
http://www.windpower.org/
Pirazzi, L. and Vigotti R., 2004. *Le vie del vento.* Franco Muzio, Roma.
DEWI, 1998. *Wind Energy Information Brochure*
IEA, 2003. *Wind Energy – Annual Report*
IEA, 2004. *Wind Energy – Annual Report*
IEA, 2005. *Wind Energy – Annual Report*
Marier D., 1981. *Wind Power – For the Homeowner*. Rodale Press, Emmaus, PA.
GWEC. Press Release 2006. Record year for wind energy: Global wind market increased by 43% in 2005

J. ENSLIN AND F. PROFUMO

POWER ELECTRONICS IN DISTRIBUTED POWER

Abstract. The Hydrogen Economy concept is a controversial and hot topic of the day. Hybrid networks, including electrical, gas and hydrogen have however a bright future. The role of power electronics as the interface between these different networks is the key technology since distributed power resources are normally connected to these different networks through a power electronic converter. These power electronic interconnections bring certainly new challenges to network operators, distributed power and power electronic manufacturers. Due to the distributed and fluctuating nature of these distributed resources, they are implemented in large numbers onto the same, sometimes remote, distribution network. These challenges include power quality issues, network stability, power balancing considerations, voltage regulation, protection protocols and unwanted islanding considerations. This paper highlights some of these issues and the role power electronics can play to make this transition between a vertically integrated energy society to a more distributed and hybrid nature.

1. INTRODUCTION

The Hydrogen Economy concept, with conversion infrastructures, storage options and networks, is a controversial, sometimes political and a hot topic of the day. The proposed hydrogen economy, shown in Fig. 1, can be seen as an integrated energy carrier and storage solution, but also as an economic and industrial gamble. As energy carrier and storage, hydrogen has many advantages (Andrews and Weiner, 2004; Gurney, 2004) and disadvantages (Bossel, 2004), but it is doubtful that hydrogen will replace electrical networks as the only major energy carrier for both stationary and mobile energy usage in the future. Hybrid networks, including electrical, gas and hydrogen have however a bright future, which is already taking some form. The role of power electronics as the interface between these different networks is the key technology, since distributed power resources are normally connected to these different networks through a power electronic converter.

These interconnections bring certainly new challenges to both network operators and power electronic manufacturers. Due to the distributed and fluctuating nature of these distributed resources, i.e. wind and photovoltaic (PV) power, they are implemented in large numbers onto the same, sometimes remote, distribution network. These challenges include power quality issues, network stability, power balancing considerations, voltage regulation, protection protocols and unwanted islanding considerations (Enslin *et al.*, 2003a, 2003b and 2004; Bolik, 2003; Povlsen, 2003; Enslin, 2003).

Figure 1. Hydrogen Economy is not only a hope for the future, but also an economic and industrial gamble.

Distributed Power (DP) may have an expected 2.5 – 5 GW/year impact on the supply of energy by 2010, with most of the DP interconnected through power electronic converters (Lovins, 2001). Concerns about reliability of the power supply, blackouts, digital power requirements, conversion efficiency, global warming, and NO_x and CO_2 emissions will accelerate this change. This paradigm shift is the forerunner of the hybrid hydrogen economy where distributed power will play an even greater role (Andrews and Weiner, 2004; Borbely and Kreider, 2001; Gurney, 2004; KEMA, 2001). The term Distributed Power is used here collectively for small to medium sized (3 kW – 100 MW) energy sources, energy storage and power conversion technologies.

The merger between electrical and natural gas networks has already taken place, with jointly owned and operated electricity/gas utilities. This paper discusses the network interconnection challenges facing this promising and potentially profitable technology.

2. HYBRID HYDROGEN ECONOMY

Different prominent research groups have discussed different scenarios for hybrid economy.

The US Road Map has envisioned a possible hydrogen future for the USA with 100 million hydrogen-powered cars and 25 million homes supplied with electricity produced through fuel cells. Both applications would entail the production of 40 million tons of hydrogen per year, achieved in a decentralized manner (via 1,000,000 electrolyzers and 67,000 reformers), through a centralized network of several hundred production plants or by adopting a mixed approach.

The scenario proposed by HyNet, the European Hydrogen Network, at the Hydrogen and Fuel Cell Technology Platform (Launch Conference on January 21st 2004 in Brussels), gives a more gradual penetration of hydrogen-powered automobiles, which would reach 5% in 2020, equivalent to 9 millions vehicles in Europe.

It is anticipated that this hydrogen process will be a gradual development process over several decades with a more hybrid character as shown in Fig. 2. Electrical networks will gradually transform from a concentrated to a distributed nature. Hydrogen, although with a low energy density, may take the role of energy storage for peak shaving and interconnecting fluctuating power sources like large off-shore wind parks to energy infrastructures, solving some of the main interconnection problems of wind and other renewable energy sources.

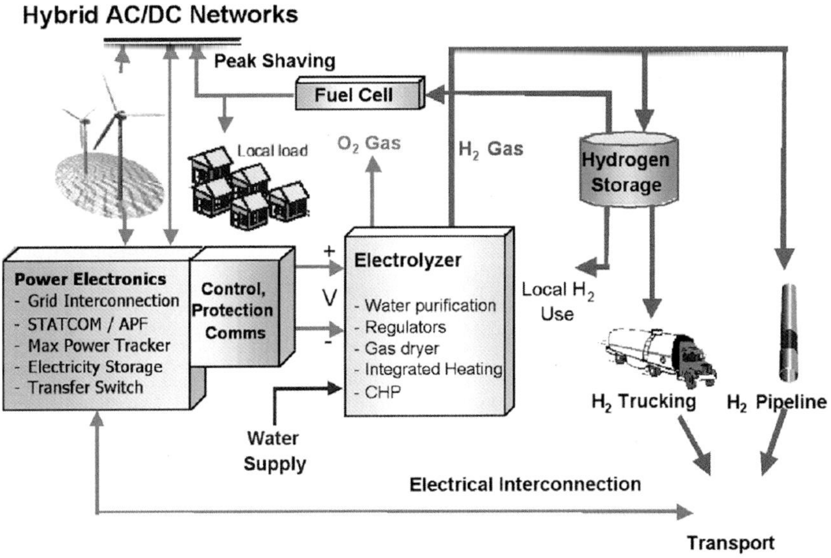

AC = Alternating Current - APF = Active Power File r - CHP = Combined Heating and Power- DC = Direct Current - STATCOM = Static Compensator

Figure 2. Hybrid Electrical/Gas/Hydrogen Network and Economy.

It is clear that power electronics will play a key role in interconnecting different renewable energy sources, fuel-cell and distributed generators, and Combined Heating and Power (CHP) systems. For power quality mitigation and power system stability, electricity storage will still play an important role with hydrogen storage playing more a longer-term storage role.

It is also clear that the mobile application of hydrogen, hydrogen and hybrid cars, will interface with this hybrid network concept. Hybrid cars all have power electronic converters that can interface with electrical and hydrogen/natural gas networks through a fuel cell and reformer (FreedomCAR & HyperCar concepts). This may well be the most important "distributed generator", producing electricity and heat in residential houses during evening hours with an improved fuel-cell membrane technology. This concept may be a gradual process with a 20 – 30 year time frame, but we are already seeing the evolution of the vertically integrated power system to the horizontal and distributed power concepts, with power electronics playing a leading role in this process.

3. ENERGY-WEB CONCEPT

Electrical networks are currently evolving into more hybrid networks, including Alternating Current (AC) and Direct Current (DC), with energy storage for individual households, office buildings, industry and utility feeders to interface these different power concepts. This energy-web concept is depicted in Fig. 3, and is also the structure for a possible future hydrogen-integrated society, combining the energy market, information data links, hybrid ac&dc electrical networks with gas, and hydrogen infrastructures (KEMA, 2001).

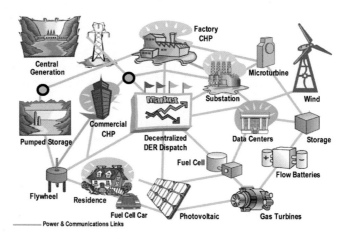

Figure 3. Presentation of the Energy-Web Concept (CHP = Combined Heating and Power).

Emerging DP technologies are available or have matured for site-specific applications in power quality, power reliability, CHP, transmission & distribution network support and price – quality management (Willis and Scott, 2000; Lovins, 2001; KEMA, 2001).

Current and continuously improving options are: aero-derivative gas turbines (25-60 MW and 40+% efficiency), small 1-5 MW industrial gas turbines (~ 40% efficiency), diesel and internal combustion engines (0.3-2 MW, 36-60% efficiency) and recuperated small gas turbines (efficiencies of 34-43%).

Emerging trends in larger and small-scale generation are: hybrid fuel cell (0,4-20 MW and 65-70% efficiency), small fuel cells (1-300 kW and 30-50% efficiency), mini-gas turbines (30-300 kW and 25-30% efficiency), micro CHP (1-10kW and 30%+ efficiency) and automotive fuel-cell systems (30-60 kW) pluggable to the distribution network when needed.

It is envisioned that most of this equipment, accompanied by flexible and cost-effective energy storage, will be connected to an electrical network via power electronics in the next 20 years. Renewable sources are intermittent, meaning that the load or capacity factor is much less than one, typically 0.25 – 0.35 for wind farms and 0.10 for photovoltaic systems. This implies that renewable energy sources might deliver much higher peak power at certain times of the day, resulting in either an over-designed network or network related power quality and stability problems. Most studies confirm that a renewable energy contribution of 10 – 15% can easily be absorbed in the electricity network without major structural changes. Higher penetration levels require additional control, storage or other means of controlling the large power fluctuations.

The power electronic converter is, in most cases, the enabling technology for these new energy related themes, and is the interface between the energy sources, the electrical network and the load. In most cases, power electronics also add value in terms of power quality enhancement, voltage regulation, reactive power regulation, energy storage interface, metering, protection and application flexibility (Enslin *et al.*, 2003a; Bhowmik *et al.*, 1999; Courault, 2002).

Detailed technical and non-technical challenges still need to be resolved. Some of the main challenges are summarized below:

- The power and data interfaces and interconnections for the Energy-Web should be improved.
- The cost-effectiveness, flexibility and efficiency of power electronics for DP units in the Energy-Web should be improved.
- Flexible and cost-effective energy storage systems should be developed and integrated into the Energy-Web.
- The complex economics of DP should be managed and supported by the business sector.
- Standards and regulations should be tightened since they are the key driving force to limit these interconnection problems.

- Public and utility acceptance for this diverse and wide variety of energy sources and services are paramount.
- The true life-cycle costs of DP should be calculated accurately and communicated so that the consumer understands the value of DP.

These interconnection issues are sensitized below with the aid of two distinct cases. The one being the interconnection of large scale PV and the other interconnection issues of wind power to electrical networks.

4. CASE A – INTERCONNECTION OF LARGE-SCALE PV

Currently, wind and photovoltaic distributed power systems are connected to the network on a large scale. KEMA was recently involved with projects concerning possible power quality problems in renewable suburbs with 200 to 500 homes with roof-mounted photovoltaic (PV) arrays (Enslin *et al.*, 2003b; Enslin and Heskes, 2003). Relatively few PV inverter-network interaction problems are reported in the literature and are mainly associated with the unwanted islanding operation and voltage regulation problems with the power feedback of large amounts of solar energy (Povlsen, 2003). Problems associated with generated harmonics and possible network resonance are seldom investigated (Enslin *et al.*, 2003b). Currently, some newly developed Institute of Electrical and Electronic Engineers (IEEE) standards (IEEE 929-2000), (IEEE 1547) and IEC standards (IEC 61727) indicate that it is expected that interconnection problems may increase in the future.

Several governments and utilities worldwide promote the use of renewable sources of DP generation, with subsidies and customer benefit programs. These include "Green" suburbs where roof-mounted PV arrays are installed on most of the roofs of individual homes, apartments and communal buildings. An example of such a project is the Dutch Nieuwland project, near Amersfoort, where, in total, 12,000 m^2 of PV arrays have been installed on 500 homes. In total 1 GWh of renewable energy is generated annually by this project. Another recent addition includes PV arrays connected on most of the roofs of the houses in Vroonermeer. A photograph of this suburb is contained in Fig. 4. As a first phase, a total of 197 homes with PV arrays and inverters were installed. This network was studied for possible PV inverter interaction issues (Enslin *et al.*, 2003b; Enslin and Heskes, 2003).

Most of these 200 – 500 homes have 1 – 3 kW inverters connected to the 220 V distribution feeders. Typically, 30 homes are on a single feeder, all supplied from a 10 kV transformer.

Figure 4. Example of Roof mounted PV suburb at Vroonermeer-Zuid.

4.1. Power Quality Problems with PV Inverters

Measurements in the networks that have a high percentage of PV generation showed that PV inverters, under certain circumstances, switched off undesirably or increased their production of harmonics significantly. This can result that the Point of Common Coupling (PCC) power quality standards are exceeded. This can happen even though all PV inverters individually satisfy the inverter power quality standards.

The PV inverters for these projects are normally based on single-phase, self-commutated voltage-source converters in the 1 – 3 kW power range. These PV inverters consist of different typologies. In practice, switching frequencies of 20 – 500 kHz are used in different power stages utilizing power switches (MOS Field Effect Transistors – MOSFETs – and Insulated Gate Bipolar Transistor – IGBTs) as the switching elements for PV inverters. For safety reasons and some standard requirements, an isolation transformer is normally required.

For all these inverter types, the AC output current is characterized by the current-feedback control loop. These inverters are capable of self-generating a 50 Hz sinusoidal output current based on internal look-up tables and only have to be synchronized with the supply voltage. Often a Phase-Locked Loop (PLL) control technique is used to detect the voltage zero-crossings.

Some inverters combine the reference source and the synchronization with the supply voltage by using the waveform shape of the supply voltage as a reference source. However, if this voltage is polluted, the reference source will also be polluted, with the result that the inverter's current regulator pollutes its own output current accordingly. Filtering of the pollution using such a controller is difficult. This kind of inverter has resistor-like characteristics, and since the inverter is feeding its current back into the network, the inverter behaves like a negative resistor. Therefore, this type of controller for PV inverters should be avoided.

4.2. Analysis of Harmonic Pollution

In order to establish the range of the natural frequencies in the low-voltage network, values for the network inductance and capacitance (cable and transformer) of a low voltage network, as well as the typical values for the household and PV-inverter input capacitance at each connection should be calculated (Enslin, 2003).

The resonance phenomenon for networks with large numbers of PV inverters can be divided into the following types:

1. *Parallel resonance* of the parallel network capacitance (PV Inverter, household and cable) and the supply inductance (transformer leakage and cable), resulting from distortion generated internally. In this case, the PV inverter can be assumed to be the generating harmonic source. The impedance at the resonance is high, resulting in higher voltage distortion either at the PCC or at the location where the PV inverter and household load are connected.
2. *Series resonance* of the network capacitance and the supply reactance, resulting from either externally generated or injected distortion. In this case, the background supply voltage distortion is the mechanism. Here the impedance at the resonance is low, resulting in higher current distortion through the load and PV inverter capacitor.

In practice, these two phenomena are linked in one circuit, resulting in increases in both voltage and current distortions.

If one of the harmonics generated by the PV inverter (parallel resonance mechanism) corresponds with the parallel resonance frequency, very high resonance voltages (damped only by the associated network resistances) will occur on the network voltage at the PCC. This may have operational effects on the PV inverter and other equipment connected to the PCC. Furthermore, this resonance can be more severe if the power network is weak, which results in a lower frequency parallel resonance.

When one of the harmonics present in the network background distortion (series resonance mechanism) corresponds with the series resonance frequency, very high resonance currents will flow in the network, damped only by the associated network resistance. The load is parallel to the network resistance for a parallel resonance and can be ignored since it is much higher than the network resistance.

5. CASE B: INTERCONNECTION OF WIND ENERGY

The other distributed power source well integrated into electrical networks and with a fast growing tendency world-wide is wind power (Slootweg and Kling, 2003), (http://www.windmonitor.nl/). Up to a few years ago, wind power did not make a large impact on network interconnection issues. This was mainly due to the fact that wind power plants were mainly sparsely interconnected to the network. Recently, the number of turbines, power ratings of the individual turbines and the power electronics interface grew rapidly (Søbrink, 1998; Heier, 1998; Bolik, 2003; Søbrink et al., 2002).

Attitudes towards offshore wind energy have changed significantly in the last few years. The European Union's White Paper proposes that 12% of energy within the EU should be provided by renewables by the year 2010, with a possible installed wind capacity of 40 GW. It is unlikely, with current technology, that all this wind power can be accommodated onshore. Although Europe does have major offshore wind resources, the development of large-scale offshore wind parks faces fundamental technical and financial constraints with regards to the connection of wind farms to the electrical network. These challenges include power quality, active and reactive power flow, infrastructure, network stability, cost recovery and profitability (KEMA, 2002). It is also widely accepted that the hydrogen economy without wind energy does not make that much sense.

Some of these large wind power projects cannot be classified as "distributed power", but on the high voltage network similar interconnection problems are experienced, including network stability, power balancing, power quality, reactive power demand, etc.

Another example is the planned large offshore wind parks on the North Sea. KEMA has done for NOVEM (KEMA, 2002; Jansen et al., 2004) an interconnection study identifying the problems and possible solutions for the connection of 6000 MW of wind power by 2020. Some of these interconnection issues and solutions are discussed here due to the common phenomenon.

5.1. Impacts of Constant and Variable Speed Wind Generators

Wind power impacts the network in the following way (Søbrink, 1998; Heier, 1998; Slootweg and Kling, 2003):

- Localized voltage regulation.
- Protection schemes, fault levels, and switchgear ratings.
- Harmonic distortion.
- Flicker.

These issues are similar for other forms of distributed power. Of particular interest to wind generators is the flicker, due to the specific stochastic nature of wind and the so-called tower shadowing effect (Enslin et al., 2003a and 2004).

In general wind turbines without power electronics contribute to fault levels and may cause extensive flicker problems. On the other hand, turbines with power electronics may lead to harmonic distortion, which may be amplified if a system resonance is occurring at one of the generated harmonics, similar to the PV case.

Wind power also has some impacts on the system level (Søbrink, 1998; Heier, 1998; Slootweg and Kling, 2003):

- Power system dynamics and stability.
- Reactive power and voltage control on system level.
- Frequency control and dispatch of conventional generating units.

The impact on the dynamics and stability of a power system is mainly caused by the fact that wind turbines do not use conventional synchronous generators. Constant speed wind turbines (without power electronics) can lead to voltage collapse after a system fault or a trip of a nearby generator. They consume large amounts of reactive power during the fault, which impedes on the voltage restoration after the fault. Solutions to prevent voltage collapse after faults include dynamic reactive power sources, i.e. STATCOMs and Static Var Compensators (SVCs) (Søbrink, 1998; Enslin and Van Wyk, 1992; Bhowmik et al., 1999; Courault, 2002).

Variable speed wind turbines (with power electronics) have hardly a risk of voltage collapse. They can resynchronize immediately after a fault, do not contribute to the fault level and do not consume reactive power. A drawback of the present generation of variable speed wind turbines is, however, that they disconnect immediately after a fault occurs. Newly developed wind power standards (Bolik, 2003) specify a precise ride-through capability for the turbine to stay connected after a network fault or deep sag, in order to minimize this problem. Most manufacturers of power electronic converters and wind turbines have responded well to this new requirement.

5.2. Mitigation Using Integrated Storage Options

An energy storage system integrated with high power electronics forms the ideal mitigation device to solve the technical interconnection problems (Søbrink, 1998; Heier, 1998; Hingorani and Gyugyi, 2000; Enslin et al., 2001; http://www.electricitystorage.org/). Such an energy storage system was considered in order to compensate the fluctuating output of 6000 MW of wind power to a constant output power (Enslin et al., 2003a; Enslin et al., 2004). If the total output of the wind farm is balanced using a large-scale energy storage plant, a capacity of 2,500 MW would be required. Such a storage plant can only realistically be made from large-scale hydro storage at this stage. This is an unpractical option for the Dutch system, since no large-scale hydro dams, space or elevations are available.

If such a storage system is nevertheless to be considered, the flow-battery may be a feasible storage device in a 10 – 20 years time-frame. An investigation into the size of such a device was considered using the current Regenesys® (http://www.electricitystorage.org/) technology. As stated, in order to balance the power flow of the 6,000 MW wind farm, a 2,500 MW battery plant will be required, with an energy storage capacity of 62 GWh. The size of such an installation would be in the order of 1x1 km and would have a Net Present cost Value (NPV) over 30 years of M€ 6,000, using current technology and prices. Balancing this against the expected savings in network upgrades (between M€ 350 – 650), it is obvious that this is by far not feasible at this moment. It is thus clear that a more integrated and flexible approach should be considered for wind park balancing and energy storage usage.

Storage can be feasible if applied in several applications in parallel (Enslin et al., 2003a and 2004). Looking into the application and feasibility of such a storage system for wind farms, these storage devices should therefore be used for as many functions as possible. The following storage applications, in the broad sense, are relevant to wind parks, power producers, transmission and distribution networks and large customers and may hence be considered: wind energy balancing; energy trading on Amsterdam Power Exchange (APX); peak power shaving; spinning reserve; black-start functionality; reduction of number of start stop of power units; delay in investment of transport capacity; network stability; power quality; voltage regulation; reactive power support; etc.

6. SUMMARY

Distributed power will change the shape of power systems in the future. DP technologies will play an important role in developing the hybrid hydrogen society. It can also be concluded that the power electronic converter is, in most cases, the enabling technology for these new energy-related themes. Power electronics add value in terms of power quality enhancement, voltage regulation, reactive power regulation, storage interface, metering, protection and flexibility.

However, short/medium term challenges, especially in terms of the large-scale interconnection of distributed power onto existing networks can hinder this development. Islanding, voltage regulation, parallel and series resonance phenomena in the network, network stability, protection, reactive power requirements and fluctuating power at the interconnection point, are some of the issues that need to be addressed. When designing power electronic interfaces, this phenomenon should be considered and the interface should be designed and used to mitigate these problems and not amplify them. Recent interconnection standards and guidelines are already making impacts on the process of mitigating these problems.

Large-scale wind park projects require a different integration approach from those used for smaller wind farms. Key wind farm technologies, AC and DC transmission systems, network load-flow and dynamic stability mitigation and storage technologies, can ensure there is minimal impact on the network. Energy storage is not an economically viable solution with the current technology in place, if only one function is addressed. Several functions should be integrated into the functionality of the storage system integrated with a STATCOM.

ACKNOWLEDGMENT

The authors gratefully acknowledge the contributions from the research group at Piemonte H_2 Laboratory (HySy_Lab), in the Environment Park, Turin, Italy.

Johan H.R. Enslin
KEMA Inc. TDC, Raleigh, NC, USA

Francesco Profumo
Politecnico di Torino, Turin, Italy

REFERENCES

Andrews C.J. and Weiner S.A., 2004. Visions of a hydrogen future, *IEEE Power & Energy Magazine*, Vol. 2, No. 2, pp. 26-34, March/April 2004.

Bhowmik S., Spée R. and Enslin J.H.R., 1999. Performance Optimization for Doubly-Fed Wind Power Generation systems, *IEEE Transactions on Industry Application*, Vol. 35, No. 4, pp. 949-958, July/August 1999.

Bolik S.M., 2003. Grid requirements challenges for wind turbines, 4th *International Workshop on Large-Scale Integration of Wind Power and Transmission Networks for Offshore Wind Farms*, Billund, Denmark, 20 – 21 Oct. 2003.

Borbely A-M. and Kreider J.F., (ed.), 2001. *Distributed Generation: The Power Paradigm for the New Millennium*, CRC Press, ISBN 0-8493-0074-6.

Bossel U., 2004. Hydrogen: Why its future in a sustainable energy economy will be bleak, not bright, *Proceedings of Renewable Energy World*, James & James Ltd, pp. 155-159, March/April 2004.

Courault J., 2002. Energy Collection on Off-shore Wind Farm, *ALSTOM Transmission & Distribution HVDC & EQCS for SRBE Meeting*, Liege, Feb. 2002.

Dunn S., 2000. Micropower: the next electrical era, *Worldwatch paper 151*, Worldwatch Institute, Washington DC, July 2000.

EN 50160, 'Voltage characteristics of electricity supplied by public distribution systems' November 1994, with amendments prAA, prAB and prAC, June 1998.

Enslin J.H.R., Jansen C.P.J. and Bauer P., 2004. In store for the future? Interconnection and energy storage for offshore wind farms, *Proceedings of Renewable Energy World*, James & James Ltd, 104-113, Jan/Feb 2004.

Enslin J.H.R., Jansen C.P.J. and Bauer P., 2003a. Integrated approach to network stability and wind energy technology for On-shore and Offshore Applications, *PCIM-2003 (Europe)*, Nuremberg, Germany, 20 – 22 May 2003a.

Enslin J.H.R., Hulshorst W.T.J., Atmadji A.M.S., Heskes P.J.M., Kotsopoulos A., Cobben J.F.G. and Van der Sluijs P., 2003b: "Harmonic Interaction between Large Numbers of Photovoltaic Inverters and the Distribution Network", in *Proc. IEEE Power Tech 2003*, Bologna, Italy, June 23 – 26, 2003b.

Enslin J.H.R. and Heskes P.J.M., 2003. Harmonic Interaction between a Large Number of Distributed Power Inverters and the Distribution Network in *Proceedings 34th IEEE Power Electronics Specialists Conference*, Acapulco, Mexico, June 15 – 19, 2003.

Enslin J.H.R., 2003. Interconnection of Distributed Power Inverters with the Distribution Network, *IEEE PELS Newsletter*, Vol. 15, No. 4, pp. 7-10, October 2003.

Enslin J.H.R., Knijp J., Groeman J. and Thijssen G., 2001. Supply Quality, Reliability and Storage, *Electricity Storage*, UK, 30 – 31 January 2001.

Enslin J.H.R. and Van Wyk J.D., 1992. A study of a wind power converter with microcomputer based maximal power control utilising an over-synchronous electronic Scherbius cascade, *Proceedings of Renewable Energy*, Pergamon Press, London, Vol. 2, No. 6, 551-562, 1992.

Gurney J.H., 2004. Building a case for hydrogen economy, *IEEE Power & Energy Magazine*, Vol. 2, No. 2, pp. 35-37, March/April 2004.

Heier S., 1998. *Grid integration of Wind Energy Conversion Systems*, John Wiley & Sons.

Hingorani N.G. and Gyugyi L., 2000. *Understanding FACTS*, IEEE Press, New York.

IEC 61727: Characteristics of the utility interface for photovoltaic (PV) systems, 2002.

IEC 61400-21: Measurement and assessment of power quality characteristics of grid connected wind turbines, 2001.

IEEE 1547: Standard for Interconnecting Distributed Resources with Electric Power Systems, IEEE Standard 1547-2003.

IEEE 929-2000 Recommended Practice for Utility Interface of Photovoltaic (PV) Systems, IEEE Standard 929-2000.

Jansen C.P.J., de Groot R.A.C.T., Slootweg J.G., Enslin J.H.R. and King H.L., 2004. *Considerations to the electrical network interaction of 6,000 MW off shore wind parks in the Netherlands in 2020*, Conf. Rec. Cigrè 2004.

KEMA: Survey of the integration of 6,000 MW offshore wind power in 2020 in the electrical grid of The Netherlands, Report on behalf of Novem, KEMA T&D Consulting, Arnhem, Laboratorium voor Elektriciteitsvoorziening, TU Delft, Twijnstra Gudde, 30 November 2002, p. 255.

KEMA: Electricity Technology Roadmap for the Netherlands in a North-West European Perspective: Technology for the Sustainable Society, September 2001, p. 260.

Lovins A., 2001. Small Is Profitable: The Hidden Economic Benefits of Making Electrical Resources the Right Size Small is Beautiful, *Keynote Address: Distributed Power 2001*, 16 – 18 May 2001, Nice, France.

Povlsen A.F., 2003. Distributed power using PV: Challenges for the grid, *Renewable ENERGY World*, Vol. 6, No. 2, 62-73, March-April 2003.

Slootweg J.G. and Kling W.L., 2003. Is the answer blowing in the wind? The current status of wind as a renewable energy source and its power system integration issues, *IEEE Power & Energy Magazine*, Vol. 1, No. 6, pp. 26-33, November/December 2003.

Søbrink K.H., Belhomme R., Woodford D., Abildgaard H. and Joncquel E., 2002. *The challenge of integrating large-scale offshore wind farms into power systems*, Paper 14-204, CIGRé-2002, Paris.

Søbrink K.H., (ed.), 1998. "Power Quality Improvements of Wind farms", *EU Report* ISBN No. 87-90707-05-2, June 1998.

US Dept. of Energy: "Workshop on Electrolysis Production of Hydrogen from Wind and Hydropower", Sept. 9, 2003.

Willis H.L. and Scott W.G., 2000. *Distributed Power Generation*, Marcel Dekker Inc., ISBN 0-8247-0336-7, New York.

http://europa.eu.int/
http://www.eere.energy.gov/
http://www.electricitystorage.org/
http://www.hypercar.com/
http://www.hysylab.com/
http://www.kema.com/
http://www.windmonitor.nl/

V. ENERGY EFFICIENCY AND RENEWABLES

2. ECO-BUILDING

L. SCHIBUOLA

ENERGY OPTIMIZATION OF A BUILDING-PLANT SYSTEM

Abstract. A fundamental contribution to solve energy issues can come from the improvement of energy conversion efficiency and from the reduction of energy waste in buildings, as the energy consumption of such structures is a considerable quota of the total energy demand. The paper shows that the energy optimization of a building-plant system starts from a correct architectural and plant design. The opportunities offered by the introduction of high efficiency solutions are then presented by referring to experimental results of some building-plant systems in some structures in the Venetian area.

1. INTRODUCTION

The problem of energy resources came to everyone's attention in the 1970s in terms of a coming exhaustion of fossil fuels. Luckily, the strong increasing cost of energy reduced the urgency of the problem since it made it profitable to use other types of energy and to exploit difficult-to-access deposits (e.g. off-shore deposits). However, nowadays the hyperbolic increment of energy demand caused by population boom, modern standards of life and industrial development in emerging countries, reopens the energy problem in terms of a limited availability of fossil fuels. In addition the production of energy from fossil fuels is accompanied by gas emissions in the atmosphere, unavoidable in the case of CO_2. It is therefore necessary to reduce the use of traditional energy resources. The first possibility is the spreading of renewable energy resources such as solar energy, wind, and geothermal energy. However, the actual level of technology and especially the costs of such resources restrain their use. Therefore, the strategy of a consistent reduction of fuel consumption obtained by both a reduced energy demand and an improvement of energy conversion efficiency is fundamental. Everyone has to be convinced that at present time, the greatest energy deposit is in fact the reduction of energy waste.

In modern countries energy consumption in buildings represents a considerable quota of the total energy demand (about 30% in Italy). For this reason, energy use optimization in buildings represents an important contribution to correct energy policies. Energy needs are: heating and air conditioning, lighting, hot water production and appliances. Air conditioning and lighting energy demand are strictly linked to the design of the building-plant system. Therefore the optimization begins with the design of the new building.

2. OPTIMIZATION OF A BUILDING DESIGN

Today, the modern softwares available on the market allow foreseeing the effects of different design options; therefore, they are prodigious tools for architects and engineers. Among these programs, we can mention for example ECOTECT, Energy Plus, and Design Builder. The analysis starts from the site in order to identify the best location for new buildings.

This paper will take into consideration the case of five residential tower buildings realized near Brescia, a city in the North-western part of Italy.

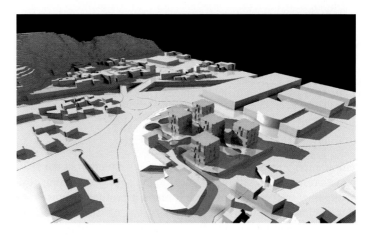

Figure 1. Model of a building area designed for a computer analysis.

Fig. 1 represents the model of a building area designed for the use of ECOTECT software.

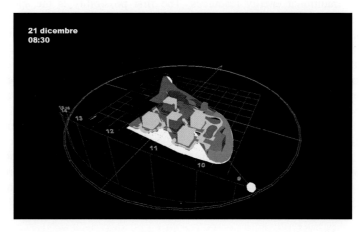

Figure 2. Solar analysis during winter solstice (December 21st, 8.30 am).

In the first phase of the design, the software allows the analysis of the solar exposure in terms of interactions between new and existing buildings. This analysis can be extended during the whole year with the calculation and the representation of the sun's contribution and the shadow evolution in continuous. It is possible to have a precise evaluation of the sun's influence on the building. Fig. 2 and 3 report the solar effect during the two solstices.

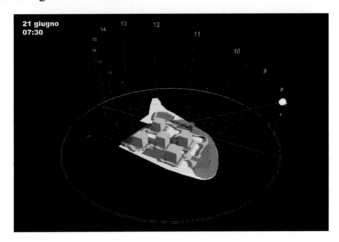

Figure 3. Solar analysis during summer solstice (June 21st, 7.30 am).

Of course the solar contribution must be maximized in winter and minimized in summer. This study is also important to establish the best sites for gardens and pedestrian crossings. The wind analysis in terms of speed and direction in various seasons (Fig. 4) gives useful information as well.

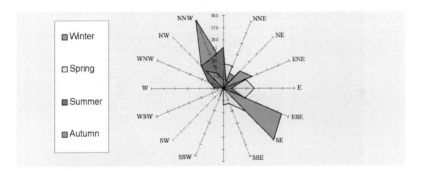

Figure 4. Wind direction, annual percentage (%).

In the case presented, the analysis suggested the realization of a wind barrier to protect the residential zone in winter. The barrier is partially covered with grass and trees. It also serves as a sound barrier to protect the residences from the road near by. In Fig. 5, a section of the barrier is reported. Fig. 6 shows the position of the barrier near the main road.

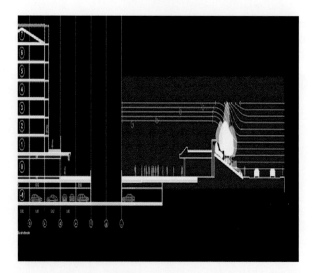

Figure 5. The wind and sound barrier.

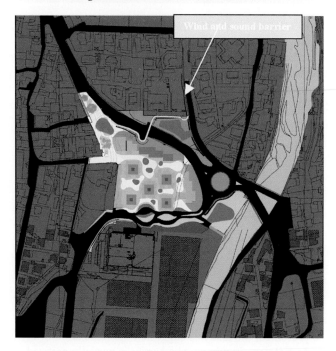

Figure 6. Setting of the barrier near the main road.

For each building, the solar analysis is fundamental in order to identify the effect on the orientation and shape of the windows in terms of solar energy contribution and natural lighting (Fig. 7).

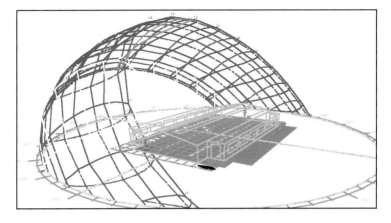

Figure 7. Solar analysis of a single building during the whole year.

Modern simulation codes give precise information on the results of the design on heating and cooling loads and internal conditions. This allows a correct choice of construction materials optimized for each kind of building. In the example in Fig. 8, the different effects of heavy or light materials are presented. In the first case, the loads through the walls are reduced and delayed.

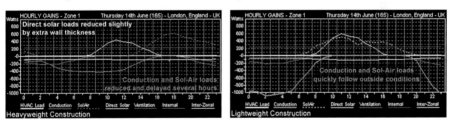

Figure 8. Effect of material thermal inertia.

Today, with the introduction of a new technology for the envelope as double skin facade, this analysis is very important. We can have a precise knowledge of the performances of different systems in order to optimize the choice in each particular application case. For example the performances of two different kinds of façades presented in Fig. 9 can be compared. The calculation of the internal temperature distribution is also fundamental to foresee the thermal comfort conditions especially in presence of radiant systems and extended glazing surfaces.

Figure 9. Different double skin façades.

3. ENERGY CONVERSION EFFICIENCY

Once the building is optimized, we can consider the plant and the energy production. The aim is to maximize the energy conversion efficiency. Cogeneration is the simultaneous production of electricity and heat. The example in Fig. 10 is referred to a case of cogeneration by internal combustion engines typical in medium and small sized cogeneration engines in building applications. The heat recovery from engine cooling and exhausted gases is an alternative to the use of traditional boilers to produce the same amount of heat.

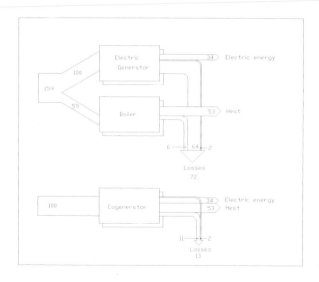

Figure 10. Cogeneration vs independent generation.

In the previous example, the global efficiency is 0.87 instead of 0.55 with the traditional independent production of electricity and heat, with an efficiency increment of 58%. It is clear that a generalized use of cogeneration can really give a fundamental contribution to reduce energy waste and gas emissions. Cogeneration can be used in district heating or commercial and office buildings. Microcogeneration can be used in blocks of flats or even small sporting facilities such as swimming pools. In summer, heat can be used to produce cool air with an absorption machine or even to control internal humidity in buildings by supplying desiccant systems (Schibuola 2001, 2003).

The increase of energy conversion efficiency also concerns the production of cool air and heat. The same inverse cycle machine (Fig. 11) transfers heat from a lower thermal level (absorbed in the evaporator) to a higher thermal level (given in the condenser).

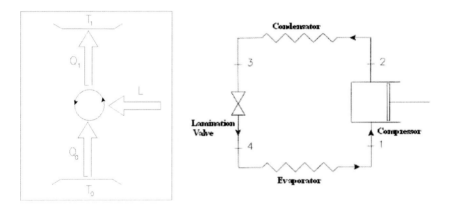

Figure 11. Inverse cycle machine.

In order to reduce heat at lower temperature levels a refrigeration machine is required, while to produce heat at higher temperatures we will use a heat pump. The efficiency, here called Coefficient of Performance (*COP*), strictly depends on thermal levels at condenser and evaporator.

Fig. 12 shows the values of *COP* as a function of condenser and evaporator temperatures. In the figure we have the comparison between the values of a real machine available on the market and the highest possible values (Carnot cycle).

In both cases, the *COP* increases with the reduction of the condenser temperature and the increment of the evaporation temperature. As for the heat pump, the curves are parametric with the condenser temperature and we can note that the *COP* increases strongly with the evaporation temperature. As far as the refrigeration machine is concerned, the *COP* increases with lower condenser temperatures. Therefore in the case of the heat pump, we always should use cold sources with high thermal levels – when available – to supply heat to the evaporator. High levels of efficiencies (*COP*) can be achieved by using water from the sea, rivers or exhausted hot fluids. In the same way, in summer the condenser of refrigeration machines must be cooled with external fluids at low temperatures. Excellent *COP* can be reached by using water from the sea, rivers or ground water instead of traditional cooling through external air or evaporative towers.

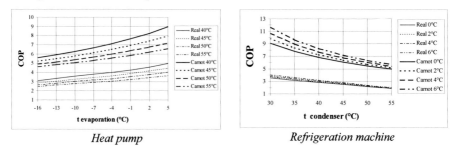

Heat pump   Refrigeration machine

Figure 12. COP as a function of condenser and evaporator temperatures.

Since refrigeration machines are usually sized for peak loads, they normally do not work at full capacity during the whole year; they work in part load conditions. Recently it has been noticed that seasonal performances of refrigeration machines in part load have increased. New standards to calculate the part load have been introduced in Italy and they are now being proposed to the European Union and also for ISO standard (Schibuola, 2000). The real performance coefficient in part load is obtained by a correction factor, the part load factor (*PLF*) multiplied by the coefficient of performances (COP_{full}) at full load, as indicated in (1).

$$COP_{real} = COP_{full} \cdot PLF \qquad (1)$$

Fig. 13 shows the trend of the *PLF* as a function of the load for an on-off control refrigeration machine (a small split air conditioner). In this case we can note a strong reduction of *PLF* and therefore a reduction of efficiency at lower loads. Through an electronic device, an inverter, we can modulate the speed of the compressor and the capacity of the machine. In this case the trend of the *PLF* is different. We can note an increment of *PLF* and therefore an increase of efficiency, in the central part of Fig. 14 (Bettanini *et al.*, 2001). The Italian standard allows a precise calculation of the seasonal mean efficiencies. In Fig. 15 the performances of typical residential and office buildings in different European cities have been considered. The comparison of refrigeration machines with an on-off or inverter control shows an increment of 25-30% of efficiency as for the inverter control compared to an on-off control (Bettanini *et al.*, 2003). New standards can give precise information to improve control systems and efficiency (Schibuola *et al.*, 2004). In the world there are millions of small split air conditioning systems. An increase of 30% of their efficiency can give a fundamental contribution to energy saving.

ENERGY OPTIMIZATION OF BUILDING-PLANT SYSTEM

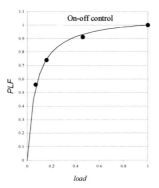

Figure 13. PLF as a function of the load for an on-off control machine.

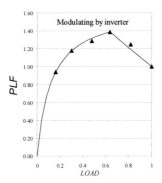

Figure 14. PLF as a function of the load for a modulating control machine.

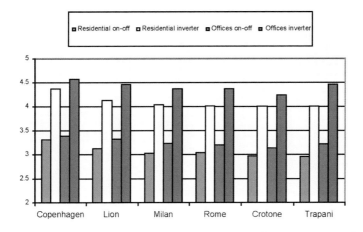

Figure 15. A comparison of seasonal efficiencies on-off vs inverter control.

4. EXAMPLES OF VERY EFFICIENT PLANTS

In this paragraph, three examples of very efficient plants in buildings located near Venice are presented. Their real performances have been measured with long period monitoring.

The first case is a Warner Bros. multiplex cinema situated in Marcon (about 15 km from Venice). Each room has an independent air handling unit sited on the top of the building as shown in Fig. 16 (Caberlotto et al., 2001, Gastaldello et al., 2003).

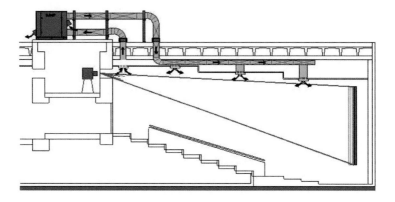

Figure 16. Section of a cinema with an air handling unit.

This special unit presents a combination of high efficient solutions: a strong heat recovery between inlet-outlet air rates, a reversible heat pump with inverter and exhaust air as source/sink, fan speed modulating control, and external air ventilation rate strictly controlled by CO_2 sensors.

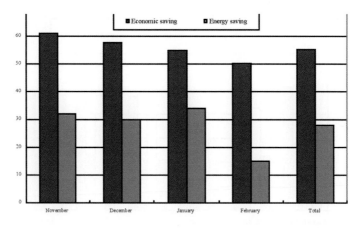

Figure 17. Saving percentage comparison: innovative vs traditional solution.

Fig. 17 shows the saving percentage of this innovative system compared to a traditional system with boilers and chillers during the heating period.

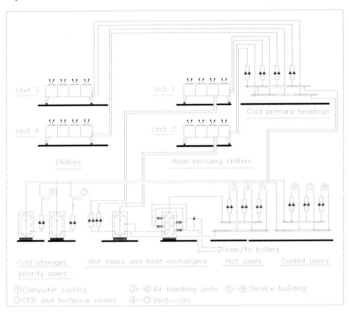

Figure 18. Sketch of the plant: total heat recovery from chillers.

The monitoring reports a total energy saving of 28% and an economic saving of 55% thanks to low electricity prices at night and in week-ends when the cinema normally works.

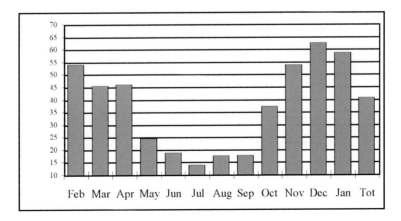

Figure 19. Cost saving with total recovery vs traditional boilers.

The second case is a data elaboration centre in Marghera (a municipality of Venice). Here the elaboration equipment requires a strong cooling also in winter (Schibuola, 1999). A total heat recovery from the chillers' condensers has been planned to heat all the buildings of the centre (Fig. 18). This way the use of traditional boilers to produce heat is not necessary. Energy consumption and heating costs are completely avoided. The monitoring shows a cost saving of nearly 40% if referred to annual heating/conditioning costs of the centre as reported in Fig. 19.

Figure 20. The bank service centre.

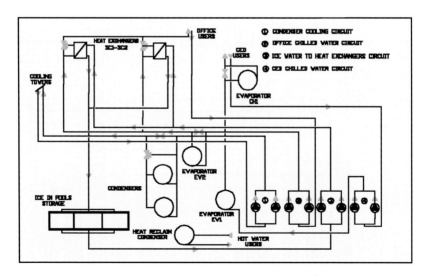

Figure 21. Sketch of the plant with ice in pools storage.

Fig. 21 shows the sketch of the plant of our last case, an office building hosting a bank service centre (Fig. 20) in Venice (Schibuola, 1995).

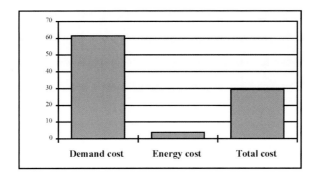

Figure 22. Costs saving referred to the case without storage.

This building was built with an ice storage in pools (ASHRAE 1989). Cool air is produced at night and during week-ends when electricity price is lower. Monitoring demonstrates a total annual saving on electric costs of about 30% as reported in Fig. 22 (Schibuola, 1998).

These case studies show that a particular attention paid to energy issues during the design of buildings and the introduction of high efficiency plants permit to achieve excellent results in terms of energy saving and management cost reduction.

Luigi Schibuola
Architectural Construction Department,
IUAV University, Venice, Italy

REFERENCES

ASHRAE, 1989. *Design guide for cool thermal storage*, Atlanta 1993. ASHRAE, *Cool storage applications*, technical data bulletin, Vol. 5, No. 3, Atlanta.
Bettanini E., Romagnoni P. and Schibuola L., 2001. Seasonal Performance Calculation for Air Conditioning Equipments, proceedings, *Clima 2000 World Congress*, Napoli.
Bettanini E., Gastaldello A. and Schibuola L., 2003. Simplified models to simulate part load performances of air conditioning equipments, proceedings, *8th International IBPSA Conference*, Eindhoven.
Caberlotto M., Gastaldello A., Schibuola L., Venco S. and Zecchin R., 2001. Experimental Results of Tests Carried Out on a New Specific HVAC System for Cinemas, proceedings, *Clima 2000 World Congress*, Napoli.
Design Builder, the program homepage available on line http//www.design builder.co.uk
ECOTECT the program, homepage of Square One, available on line: http//www.squ1.com
Energy Plus, the program homepage of U.S. Department of Energy (DOE) available on line: http//www.eere.energy.gov/buildings/energyplus
Gastaldello A., Gerrard-Smith A., Schibuola L. and Venco S., 2003. Long-term monitoring of the performances of an innovative HVAC system for cinemas, proceedings, Healthy Buildings, 2003, *7th International Conference*, Singapore, Vol. 3, pp. 665-670.
Schibuola L., 1995. Thermal Storage: An Opportunity to Reduce Climatization Management Costs in Buildings, Healthy Buildings '95, *International Congress on Healthy Buildings in Mild Climate* proceedings, Milano, Vol. 3, p. 1573.
Schibuola L., 1998. Experimental Study of an Ice Storage Performance in an Office Building, *International Journal of Energy Research*, John Wiley and Sons Ltd, 1998, pp. 751-759.

Schibuola L., 1999. Experimental analysis of a condenser heat recovery in a air conditioning plant, *Energy the International Journal*, Pergamon, pp. 273-283.

Schibuola L., 2000. Heat Pump Seasonal Performance Evaluation: A Proposal for a European Standard, *Applied Thermal Engineering*, Pergamon Elsevier Science Ltd, Oxford, UK, pp. 387-398.

Schibuola L., 2001. Humidity Control by Heat Reclaim, *International Journal of Energy Research*, John Wiley and Sons Ltd, pp. 1208-1209.

Schibuola L., 2003. High efficiency desiccant systems integrated with cogeneration units, proceedings, *International Conference on Energy and the Environment*, ICEE 2003, Shangai, Vol. 1, pp. 333-338.

Schibuola L., Gastaldello A. and Venco S., 2004. Seasonal Performance Evaluation of Chillers characterized by new design solutions in air conditioning plants, proceedings, *International Conference Climamed*, Lisbona.

F. BUTERA

THE SINO-ITALIAN ENVIRONMENT & ENERGY BUILDING (SIEEB): A MODEL FOR A NEW GENERATION OF SUSTAINABLE BUILDINGS

Abstract. The Sino-Italian Environment & Energy Building (SIEEB) is an intelligent, ecological and energy-efficient building and it seen as a model for a new generation of sustainable buildings. This paper describes the integrated design procedure for the SIEEB building design and the methodologies adopted for sustainable architecture and energy saving measures by using advanced technological solutions and control strategies. The results on the building energy simulation, plant optimization and first estimation of CO_2 emission reduction potential through SIEEB are also presented.

1. INTRODUCTION

China is experiencing an extraordinary growth in its building stock. From 1991 to 2000, residential buildings were built for nearly 5 billion square meters. In only four years (1996-1999), energy consumption of the building sector jumped from 24.59% to 27.81% of the total energy consumption. It is expected that the building stock, residential and commercial, will be doubled by year 2015 (Chen, 2003). The energy structure of China is coal-based, resulting in emission of large quantities of pollutants and greenhouse gases (GHGs). It is therefore strategically important to introduce advanced environmental and energy technologies into this field and to promote the construction of green energy-saving buildings.

The Italian Ministry for the Environment and Territory, in cooperation with the Chinese Ministry of Science and Technology, promoted the construction of a new-generation of sustainable building, the Sino-Italy Environment & Energy Building (SIEEB) in the campus of the University of Tsinghua in Beijing. SIEEB is technologically advanced, efficient from the point of view of the environment and energy consumption, and intended for offices, laboratories, classrooms, an exhibition area on Italian technology, and a conference hall with a total floor space of 20,000 m².

The SIEEB is also regarded as a platform to develop the bilateral long-term cooperation in the environment and energy fields, and a model case for showing the CO_2 emission reduction potential in the building sector in China.

The building design was carried out by the Department of Building and Environment Science and Technology (BEST) of the Politecnico di Milano, in cooperation with University of Tsinghua, MCA Mario Cucinella Architects and China Architecture Design & Research Group, in a collaborative experience among consultants, researchers and architects. The integrated design process of the SIEEB is a most distinctive part of the project and a key issue for green buildings.

In the preliminary design process, a number of appropriate shapes were considered and a feasibility analysis was carried out to check how the building was able to cope with all the requirements in terms of available area, specific building volume and space distribution. The resulting shapes were then analyzed in terms of their solar performance. Using the shape analysis, the best shape was developed with the aim of maximizing solar gains in winter and minimizing them in summer. Further, the designing of the SIEEB building was carried out on the basis of various advanced technological solutions and control strategies which include sun shading, radiant heating and cooling, displacement ventilation, efficient artificial and natural lighting etc.

This paper describes the integrated design procedure for the SIEEB building design and the methodologies adopted for sustainable architecture and energy saving measures by using advanced technological solutions and control strategies. The results on the building energy simulation, plant optimization and first estimation of CO_2 emission reduction potential through the SIEEB are also presented.

More detailed information about the building design can be found in Butera *et al.* (2003) and Adhikari *et al.* (2004).

2. BUILDING DESIGN PROCESS

2.1. Building Shape

The SIEEB building shape derives from the analysis of the site and the specific climatic conditions of Beijing. Located in a dense urban context, surrounded by some high-rise buildings, the building optimizes the need for solar energy in winter and for solar protection in summer. The shape of the building evolves from a series of tests and simulations.

2.2. Advanced Technologies and Control

The envelope characteristics derive from a series of simulations of the thermal behaviour of the building, optimizing energy and architectural factors. Building energy analysis, carried out by means of detailed computer simulations, showed that, in order to minimize CO_2 emissions, the key issue was electricity consumption, mainly because of the highly polluting electricity production and distribution system in China. The envelope characteristics were defined on the basis of the advanced technologies for energy savings.

From such evidence the following design strategies were considered:

- to maximize natural lighting, for minimizing the need of artificial lighting
- to minimize the electricity demand of the heating and cooling systems
- to cover as much as possible the electricity demand of the building by means of cleaner production systems

2.2.1. Building Envelope

The building was designed for optimising energy performance thanks to a dynamic structure that modifies itself in response to weather and light conditions, both internal and external. Form and function are integrated in order to minimize environmental impact (Fig. 1).

The envelope characteristics were defined on the basis of the advanced technologies for energy saving. On the facade facing the sun, a system of semi-reflecting glazed louvers move in relation to the sun's position, deflecting the rays onto the ceiling of the spaces behind so that light penetrates deep into the building (Fig. 2). The louvers also reflect solar radiation in summer and let it pass through in winter. Artificial lighting is based on high efficiency lamps and fittings, controlled by a dimming system capable to adjust the lamps power to the actual local lighting needs, in combination with the natural light contribution. The geometrical positions of the lamps are optimized too. A presence-control system switches off the lights in empty rooms. The integration of the envelope components chosen and the control systems will reduce by several times the electric energy consumption for lighting.

In the east and west facades a horizontal element, a light shelf, diffuses the light onto the ceiling, and internal reflecting blinds control direct sunlight.

A large surface of photovoltaic cells completes the shell.

Figure 1. SIEEB Building Envelope.

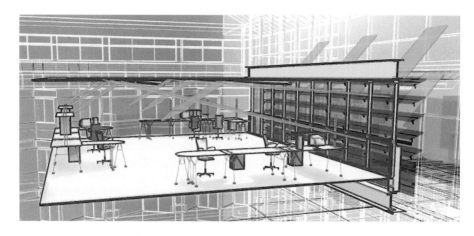

Figure 2. Natural lighting enhancement.

2.2.2. Heating Ventilation Air Conditioning (HVAC) and Combined Heat and Power (CHP) Systems

Thermal comfort conditions are provided by primary air, distributed by means of a displacement ventilation system, and radiant ceiling system. This combination minimizes electricity consumption in pumps and fans. Light weight radiant ceilings allow for reduced energy consumption; moreover, the presence sensors, coupled with CO_2 sensors, can modulate both the air flow and the ceiling temperature when few or no people are in the room, thus avoiding useless energy consumption. In summer nights, cooling takes place.

The CHP system is the core of the energy system of the building. It consists of gas motors coupled with electrical generators to produce most of the electricity required. The waste heat from engines is used for heating in winter and cooling in summer, by means of absorption chillers, and for hot water production throughout the year (Fig. 3). In order to minimise the electricity exchanges with the grid, the system is controlled in such a way that neither the electricity production exceeds the building's demand nor the waste heat produced exceeds the heating or cooling demand. This means that sometimes, when thermal loads are low, electricity production is not sufficient and some power has to be taken from the grid. Some other times the cooling loads – that are higher than the heating ones – are so high that too much electricity would be produced; in this case, the excess electricity is diverted to compression chillers, slightly reducing, at the same time, the power of the engines. A sophisticated, intelligent control system manages the plant.

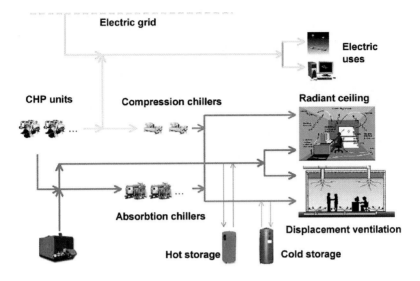

Figure 3. SIEEB HVAC and CHP Systems.

3. BUILDING ENERGY SIMULATIONS

The energy simulations of the SIEEB were carried out using DOE 2.1 building energy simulation programme, developed by the Lawrence Berkley Laboratory (1980). They showed that, because of the cleaner electricity produced and the low energy design, the amount of CO_2 emissions of the SIEEB will be far lower than in present Chinese commercial building stock, as shown in Fig. 4, where the CO_2 emissions of the final design are compared with those of the same building in which the envelope and the HVAC technologies are the ones currently used in Chinese commercial buildings.

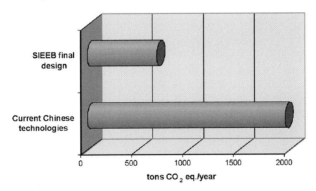

Figure 4. Comparison of CO_2 emissions.

4. CONCLUSIONS

The SIEEB design process, resulting from an interaction among architectural and technological issues, is a model of sustainable building design and shows that, in order to achieve the planned results, it is necessary to use advanced design tools and highly qualified skills. The SIEEB experience shows also that a new generation of low energy and low emission buildings – in which function and form, current practice and innovation are integrated in order to achieve very low CO_2 emissions and ecological design combined with high functional and comfort standards – is a concrete possibility.

Federico Butera
Department of Building Environment Science & Technology,
Politecnico di Milano, Italy

REFERENCES

Adhikari R.S., Aste N., Beneventano U., Butera F., Caputo P., Ferrari F. and Oliaro P., 2004. Advanced Technological Solutions for Building Load Reduction in the Sino-Italy Environment and Energy Building (SIEEB). *Proc. Eurosun-2004*, 2/514-2/521.

Butera F., Ferrari S., Aste N., Caputo P., Oliaro P., Beneventano U. and Adhikari R.S., 2003. Ecological design procedures for Sino-Italian Environment and Energy Building: Results of Ist Phase on the Shape Analysis', *Proc. PLEA-2003 Conference*, Santiago, Chile, November 2003.

Chen J., 2003. *Sustainable Buildings: the Chinese Perspective, Challenges and Opportunities*, Presented at the COP-9 Conference, December 1-12, 2003, Milan, Italy.

DOE-2 Manuals (Version 2.1), 1980. US National Technical Information Service, Department of Commerce, Springfield, Virginia, USA.

V. ENERGY EFFICIENCY AND RENEWABLES

3. ENERGY POLICY

M. PAVAN

NEW POLICY SCHEMES TO PROMOTE END-USE ENERGY EFFICIENCY IN THE EUROPEAN UNION

Theory and practice

Abstract. Uniquely among energy policy options, energy efficiency can tackle all the major energy policy goals, i.e.: security of supply; pollution impacts; reducing the costs of the services derived from energy for equity as well as international competitiveness purposes. Regulation has a critical role to play in reducing the barriers to the development of a market for energy efficiency products and services. Traditional policy approaches to the promotion of end-use energy efficiency include "command and control" type of regulation and various forms of economic incentives. In a liberalized market framework new types of regulatory approaches have to be defined to promote end-use energy efficiency. The paper discusses recent developments in the European Union related to policy schemes aiming at promoting investments in end-use energy efficient technologies by combining traditional tools with more innovative market-based ones. The discussion is developed from both a theoretical and an empirical point of view.

1. THE ROLE OF END-USE ENERGY EFFICIENCY IN A BALANCED ENERGY POLICY

Energy efficiency projects cover measures that yield a reduction of the level of "energy intensity" of a product or a service (i.e. the amount of energy needed for a given unit of output) through the substitution of the capital stock or, to a lesser extent, through changes in management/maintenance practices. These measures result in producing the same or better levels of amenities (e.g. lighting, space conditioning, motor drive power, etc.) using less energy, they are generally long lasting and save energy across different time periods.

Uniquely, among energy policy options, energy efficiency can tackle all the major energy policy goals (i.e.: security of supply) by reducing the demand for imported energy; reducing polluting emissions; reducing the costs of the services derived from energy in such a way that comfortable households are affordable for all and business international competitiveness is improved. This makes energy efficiency a key component of any sustainable energy policy.

2. POLICY APPROACHES TO THE PROMOTION OF END-USE ENERGY EFFICIENCY

Together with the above-mentioned range of valuable public benefits, investments in end-use energy efficiency deliver private benefits in terms of lower energy bills (coupled with equal or even higher levels of energy service) and possibly higher levels of comfort (coupled with comparable energy bills).

The unit costs of energy efficiency measures (i.e. the costs per unit of energy saved as a result of these measures) are very often lower than the costs of producing the same unit of energy. The saved energy is in most cases sufficiently valuable to repay the cost of investment in a reasonable length of time and to cover interest charges. In the European Union (EU) the average cost of saving a unit of (off-peak) electricity in the domestic sector is around 2.6 € cents, against an average (off-peak) price for delivered electricity around 3.9 c€/kWh. Similar gaps between the cost of savings and the price of delivered energy exist for the other energy carriers (European Commission, 2003).

Despite this, the market for energy efficiency products and services finds it hard to develop. The more common obstacles to this development include, but are not limited to: the lack of knowledge on the cost-effectiveness, returns and risks of investments in energy end-use efficiency; the lack of visibility of savings potential; the limited access to capital; market prices which do not reflect the social costs of energy production and consumption (so-called "externalities"); the so-called "investor-user dilemma", where owners of, for example, residential buildings and offices tend to minimize investment costs in efficient energy-using technology since the resulting higher running and energy costs from using less efficient technology will not be paid by them but by their tenants, renters or other users; the fact that the fees for suppliers of energy end-use technologies, as well as for installation engineers, builders and architects are usually proportional to the total investment cost and are not based on the energy performance.

The role of public policy is to reduce or overcome these barriers in order to foster the market penetration of energy efficiency products and services.

Traditional policy approaches to the promotion of end-use energy efficiency in Europe embrace "command and control" (CAC) regulation and various forms of economic instruments. The first 'category' of policy tools includes, for example, compulsory efficiency product standards, minimum efficiency requirements (possibly coupled with energy efficiency labels) and building codes. The latter includes various forms of taxes, subsidies and incentives (such as feed-in tariffs).

Both approaches have advantages and drawbacks. One of the major differences between the two is that while 'pure' command and control regulation fixes compulsory quantitative targets but is unable to control the cost at which these targets will be achieved, economic instruments set an overall cost target but are unable to ensure certainty over the energy efficiency improvements that will be achieved at this cost.

Integrated policy packages that combine 'command and control' elements with market-based ones (MB), such as certificates trading, have been implemented in Europe to promote electricity from renewable energy sources (i.e. green certificates schemes) and to cut harmful emissions (i.e. emissions trading schemes). Certificates trading is expected to deliver economic efficiency benefits both by equalizing the marginal costs spent on complying with a target (so-called static efficiency), and by encouraging technological development (so-called dynamic efficiency). The economic textbook argument is that integrated 'CAC plus MB approaches' may combine the certainty of results of "command and control" regulation with the economic efficiency of market-based instruments.

Recently, the scope for the implementation of 'integrated CAC plus MB' policy packages to the promotion of end-use energy efficiency has become an issue of debate in a number of EU member States and within EU institutions. The implementation of similar approaches to foster the market penetration of energy efficient end-use technologies poses a number of technical challenges and faces a number of trade-offs, only some of which are common to other more widespread certificates trading schemes. The extent to which these are absolute trade-offs is strictly connected to some fundamental design choices of the different components of the scheme.

In the following sections we introduce the basic elements and major design features of integrated 'CAC plus MB' policy mechanisms recently introduced or currently under consideration at the European level. The discussion is developed from both a theoretical and an empirical point of view, on the basis of the analysis of recent experiences with some or all the components of this policy kit in certain EU member States, namely: Italy, United Kingdom, France and Belgium (i.e. the Region of Flanders).

3. ENERGY SAVING OBLIGATIONS

In the last few years, a number of EU member States have introduced mandatory energy saving targets at the national level. The UK introduced the first obligation in 1994, Italy in 2002 (then post-postponed to 2005), the Flemish Region of Belgium in 2003, while France has approved the introduction of a similar obligation at the national level starting from 2006.

Although referred to as "energy saving targets" or "energy saving obligations" hereafter, these obligations require energy consumption to be reduced by a certain amount over a certain period of time via the adoption of energy efficient technologies that deliver the *same level of energy service(s)* (e.g. lighting, ambient temperature, production level) compared to the less efficient technologies that they substitute. In other words, the savings have to result from investments in energy efficient technologies rather than from mere curtailments of the energy usage.

Apart from differences related to the size of the targets, the obligations differ with respect to a number of other characteristics: the metric, the time horizon for the obligation, as well as the length of the compliance period.

The size of the energy savings target needs to be based on a sound technical-economic analysis although, in the end, it is often the outcome of negotiation among all the interested parties.

The choice of the metric is linked to the policy goal(s) driving the introduction of the system. For example, if the prevailing policy aim is energy security, then the target is more likely to be set in terms of tons of oil equivalent saved; if it is the reliability of the electricity system, then the target is set in terms of kWh saved; if it is air pollution, for example, the national target is fixed in terms of primary energy savings, while in the Flemish and French schemes the obligation is set in TWh saved. In the current UK Energy Efficiency Commitment scheme (EEC) the saving target is set in terms of fuel standardized TWh, i.e. each energy source or carrier is weighted according to its CO_2 emission factor.

Further differences among the various schemes are related to the time horizon and the compliance period of the obligation. Theoretically, a longer time horizon ensures adequate returns for investors in energy efficiency projects, thus encouraging investments. A multi-year compliance period allows obliged parties a larger flexibility in designing their compliance strategy, but may require periodic monitoring of the progress towards the target during the compliance period, in order to allow for policy adjustments if this progress is not adequate[1].

In Italy, annual targets are set for the period 2005-2009 and increase in stringency each year. The mechanism is planned to deliver energy savings equivalent to 5.8 millions toe in the five-year target period. Electricity savings could amount up to 14 TWh, while natural gas savings could total 3.3 billions cubic meters. Targets for the post-2009 period will be fixed by the Government by the end of the year 2006. In Great Britain, the EEC runs in 3-year cycles from 2002 to 2011. It replaced the previously existing Energy Efficiency Standards of Performance (EESOP, 1994-2002). In the EEC-1 the overall saving target was 62 fuel standardized TWh; in the EEC-2 the target has been increased to 130 TWh. In the French scheme the aggregate target for the first three years (2006-2008) is 54 TWh (in final energy). In the Flemish region of Belgium the target was 381 GWh of primary energy for 2003, 551 GWh for 2004 and 579 GWh for 2005.

One very important feature of the energy savings obligation is whether the target is set in terms of cumulated savings over the lifetime of energy saving projects or in terms of the savings produced each year. In both cases the 'reference lifetime' can be shorter that the physical (i.e. actual) lifetime of the projects. It is clear that the choices made with respect to these issues deeply influence the stringency of the energy efficiency target. Both in the UK and in the French scheme energy savings are cumulated over the physical life of the energy efficiency measures (as measured by default factors), with energy savings in future years discounted with different discount rate (4% in France, 3.5% in the UK). In the Flemish scheme no discounting is applied to lifetime energy savings. In Italy only savings generated during the year are accounted for towards the annual target, and the crediting lifetime is currently fixed at five years for all project types except for measures that impact on energy consumption for heating and air conditioning, that have a crediting lifetime of 8 years.

The obligation can be banded or un-banded, that is to say it may include one or more constraints to achieve a certain share of the target by means of specific types of projects, i.e. technologies, end-uses or target groups. For example, in the UK EEC scheme suppliers must achieve at least half of their energy savings in households on income-related benefits and tax credits. In the Flemish scheme separate kWh saving targets are set for low voltage clients (< 1kV, mainly residential) and high voltage ones (> 1kV). In the Italian scheme at least half of the target set for the electricity and natural gas sector has to be achieved via reductions in electricity and natural gas consumption respectively (so-called "50% constraint"). Although these are legitimate policy goals, if the energy saving obligation is combined with a certificates trading mechanism, then a banded obligation reduces the scope for cost savings, restricts the fungibility of certificates and increases the related administrative costs, since it requires certificates to be labelled (e.g. by energy carrier/technology/consumer group).

A further differentiation between the different schemes in operation concerns the choice of the target group, that is the organizations and/or individuals on whom the obligation is to be imposed. In general this choice is based on the characteristics of the energy market(s) in the country. In the Italian scheme national targets are currently apportioned among those electricity and natural gas distributors that served at least 100,000 final customers in 2001, linearly on the basis of their market share (measured in quantity of electricity/natural gas distributed to final customers). Rules for distributors under the above-mentioned threshold are expected to be fixed by the Government by the end of 2006. In the current UK EEC the national target is allocated among electricity and natural gas suppliers with 50,000 or more domestic customers. The apportionment is linear to the number of customers served. In the Flemish scheme the obligation is imposed on electricity distributors, while in the French system obligations are set for suppliers delivering electricity, gas, domestic fuel (not for transport), cooling and heating for stationary applications with sales over 0.4 TWh/year. The aggregate target is distributed between suppliers on the basis of their market share (measured in turnover) in the household, public and commercial sectors combined (i.e. excluding industry).

3.1. Eligible projects

The projects that are considered as eligible to meet the energy saving targets differ quite significantly between the various schemes. What varies most is the scope of the obligation.

In general, the scope tends to be wider in schemes that combine the obligation with certificates trading (i.e. Italy and France), given the economic textbook argument that the potential benefits of a market-based instrument are better exploited when the diversity of the costs of meeting the target is larger. Apart from that, restrictions may apply to specific types of projects with the aim of avoiding the double counting of the benefits delivered by (and thus the over-subsidization of) projects that profit from other public incentives. Minimum size thresholds for projects may be imposed, mainly to reduce administrative costs.

The Italian system is open-ended, i.e. any type of end-use energy efficiency project qualifies for the issuing of white certificates[2]. Only "hard measures" (i.e. investments in energy efficiency technologies) are included in the system. Soft measures such as behavioural changes are not eligible. Information campaigns are eligible only if add-on and paired to specific 'hard' measures. This restriction stems from the fact that the energy saving impact of these campaigns is difficult to quantify (see section 5). Other restrictions apply to end-use projects that already benefit from other forms of public incentives, for example: small photovoltaic systems that benefit from feed-in tariffs and district heating based on cogeneration plants that qualify for green certificates. In order to be eligible, projects must reach a minimum size in terms of delivered energy savings, which is currently differentiated according to project type and project developer (i.e. obliged versus non-obliged parties).

In the French scheme energy savings can be made in any sector and with any energy source or carrier, with the exception of plants under the European emissions trading scheme and projects that result in a mere fuel-switching. A minimum project size is imposed. Individual projects may be implemented by the suppliers themselves or subcontracted to a wide range of other actors. The UK EEC projects can be related to electricity, gas, coal, oil and LPG use in the domestic sector. They may be implemented by suppliers themselves or subcontracted to energy service companies or other actors. In the Flemish scheme eligible actions refer to residential and non energy-intensive industry and service and can involve saving fuel from any sources.

Early actions (i.e. projects realized prior to the entering into force of the obligation) can be considered as eligible when graduality in the implementation of the obligation is one policy aim. Allowing for early actions inevitably reduces the stringency of the target(s), since it lessens the incentives to develop new projects, at least in the first phases of the mechanism. In Italy the energy savings produced by projects developed in the four years prior to the coming into force of the obligation (i.e. 2001-2004) are considered to be eligible, subject to the Regulator's approval[3]. In the proposed Directive on energy services (COM(2003)739) measures taken as early as 1991 qualify to meet the energy saving targets set for member States.

4. THE WHITE CERTIFICATES MARKET

The idea to combine energy saving obligations with energy efficiency certificates trading is currently debated both at the academic and at the policy and decision-making levels in a number of EU member States and within EU institutions. Some of the countries where energy saving targets have been in place for some time are considering the scope of adding a market-based component to the policy package in order to promote the cost effectiveness in meeting these targets, while other countries that are currently implementing different policy instruments to foster the market development of energy efficient technologies are investigating the pros and cons of introducing this integrated policy scheme. In the proposed Directive on energy services the so-called "tradable white certificates" are specifically mentioned as one of the tools that member States may choose to use in order to comply with the energy saving targets imposed on them. The Proposal leaves the option to the European Commission to recommend later on the introduction of this market approach.

To date Italy is the only country worldwide to have a fully-fledge tradable energy efficiency certificates scheme. Obliged distributors have three basic options to comply with their target: a) they can implement end-use energy efficiency projects; b) they can develop projects jointly with third parties, such as energy service providers, energy service companies, product manufacturers and installers; c) they can buy certificates called "energy efficiency certificates" or "white certificates" that attest that the corresponding amount of energy has been saved by third parties.

In July 2005 the French Government has approved the introduction of a national white certificates scheme, which became operational in 2006.

In the UK there is no certificate trading in the strict sense of the word. The energy efficiency programs of individual suppliers are monitored and approved by the Regulator, but no certificates are issued. Suppliers can trade their targets, in such a way that one supplier takes responsibility for a share of another supplier's obligation in exchange for payment. Suppliers can also trade their performance under the scheme, so that one supplier sells fuel standardized kWh of achieved energy benefits to another supplier, who uses this towards his own EEC target. Both forms of trade require approval from the Regulator (OFGEM). To date no trading has been concluded. Work is ongoing to study the pros and cons of introducing a white certificates market.

The major features of a tradable white certificates scheme concern: which parties can be awarded certificates (the so-called 'eligible parties'); who issues the certificates; what is the metric and the unit value of one certificate; how many types of certificates are issued and which is their temporal validity; how are certificates traded and by whom (apart from eligible parties, if any). Some of these features inevitably come from the design choices made with reference to other elements of the policy package, while others require further regulatory decisions.

The metric of a certificate, for example, is inevitably the metric of the energy savings obligation. On the contrary, the certificate can have different unit values. This value may have implications on the number of parties that can offer certificates for sale (unless other restrictions apply) and, ultimately, on the liquidity of the certificates market. In the Italian scheme certificates are expressed in primary energy saved and the unit is 1 toe.

The number of certificate types that are issued and traded in the market is a further attribute that comes from the design choices related to other elements of the integrated policy tool-kit. In particular, a banded energy savings obligation requires more than one type of certificate, in order to implement the restrictions in place. In the Italian system three types of certificate are issued to implement the "50% constraint" (cf. section 3): certificates for electricity savings, certificates for natural gas savings and certificates for savings in other energy sources or energy carriers. The three types of certificate are only partially fungible.

In order to allow obligated parties some degree of inter-temporal flexibility in meeting their annual target, banking and borrowing of certificates may be authorized. In the Italian scheme, banking is allowed throughout the 2005-2009 period, i.e. certificates issued in any of these five years can be used to meet the targets set for the period. Banking is expected to increase the scope for cost savings and help reduce the risk of price volatility, e.g. the risk of price spikes/falls in periods of excess demand/supply, which is particularly high with annual energy saving targets. Borrowing is not allowed because of concerns about long-term compliance.

Both banking and borrowing may be unrestricted or there may be restrictions imposed on the number or portion of banked/borrowed certificates that can be presented for redemption in a given period. In the Italian scheme no restriction to banking is currently applied.

As in the case of eligible projects, a wide scope of the certificates market in terms of eligible parties is expected to promote the effectiveness of the scheme by ensuring the diversity of the cost-options to meet the target, thus increasing the scope for cost-effectiveness via trading. In Italy eligible actors include: i) all electricity and natural gas distributors; ii) companies controlled by those distributors; iii) companies that operate in the energy services market that encompass, but are not limited to, energy service companies (ESCOs). In the French scheme any economic actor can develop energy savings measures and be awarded certificates.

Different choices are possible with respect to the body in charge of the issuing of certificates, depending on the institutional structure in place in the country. The choice should aim, once again, at minimizing the costs associated with this activity. In Italy the Electricity Market Operator is in charge of issuing certificates and of administering the certificates Registry. Certificates are issued following the specific authorization of the Regulator (AEEG), once the latter has completed the measurement, verification and certification of the energy savings achieved by each project (cf. section 5).

The trading of certificates can take place in a specific marketplace and/or over the counter. In addition to parties that can be awarded certificates, other parties can be allowed to trade certificates (e.g. market intermediaries). A large number of trading parties generally promotes market liquidity. Rules and procedures to administer market sessions as well as the certificates Registry need to be designed in order to guarantee the effective working of these instruments and the minimization of administration and compliance costs.

In Italy both trading in the marketplace and over-the-counter (OTC) is allowed. The spot market as well as the certificates Registry are organized and administered by the Electricity Market Operator according to rules set jointly with the Regulator. There are three different certificates markets (one for each of the three types of certificates). Any interested party can operate in the spot market and have an account in the Registry to record certificates bought and sold also via bilateral contracts, provided that he/she meets standard legal and technical requirements. Each market operator has to pay an annual fee that covers the costs borne by the Electricity Market Operator to administer the Registry and the market. Market sessions are organized at least once a month during the year, and at least once a week in the four months prior to the annual compliance check (cf. section 6). Market rules include procedures to ensure the positive conclusion of market deals both to sellers and to buyers.

In the French scheme trading occurs only via bilateral contracts.

5. MEASUREMENT AND VERIFICATION OF ENERGY SAVINGS

The measurement and verification of energy savings (M&V) is much less straightforward than e.g. the monitoring and verification of power generation from renewable sources within a green certificates scheme. This is because the quantity of energy saved cannot be directly metered, but must be estimated by comparing measured or calculated consumption with a counterfactual baseline. The credibility and success both of stand-alone energy saving obligations and of systems that combine obligations with certificates trading greatly depend upon how M&V is designed and carried out.

To this respect, the technical challenges related to the application of certificates trading to the promotion of end-use energy efficiency are similar to those posed by the implementation of project-based trading schemes such as Joint Implementation and the Clean Development Mechanism. Common to these schemes is also the inherent trade-off between M&V rules that guarantee accuracy and precision of energy savings/emissions reductions measurements, and the economic efficiency of these rules in terms of underlying compliance and administration costs.

But what is special about the M&V of energy?

5.1. Main M&V issues

As we already mentioned, energy savings cannot simply be measured at the meter, as they depend on a range of factors that are not always known in detail.

Generally speaking, energy savings have to be quantified via a comparison of energy consumption *before* (or *in the absence of*) and *after* the implementation of a project. But in some cases the 'before the project' scenario is not known, for example because historic data on energy consumption are lacking or because you are dealing with a new installation. Hence assumptions have to be made on what would have happened in the absence of the project.

In other cases the 'before the project scenario' is known, but the impact on consumption trends of variables other than those on which the energy saving project has an influence need to be net out (e.g. climatic conditions, working hours that vary in the pre- and post-project scenario). A related issue is the so-called 'rebound effect', that may reduce the energy savings from improved energy efficiency due to the likely proportion of the investment to be taken up by improved comfort i.e. increased consumption of the energy service (e.g. thermal comfort).

One of the most important elements of M&V rules for energy savings is the choice of the technological baseline, i.e. the technology that is *assumed* to be substituted for with the implementation of the project. In the case of projects that involve capital stock substitution you might want to accredit savings and to issue certificates only to end-use energy efficiency improvements achieved over and above spontaneous market trends or legal requirements. And you might make an analogous choice also in the case of projects dealing with new installations in order to ensure that you are promoting only savings that are 'additional' to what would have occurred in the absence of the scheme. A related M&V issue concerns whether and (if yes) how, you want to deal with free riders, i.e. whether you want to consider savings that would have been achieved by project participants even in the absence of the project.

Various M&V approaches and techniques can be applied to tackle the above-mentioned issues. In the Italian scheme, projects are not subject to approval before implementation, although developers may request an eligibility check. Instead, the Regulator makes an *ex-post* evaluation and certification of the savings achieved by each project and authorizes the Electricity market Operator to issue an appropriate quantity of certificates. The method of monitoring and verification takes one of three forms depending upon the type and complexity of the project, namely: a) a 'deemed savings' approach, where energy savings from particular technologies are estimated using standard parameters and no on-field measurement is required; b) an engineering approach, where energy savings are determined by a formula that depends upon one or more parameters which need to be monitored at the site; and c) a comprehensive approach, where a site-specific baseline and monitoring of energy consumption is required. 'Deemed savings' are developed by the Regulator following consultation of all interested parties and represent the prevailing methodology for projects in the household sector. The engineering approach is more applicable to projects in public and commercial buildings and, although to a much lesser extent, in industry, while the comprehensive approach is suitable for large-scale projects. Only 'additional' savings are certified, i.e. savings achieved over and above market average or legislative requirements. In the 'deemed savings' and engineering method the choice of the technological baseline is rooted in the calculations (and it is always the average market technology, provided that it meets existing legislative requirements).

Since the UK ECC is confined to small-scale projects within households, there is no monitoring of individual projects. Instead, 'deemed savings' are used to estimate the energy benefits from different types of energy efficiency measure. The 'rebound effect' is taken into account and dead-weight factors are used in order to consider the effect of investments that would be made anyway. Each supplier submits proposals to the Regulator (OFGEM), detailing the measures he is planning to undertake. OFGEM determines whether the measures qualify and uses formulae to calculate the discounted energy benefits to be attributed to them. These are assigned to the suppliers on an *ex-ante* basis. OFGEM monitors the overall progress of each supplier towards its target and audits a selection of individual schemes over the course of the programme.

In the French scheme, ADEME[4] is responsible for monitoring and verification and it is expected that all projects will be of the 'deemed savings' type. ADEME will receive report from suppliers on their energy efficiency programs, will calculate the savings, award certificates and audit a selection of individual projects. Also in this scheme measures undertaken to comply with existing legislative requirements are not eligible.

6. ENFORCEMENT MECHANISMS

Adequate compliance and enforcement mechanisms are necessary to ensure both the credibility of the energy saving obligation and the effective operation of the certificates market. The enforcement mechanism can be designed in different ways, according to different policy purposes and different national legislative frameworks.

Compliance with targets is better enforced through a financial penalty, which can be defined alternatively at the start or at the end of the compliance period. In a market-based scheme a pre-defined penalty may serve the policy goal of establishing a ceiling to the price of certificates and thus, ultimately, a cap to the overall cost to the country of meeting the energy efficiency targets. The major drawback of this option is the risk facing the Regulator to set a penalty that is too low to promote investments in energy efficiency projects, or too high to determine an effective upper limit to the total cost of the scheme. To be effective, the level of the penalty should be fixed on the basis of a thorough analysis of the marginal cost curve of saving energy in the country, as it should be greater than the average investment required to achieve the target.

Alternatively, the non-compliance penalty can be determined *ex-post*, at the end of the compliance period. In a policy scheme that combines energy saving targets with tradable certificates, this option has the advantage of allowing the Regulator to use the price signals produced by the certificates market as pieces of information to set the penalty.

The unit value of the penalty can be fixed or can be linked, for example, to the market price of certificates. Sanctions proceeds may be used for different purposes, that can be related or not to the pursuit of energy efficiency improvements. One way to use sanctions proceeds is to recycle the money collected to over-complying parties under the scheme. In the case of a pre-defined penalty, this reduces the risk that the penalty acts as a sort of 'reference price' for the trading of certificates, thus altering the effective functioning of the market. More generally, this way of using the revenues collected from penalties enforces the effect of the penalty itself by increasing the opportunity costs of non-compliance.

In the Italian scheme, compliance with targets is assessed by the Regulator[5], who is also in charge of determining the penalty for non-compliance. A two year grace period[6] applies if the share of the target not fulfilled by the obliged distributor is equal or higher than the ratio between the amount of certificates issued during the compliance period and the total (allocated) national target for that period. This ratio is computed by the Regulator each year and acts as a sort of 'benchmark' in the assessment of compliance with the obligations, as it is meant to represent the concrete possibility for obliged distributors to meet their target (via the purchase of certificates). Although appealing in principle, this benchmark is not straightforward to calculate, mainly as a result of the possibility to bank certificates and of the existence of three certificates markets that are only partially fungible.

As for the level of the penalty, according to the criteria set by the Government, it has to be proportional and in any case greater than the investments required to compensate the non-compliance. This criterion has to be applied in the framework of the more general criteria that preside over the definition of financial penalties in the country. These criteria call for a case-by-case assessment of the non-compliance and of the related penalty, and prevent the Regulator from defining the value of the penalty totally *ex-ante*. As a result the sanction does not act as a reference price for the trading of certificates which, in turn, guarantees (at least to this respect) that the certificates market send correct signals on the real cost of saving energy. As a general guideline, the Regulator has indicated that the unit value of the penalty (expressed in terms of €/toe not saved) will be defined, *inter alia*, with reference to the average price of certificates traded in the marketplace and to the average cost of saving energy in the country. Sanctions proceeds used to finance information programs on end-use energy efficiency.

In the UK EEC the Regulator has the power to consider whether it is correct to set a sanction for non-compliance; however, there is no specific guidance on how the sanction would be calculated. The legislation makes general reference to the qualifications for electricity and gas supply licences, with the implication that these could be removed.

In the French scheme, a penalty of 2 c€/kWh is envisaged, with an estimated average cost of saving one unit of electricity around the same value.

In the Flemish scheme there is a penalty of 10 c€/kWh of shortfall, which can not be passed along in tariffs.

7. THE RECOVERY OF COSTS

The way in which the parties that are subject to the energy saving obligation may recover the costs borne to comply with their target is strictly dependent upon the structure of the market in which they operate.

In the UK, since the electricity and gas supply markets are liberalized, there are no requirements on cost recovery. Suppliers may cover the costs of the scheme through any means they choose, and typically share the costs of each investment with either consumers themselves or third parties such as housing associations. Although there is no explicit levy to finance the scheme, the cost of "doing business" in the EEC2 is estimated to amount to 13 £/customer/fuel.

Most of the suppliers under the French scheme are regulated monopolies and cost recovery is allowed to a maximum of a 0.5% increase in unit tariffs.

In the Flemish scheme programs costs are incorporated into electric tariffs.

In the Italian scheme, the cost-recovery mechanism is designed, updated and administered by the Regulator (who has the responsibility of setting tariffs in the non-competitive segments of the electricity and natural gas market). The system only applies to costs related to electricity and natural gas savings, up to the occurrence of the distributor's target. The Regulator determines an average standard cost per unit of primary energy saved. Currently the average standard cost is set at 100 €/toe, i.e. roughly 2,2 c€/kWh saved and 8,2 c€/cubic meter of natural gas saved. The costs of purchasing certificates from third parties are included in the system.

8. SUMMARY AND CONCLUSIONS

Energy efficiency is an important cornerstone of a sustainable energy policy. Various policy approaches and tools can be applied to overcome the barriers that prevent the development of a competitive market for energy efficient products and services.

The paper has examined recent developments in the European Union based on energy saving obligations possibly coupled with market-based tools such as certificates trading. The discussion has focused both on the theory and on the practice, via a comprehensive analysis of the ongoing and planned experiences in Europe. The operational schemes described above are quite diverse.

Unquestionably the introduction of a certificates trading mechanism driven by an energy efficiency obligation has a number of advantages over more 'traditional' approaches to foster the market penetration of efficient end-use energy technologies and energy services:

a) it secures that a predefined and certain quantitative target is achieved;
b) it enables least cost solutions to be developed therefore limiting the overall costs of meeting this target;
c) it is consistent with a liberalized market framework.

On the other hand, the effectiveness of such a policy package in delivering the expected results crucially depends on a number of factors such as: the actual technical-economic potential for savings, the number of actors involved in the market, their diversity in terms of technological and cost options, and the degree of complexity of the rules shaping the mechanism.

The ongoing experiences at the European level will help understanding which design choices are best in which context, as well as the extent to which the theoretical advantages of an 'integrated CAC plus MB' approach actually deliver the expected results in terms of both effectiveness and cost-efficiency.

Marcella Pavan
Autorità per l'energia elettrica e il gas, Italy
Italian Regulatory Authority for Electricity and Gas

The opinions expressed in the paper are those of the author, and do not necessarily reflect the position of the Autorità per l'energia elettrica e il gas.

NOTES

[1] In addition, in a market-based scheme, a system with periodic assessments of compliance is also likely to generate more trading activity than one with a single assessment of compliance at the end of the scheme.

[2] Projects aiming at increasing the energy efficiency of power generation are explicitly excluded from the system. Small photovoltaic systems (below 20 kV) are the only exception to this rule.

[3] In the Italian case, this choice is likely to be linked also to the fact that the system was initially designed to enter into force in 2002, but was then deferred to allow adequate time for the design of the technical and economic regulation needed to make it work.

[4] Agence de l'Environnement et de la Maîtrise de l'Energie.

[5] At the end of each year, obliged distributors have to surrender a number of certificates equal in volume to their annual energy savings target, taking into account the "50% constraint".

[6] In other words, the distributor that falls short of target must make up the shortfall in the two subsequent years.

REFERENCES

Baumol W.J. and W.E. Oates, 1988. *The Theory of Environmental Policy*, 2nd edition ed. Cambridge University Press: Cambridge.

Baudry P. and S. Monjon, 2005. The French Energy Law and white certificates system, *RECS open seminar*, 25 September 2005, Copenhagen.

Bowie R. and H. Malvik, 2005. "Measuring savings target fulfilment in the proposed Directive on energy end-use efficiency and energy services (COM(2003)0739)". In *Proceedings of the 2005 ECEEE Summer study*, European Council for an Energy Efficient Economy (ECEEE).

Bertoldi P., S. Rezessy and M.J. Bürer, 2005. "Will emission trading promote end-use energy efficiency and renewable energy projects?", in *Proceedings of the 2005 ACEEE Summer Study on Energy Efficiency in Industry*, Washington D.C.

Collys A., 2005. "The Flanders (BE) regional utility obligation". Presented at the *Workshop on Bottom-up Measurement and Verification of Energy Efficiency Improvements: National and Regional Examples*, arranged by European Commission DG TREN, European Parliament and ECEEE. Brussels, March.

Coase R.H., 1960. "The problem of social cost", in: *The Journal of Law and Economics*, 1960, III: pp. 1-44.

European Commission, 2003. *Proposal for a Directive of the European Parliament and of the Council on energy end-use efficiency and energy services*, COM(2003)739 final, European Commission, DG TREN, Brussels.

European Commission, 2005. *Green Paper on Energy Efficiency. Doing more with less. COM(2005) 265 final of 22 June 2005*, European Commission, DG TREN, Brussels.

Langniss O. and B. Praetorius, 2003. "How much market do market-based instruments create? An analysis for the case of "white" certificates", In *Proceedings of the 2003 ECEEE Summer study*, European Council for an Energy Efficient Economy (ECEEE).

Lees E., 2005. *Summary of Workshop on Bottom-up Measurement and Verification of Energy Efficiency Improvements: National and Regional Examples*, arranged by European Commission DG TREN, European Parliament and ECEEE. Brussels, March.

Lees E., 2004. "The UK Energy Efficiency Commitment". Presented at the Expert Seminar on Measurement and verification in the European Commission's Proposal for a Directive on Energy Efficiency and Energy Services, organized by the European Commission, the European Parliament and ECEEE. Brussels, September.

Pavan M., 2002. "What's up in Italy? Market liberalization, tariff regulation and incentives to promote energy efficiency in end-use sectors". In *Proceedings of the 2002 ACEEE Summer Study on Energy Efficiency in Buildings*, 5.259-5.270. American Council for an Energy Efficiency Economy, Washington D.C.

Pavan M., 2003. "Expectation from Italy: suggestions to focus the project proposal". Presented at the *Workshop on organizing a Task on white certificates under the IEA-DSM Implementing agreement*. CESI, Milan.

Pavan M., 2005. "Italian Energy Efficiency Obligation and White Certificates: Measurement and Evaluation". Presented at the *Workshop on Bottom-up Measurement and Verification of Energy Efficiency Improvements: National and Regional Examples*, arranged by European Commission DG TREN, European Parliament and ECEEE, Brussels, March.

Pavan M., 2006. "New trends in energy regulation: the integration of command and control approaches, tariff regulation and artificial markets for demand-side resources". In *Proceedings of the 2006 International Association of Energy Economists (IAEE) Conference*.

Quirion P., 2005. "Distributional impacts of energy-efficiency certificates vs. taxes and standards". In *Proceedings of the 2005 ECEEE Summer study*, European Council for an Energy Efficient Economy (ECEEE).

Schipper L., 2000. "On the Rebound: The Interaction of Energy Efficiency, Energy Use and Economic Activity". *Energy policy*, 28(6-7).

Wuppertal institute, ACE, ADEME, ARMINES, CCE, DEA, Energy piano, EEE, ESI, InterRegies, Lund University, and Politecnico di Milano (2000) *Completing the market for least-cost services. Strengthening energy efficiency in changing European electricity and gas markets*. Wuppertal Institute for Climate, Environment, Energy. A study under the SAVE Program.

G. PIREDDU

SUBSIDIES AND MARKET MECHANISMS IN ENERGY POLICY

Abstract. The promotion of renewable energy sources is seen as a good alternative to thermal produced electricity in the light of climate change and the Kyoto Protocol to reduce greenhouse gases. This chapter focuses on incentive mechanisms designed to sustain renewable energy technologies in a liberalized and competitive electricity market. A series of fundamental reforms changed the shape of energy markets, particularly the electricity market. Electricity produced from renewable energy sources has to compete in a liberalized scenario because, owning up to higher generation costs, it needs some type of incentives. In this chapter we describe and compare the main incentive schemes for renewable energy sources in order to fulfil environmental goals in a competitive electricity market. Tradable green certificates and tradable pollution permits are the incentive schemes that are consistent with free market rules.

1. INTRODUCTION

The development of renewable energy technologies (RETs) has received political support within the European Union after the oil shocks of the 20th century. Two main issues have increasingly dominated the central aim of the European Commission's energy policy: environmental constraints and energy market liberalization. The promotion of electricity from renewable energy sources (RES-E) is mainly seen as a good alternative to thermal produced electricity in the light of climate change and Kyoto Protocol strategies to reduce greenhouse gases, most notably CO_2 emissions. The European Union (EU) adopted a Directive on the promotion of renewable electricity in member States in September 2001 (Directive 2001/77/EC). The EU has set itself a target to increase electricity produced from RES-E to 22% of its total electricity generation by 2010. Member States must also take appropriate steps to encourage consumption of green electricity to meet the levels indicated in the Directive. At the beginning of 2007 the European Commission set each Member State the so-called 20-20-20 target to meet in 2020: increasing energy efficiency by 20%, consuming at least 20% of renewable energy, and abating greenhouse gases by 20%.

Increasing the share of RETs in the energy balance will enhance sustainability and improve energy supply security by reducing energy dependence on oil and natural gas imports. To realize environmental goals in liberalizing energy markets the implementation of incentive schemes should be established in accordance with free market rules. The expected result is a strong development of the amount of energy generated from RETs such as wind, solar, geothermal, energy from waste and small scale hydro-electric plants.

Most of current incentive schemes, designed in the pre-liberalization era, are not able to fulfil free market principles. The strategy of liberalization and the creation of a single electricity market in the European Union have to balance opposite targets (COMM(2006) 105 final): on one side RETs should be established and operated at the lowest possible costs, on the other side mainly RETs can not compete with thermal based power without additional support. In other words, RETs support mechanism must link RES-E cost-effective with energy efficiency promotion.

This chapter focuses on incentive schemes designed to sustain RETs. In paragraph 2 we shortly describe the changing shape of the energy markets, which are experiencing a series of fundamental changes. In paragraph 3, we grasp the essentials of power revolution in order to understand the economics of a competitive electricity market. In paragraph 4, we shortly describe and discuss the main incentive schemes for RES-E as they currently exist in Europe and we conclude comparing main advantages and disadvantages.

2. CHANGING SHAPE OF ENERGY MARKETS

Three decades ago, the oil industry started to learn many of the lessons now being addressed by those electricity and gas businesses that are facing market liberalization processes. The oil market is no stranger to change. The biggest shake up came over the period 1978 to 1985 and was characterized by two separate circumstances. Following very sharp price rises over 1979, there were subsequent downward price movements continuously from 1980 until 1985. At the same time the major oil companies had changed from being generally crude long to being crude short as a consequence of nationalization reserves programmes. The need to buy crude oil at market prices from refining led to critical examination of refining and downstream profitability, whereas, the downstream market had previously been a tap to provide a route to market for upstream production. There was a move away from priced contracts for crude oil to arrangements based on clearing market prices and notably with production netbacks putting pressure on crude oil prices. As the full market liberalization takes effect in downstream trading for electricity and gas, some of the lessons learned by the oil sector are pertinent.

Acceptance of competition is changing the shape of traditional utilities. Although this is often coupled with privatization, the question of ownership is not central to the structural and behavioural change within the sector. The forces of change are liberalization, globalization and convergence.

Liberalization. Open competition, given effect by measures to promote liberalization and reinforced by customer trends and technological changes, has a profound influence on the shape of the energy market. The nature of competition in the electricity and gas markets can either be for the market or in the market. The former (competition for the market) concerns long term projects where there is a competition for the franchise or the project construction and ownership but backed by long-term off take contracts; in many cases this type of ex ante competition is requested in order to cope with the existence of natural monopoly conditions. The latter (competition in the market) is characterised by a clearing market responsive to short market forces. Competition in these markets requires that the segments of any integrated business be unbundled to ensure equal access condition to all market entrants. Competition for the market does not necessarily require such unbundling.

Globalization. Ownership patterns are becoming more and more international. The boundaries of utilities markets are swept aside. Companies who previously played a monopoly role on a regional or national stage are becoming competitors in a global arena. This trend to globalization is driven by three main factors. First, the ability to create value by operating globally. There is potential to capture and share know-how in fuel procurement, electricity contracting and financing on a world scale as well as to consolidate relationships with plant manufacturers. This means that key competitive advantage is to be gained from globalization. Second, more assets are available, due to privatization. Privatization was the spark igniting globalization. The wave of privatization continues to sweep through utility markets delivering opportunities for companies to build global positions through acquisition. Third, faced with limited growth prospects in their home market, most companies have little option but to move overseas if they want to grow earnings.

Convergence. Convergence is playing a critical role in shaping the utility markets of the future. Convergence is being felt in many different ways, shaping and transforming products, markets and trading systems. Old boundaries between products and services that shape current customer relationships are being swept away as companies seek to exploit the synergies between trade and supply of electricity and gas to the same customer.

3. ESSENTIALS OF LIBERALIZED ELECTRICITY MARKET

Before talking about RES-E promotion schemes, it should be useful to grasp the essentials of the competitive electricity market because it is the natural candidate to support renewable energies. There is fairly wide agreement about the goals that electricity sector reforms should achieve and even on the basic architecture of a model for creating competitive wholesale and retail markets to achieve these goals. It is less clear that there was broad understanding of what would have to be done to achieve these goals and how long it would take to achieve them (Joskow, 2003).

The reform goal is to create new governance arrangements for the electricity sector that will provide long-term benefits to consumers. These benefits will accrue by relying on competitive wholesale markets for power to provide better incentives for controlling capital and operating costs of new and existing generating capacity, to encourage innovation in power supply technologies, and to shift the risks of technology choice, construction costs and operating "mistakes" to suppliers and away from consumers. Retail competition would allow consumers to choose the supplier offering the price/service quality combination that best meet their needs, and competing retail suppliers would provide an enhanced array of retail service products, risk management, demand management, and new opportunities for service quality differentiation based on individual consumer preferences.

It was also widely recognized that significant portions of the total costs of electricity supply – distribution and transmission – would continue to be regulated because of natural monopoly conditions. First, regulatory mechanisms with good incentive properties would lead to lower distribution and transmission costs and this in turn would help to reduce retail electricity prices. Second, the efficiency of wholesale markets, in particular depends on a well functioning supporting transmission network and its efficient operation by a system operator. Good operating and investments incentives are important for providing an efficient network platform upon which wholesale and retail competition would proceed. This is the dream in the long run: the promise is that these reforms would lead to lower costs and lower average retail price levels reflecting those costs savings compared to the old regulated monopoly alternative, while maintaining or enhancing system reliability and achieving environmental improvement goals.

Whatever the political motivations, the basic reform for transitioning to competitive electricity markets has already been developed in theory and applied in practice in many countries. A liberalized and competitive electric system should be characterized by:

- vertical separation of competitive activities (e.g. generation, marketing and supply) from regulated activities (distribution, transmission, dispatching and system operations); functions must be separated (legal or administrative unbundling) within the same corporation; structural separation (property unbundling through divestiture) is the ideal rule;
- transmission system operator independent of the interests of the electricity industry to guarantee the grid open access to third parties – a need for transparent, predictable and non-discriminatory tariffs for access to essential transportation infrastructures (Regulated Third Party Access) – to manage the network operation, to schedule generation to meet demand and to maintain the physical parameters of the network (frequency, voltage, stability);
- power exchange bourse to facilitate economical trading opportunities among suppliers and between buyers and sellers, and operating reserve market to support requirements for real time balancing, to respond quickly and effectively to unplanned outages of transmission and generating facilities consistent with the need to maintain the above grid parameters within narrow limits;
- a regulatory authority independent of the interests of the electricity industry responsible for monitoring the market and mitigating market power, ensuring non-discriminatory access to network with power to fix or approve transport tariffs and to allocate the limited transmission capacity, setting final tariffs in order to separate power supplies and associated support services (competitive activities) from distribution and transmission services (regulated monopolies) in order to avoid discrimination, cross-subsidization and a distortion of competition.

Electricity cannot be stored economically and demand must be cleared at the same time electricity is produced, i.e. production from generating capacity available at the same time that electricity is consumed. Because demand varies widely from month to month, between day and night and between weekends and workdays, a significant amount of generating capacity operates for a relatively small number of hours during the year in order to meet peak demands. This means that generating plants that produce for a small fraction of the year can recover investments and other fixed operating costs during periods when demand and prices are at their highest levels. In other words, it means that it is necessary to maintain standby plants that can respond very quickly to both supply and demand variations.

Price determination in a competitive electricity market is shown in Fig. 1. Merit order line (MC) is drawn adding additional generating capacity starting from the lowest to the highest marginal costs. We can roughly associate the marginal cost of a thermal power plant with its energy cost. MC line is equal to the incremental cost of producing a little more or a little less energy from generating unit on the margin in the merit order. Under typical operating conditions the system marginal price (SMP) should be equal to the marginal cost of the last increment of generating capacity that just clears supply (MC line) and demand (D1 line) at each point of time; in Fig. 1 we can suppose an excess of generating capacity (K1 < Kmax). Area "A" between the SMP1 and MC line represents the total infra-marginal rent that all dispatched generating plants have, and are necessary to recover their capital costs.

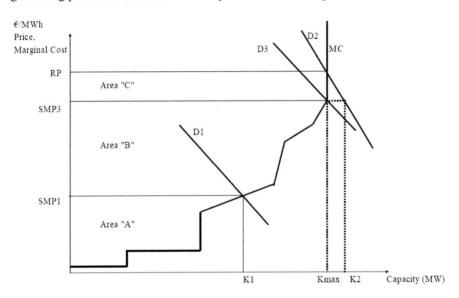

Figure. 1. Different states of power market and price determination.

We can also suppose a perfect matching between demand (D3) and the maximum installed capacity. In this case the clearing price line is SMP3 and area "B" represents an additional infra-marginal rent that all dispatched generating plants have. We can now consider a different state of the system associated with a relatively small number of hours when there would be excess of demand (K2 > Kmax) at a price that equals the marginal supply cost of the last increment of generating capacity, that can physically be made available to supply energy. In this case market must be cleared on the demand side: consumers have to offer prices up to a higher level reflecting their willingness to pay. In other words, the willingness to pay is equal to the value that consumers place on consuming less electricity as the demand is reduced to match the limited capacity made available by generators. Area

"C" between the new clearing price line (RP) and SMP3 line – at which the demanded quantity is equal to the maximum installed capacity – represents the total additional scarcity rent that all dispatched generating plants have in order to pay capital costs. This additional rent is especially important for those plants that run infrequently during the year. The amount of generating capacity that is available to the system depends on investors in generating capacity. Investors must balance the costs of additional capacity against net revenues they expect, including the "scarcity" rent.

RETs have to compete in this complex scenario. In other words, a RET plant can survive in a liberalized market if and only if clearing prices are sufficient to recover both marginal energy and capacity costs. Owning up to higher generation costs, RETs are not yet competitive within the electricity market, except for large hydropower plants. This leads to the need for some type of support in order to increase RES-E share of electricity produced.

4. SUBSIDIES VS. MARKET-BASED INCENTIVES TO PROMOTE RENEWABLE ENERGY SOURCES

There is fairly wide agreement among economists that subsidies in general have the potential to distort market. In order to reach cost-effectiveness and at the same time avoid weakening competition within the market, it is important to implement a RES-E support system that can fulfil minimum standards, such as reflecting the existing potentials of different RETs, enhancing competition between generators, encouraging RES-E suppliers to improve operation performance and technological efficiency, offering transparent information to consumers about the true costs of RES-E, introducing market based mechanisms.

The best policy measure that is in line with free market principles, from economists' point of view, is the internalization of external costs of non-RETs. The internalization of external costs determines a direct price increase of the polluting energy sources. This can be done by taxing emissions of greenhouse gases generated from those technologies or by taxing energy from which renewable sources are exempted. A tax exemption measure is specifically aimed at renewable sources, whereas taxing emissions also give an advantage to non-renewable options like energy conservation and pollution abatement.

Different types of incentive schemes are shortly described in this paragraph. Incentive scheme classification can be distinguished between generation-based (output) and investment focussed (installed capacity) schemes. This classification can be combined with other schemes, which can be designed to stimulate the supply or the demand side. Incentive schemes can also be distinguished between price and quota mechanisms. The first approach fixes the price for the sale of electricity to the grid leaving the quantity to be produced to the market demand. The second approach fixes the quantity to be produced leaving the price to be set by market forces.

4.1. Financial and Fiscal Subsidies

Financial incentives. RES-E plants are often capital-intensive technologies with relatively low running costs. Financial incentives can reduce total investment costs or interests to be paid to the investors. Subsidies can also be recognized for research and development of new and more efficient technologies. These incentives are the most feasible way to introduce non-competitive technologies into the market, but they suffer from many disadvantages because this scheme gives no incentive to operate RETs as efficiently as possible and it does not guarantee that incentivized RETs will actually be on operation.

Fiscal incentives. Numerous types of incentives can be grouped in this category, including exemption from energy taxes or corporate taxes, tax refund, reduced value added tax (VAT) rate.

Fiscal and financial incentives are schemes that not only can distort market competition but also can merely promote available capacity and not the energy actually produced.

4.2. Feed-in Tariffs

Feed-in tariffs are "administered" (i.e. non-market) guaranteed premium prices in combination with a purchase obligation by transmission/distribution or utility companies. Feed-in tariffs are designed to stimulate RES-E generators to produce electricity. Because generation costs are different across RETs, the feed-in tariffs are usually different per RET; in some cases feed-in tariffs are designed for different periods of time in order to take into account the minimum time necessary to recover investment costs. The feed-in tariff can be the only revenue for the generator or it can be a supplement to the power price, or to other subsidies.

Feed-in tariffs usually set the "price" but not the quantity to be produced. In other words, RET generators provide the grid/utility with the entire uncertain electricity production (kWh) obtained by a given installed capacity (kW). In some cases, it is also possible to provide a contracted production (kW times contracted yearly hours) at an agreed full feed-in tariff and to provide a possible production surplus exceeding the contracted production at a different feed-in tariff.

The feed-in tariff should be designed to reflect some economic targets, such as the long-run marginal production costs, including a fair rate of return (a "reasonable" profit) in order to stimulate investments in RETs. Security in revenue is the main reason of popularity for feed-in tariffs across investors because this incentive guarantees bankable projects.

The US Federal Public Utilities Regulatory Act of 1978 (well known as PURPA Act) established tariffs under which utilities had to purchase power from certain kinds of private producers known as "qualified facilities". Such tariffs were designed so that consumers were not affected by these transactions. In other words, utilities should buy power at a price equal to costs they would "avoid" by making a purchase at privately generated power (Kahn, 1988). Avoided costs could be identified with the marginal costs at expanding capacity. A feed-in tariff designed as avoided cost was also established in Italy in 1992 by the so-called Provision CIP 6/92.

RETs can take advantage from technological progress. Therefore, total installed capacity in a certain year is given by a "vintage" of RETs characterized by different production efficiencies, as well as investment and operating costs. The feed-in tariff should be set taking account of those differences in marginal and average costs according to RETs vintages. In other words, the feed-in tariff should be effective and efficient, but in practice it is very rare due to imperfect information of actual costs.

When feed-in tariffs are set too low with respect to the long-run marginal costs, the incentive is cost-efficient (i.e. consumers pay lower costs) but not effective (RES-E reduced promotion). On the opposite, when feed-in tariffs are set too high, the incentive is effective (RES-E accelerated promotion) but not cost-efficient (consumers are charged of extra profits).

Astonishing development of wind power in Germany and Spain can be also explained by generous and low-risk feed-in tariff mechanisms. There is a great risk of underestimating or more often overestimating the feed-in tariff level. For this main reason policy makers are moving to a more efficient, marked oriented scheme such as a tradable "green certificate" scheme.

4.3. Tendering Procedures

According to this scheme, a public institution invites generators to compete through tender (competition "for" the market) for a specific financial budget or capacity. Because of different RETs, there are usually separate tenders for different technologies and technological bands. Contracts and the corresponding feed-in tariffs are only awarded to the cheapest bids. In theory this scheme should stimulate competition within RETs generators and hence cost-efficiency and price reduction.

4.4. Green Pricing

The simple idea of this incentive scheme is possible when there is a willingness to pay a surplus on the electricity bill in order to promote RES-E. This surplus is *voluntarily* paid by consumers, and generators cash it in order to cover additional generation costs of "green electricity" (i.e. RES-E) with respect to thermal based power in the market. Such willingness to pay seems to be influenced by personal factors related to consumers' environmental awareness and to their certainty that electricity is really generated from RETs.

4.5. Tradable Green Certificates

A "green certificate" is a public certification that a certain quantity of electricity is actually produced from RETs. The certification provides an accounting system to register production; it authenticates the source of electricity (i.e. guarantee of origin of RES-E) and it verifies whether demand has been met. When these certificates are tradable they are denoted "tradable" green certificates (TGCs).

With respect to the supply side, RETs generators provide electricity to the grid: they are paid for the electricity sold to the power market, also called the "grey" electricity market, and receive by the authority a correspondent number of TGCs. These TGCs are financial assets. They can be sold in specialized markets established for green certificates and thereby RETs generators can realize additional revenue necessary to cover RES-E additional costs with respect to the electricity market price. It is important to point out the fact that electricity market prices in a competitive arena are structurally determined and driven by low cost fuels and technologies. In other words, RETs can compete with thermal based power thanks to this additional support, the "green certificate" revenue.

Demand for TGCs is crucial driver for stimulating supply of RES-E. Demand can be created by imposing a mandatory quota on certain actors of the electricity market, such as producers, consumers or distribution companies. Every actor is obliged to acquire every year a certain number of TGCs established by law. For example, in the Italian case, every generator must provide the grid a certain percentage of RES-E with respect to his total electricity production from fossil fuels. Therefore, a generator has two options: to install and directly run RETs or to buy TGCs in order to fulfil his obligation.

The green quota and hence the number of TGCs must be tuned and yearly changed by the regulator following long-term goals of a designed energy policy. A demand-related mandatory quota creates the need for a financial market for green certificates; the quota must be supported by a penalty in order to ensure that demand obligation is actually fulfilled. Since the TGCs market is independent of physical constraint, i.e. transmission capacity, TGCs can also be traded between EU countries thanks to the guarantee of origin certification.

It is important to understand that the price of the green certificates, in contrast to what happens in the feed-in tariffs case, is not fixed but always determined by the market forces (Morthorst et al., 2003). In a competitive scenario there is a trade-off between the "green" electricity market and the "grey" electricity market; and also between the price of the green certificates and technological improvement.

Fig. 2 shows how the market for certificates works. The clearing price level (SMP) is determined in the power "grey" market according to the rules previously explained in Fig. 1; in this example SMP level is depicted not sufficient to recover RET costs.

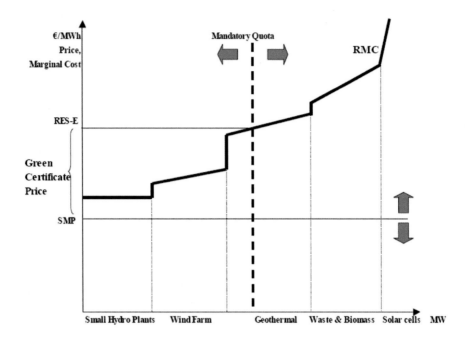

Figure 2. Price determination on the tradable green certificate market.

The requested financial incentive is determined on the TGCs markets in which a stylized merit order line (RMC) for RETs is usually drawn starting from the cheapest technology, like small hydro-power plants, to the most expensive one, like solar cells. RES-E price is set by intersection between mandatory demand and RMC line. Distances between RMC and SMP lines measure the unity incentive (i.e. TGCs price) that each specific RET should receive in order to recover costs exceeding the "grey" price determined in the power market. Fundamentals of TGCs price determination are, given RES-E technologies: the mandatory green quota and the SMP level. In Fig. 2, RETs on the right-hand side of the mandatory quota is depicted as a scenario for which the sum of "green" certificate price and the "grey" price level is not sufficient to assure economic sustainability for the most expensive drawn.

4.6. Tradable Pollution Permits

We anticipated that a solution to the problem of negative externalities generated by polluting firms, suggested by Pigou, consists of taxing the difference between the marginal private cost of production and the marginal social cost, which includes negative externality costs. In this way, the environment damage internalized in the market price reduces the demand and, consequently, the supply of polluting goods, in our case the electricity generated from fossil fuels. To determine a Pigouvian tax that leads to the optimal resource allocation, it is necessary to equalize the Pigouvian tax to the marginal value of the external damage.

Following Coase's criticism to Pigou, we know that wealth distribution depends on property rights and a Pareto-efficient resource allocation depends on wealth distribution. Whether we assume that the polluter must pay the polluted party in order to allow for increased production or whether the polluted party must pay the polluter so that the latter reduces pollution, Coase's theorem states that the optimum levels of output and pollution remain unchanged in relation to the distribution of property rights on the receptor body affected by pollution (for instance air, land and water). In other words, the same amounts of output and pollution would be produced regardless of which party owned property rights at the beginning of bargaining. Coase's theorem remains valid as long as the transaction costs associated with the bargaining between the parties are practical negligible.

A significant and practical application of Coase's theorem is the implementation of a tradable pollution permits (TPPs) scheme (Pireddu, 1996). The authority could establish the tolerated quantity of pollution or abatement to be achieved, and letting market forces determine the appropriate price level. This pricing incentive can be problematic when government will not know in advance what price level will achieve a quantity-based environmental target. In theory it may be more efficient to use a market mechanism that operates from a socially desirable quantity of pollution or abatement, and leave the market to establish the price. Such is the underlying premise of TPPs, which are characterized by two essential components: on one side the issuance of some fixed number of pollution permits; on the other side a provision for trading of these permits among polluting sources. A pollution level, which is recognized as acceptable, must bind the number of permits issued and polluters are allowed to trade them. Tradability gives rise to a distinct market for pollution rights. Therefore, polluters, following their own self-interests, either purchase these rights to pollute or abate according to the cheapest alternative. High-cost abators will have an incentive to bid for available permits, while low-cost abators will have the opposite incentive to sell their permits surplus as a consequence of the abatement strategy. The final result, from a social point of view, is a cost-effective allocation of abatement responsibility.

An important development for the implementation of the targets set under the Kyoto Protocol is the EU emissions trading scheme (EU-ETS) for CO_2 (Directive 2003/87/EC), started in January 2005 in order to gather emissions-trading experience before the commitment period of the Kyoto Protocol in 2008 with regard to its flexible mechanisms (CDM and JI). This is the largest and most ambitious system of TPPs to be introduced for environmental protection. The EU-ETS establishes a cap-and-trade system to limit CO_2 from power sector and other energy intensive industries. The main idea of the EU-ETS is to allocate CO_2 tradable allowances on industrial plants in the period 2005-2007 according to national allocation plans. The trading element allows plant owners that have reduced their emissions by more than allocated quotas, to sell their surplus allowances to other plant owners who are not able to reach their binding quotas. The overall environmental outcome is the same as if each plant uses its allowances exactly, but with the important difference that buyer and seller benefit from the flexibility offered by trading in terms of opportunity cost (differences in marginal abatement costs across industry sectors and plants, and across European countries). Permits associated with CDM and JI represent quantities of greenhouse gases and may be traded; trading allows industrialized countries to meet their commitments under the Kyoto Protocol when unable to do so by domestic actions. With the implementation of CDM and JI mechanisms in 2008 a growing international market for CO_2 emissions trading will emerge.

Emissions trading instruments show many advantages with respect to the traditional command-and-control instruments because they enable least-cost compliance with emissions limits. Companies may choose to reduce their actual emissions or buy in permits from elsewhere to cover excess emissions, whichever is less expensive. Trading should encourage innovation and investment in cleaner technology, as companies are stimulated to find cheaper ways to reduce emissions (Pireddu, 2002).

TPPs price determination is shown in Fig. 3. In Fig. 3.1 it is assumed that there are two polluters, A and B, which are characterized by different marginal abatement costs (respectively MACa and MACb): polluter A is a high-cost abator while polluter B is a low-cost abator. Supposing a regulator establishes an abatement standard level (X), a cost-effective allocation of abatement responsibilities determines different abatement units between the two polluters (respectively, 0A and 0B). This result is shown by equating marginal abatement costs, i.e. MACa = MACb.

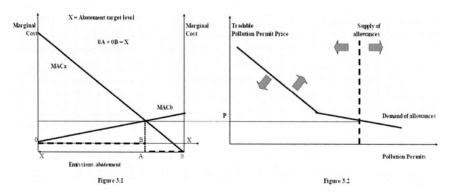

Figure 3. Price determination on the tradable pollution permits market.

Marginal abatement cost curve represents an opportunity cost with respect to the incentive to abate or to buy/sell TPPs; therefore, the marginal abatement cost curve is specular to the demand (D) for TPPs shown in Fig. 3.2. The supply curve for permits (S) is given and tuned by the regulator. TPPs price level is given by the traditional market rule, i.e. a price for which TPPs demand and supply are cleared. If the Government reduces total pollution permits over time, polluters are therefore forced to invest in pollution abatement technologies or rely on buying from a decreasing supply of pollution permits, which may escalate in price as a result of their increasing scarcity.

The strategy for a green certificate market is to ensure a planned development of RETs under liberalized market conditions, while the strategy for a tradable pollution permits is to ensure a lower level of emissions in an economically efficient way. Generators that reduce emissions under the allowed level can sell excess emission permits (i.e. allowances). In both cases, a desired share of RES-E can be obtained by setting the appropriate number of green certificates; a desired level of emissions can be obtained by issuing an appropriate number of total emission quotas. The interdependence between TGCs, TPPs and power market is represented in Fig. 4.

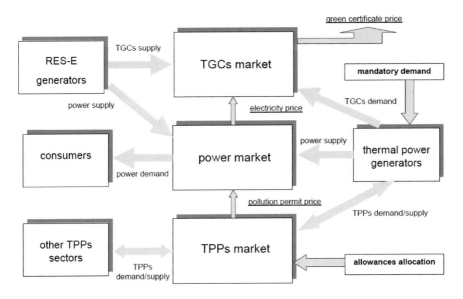

Figure 4. Power market, green certificate market and pollution permits market interdependences.

Consumers purchase power on the power market, and "green" and thermal generators deliver power to the power market: demand and supply determines power price. Green generators provide the green certificate market with certificates equal to the amount of power they produce and thermal generators are obliged to demand a certain quota of their thermal production in terms of green certificates: demand and supply determines green certificate price. Thermal generators are also obliged to have a number of allowances corresponding to the amount of emissions caused by their thermal production: if they hold a number of allowances greater than their obligation, they can sell the surplus on the pollution permits market otherwise they must cover the deficit. At the same time, other industries obliged to have emissions permits can be either seller or buyer of allowances on the same pollution permit market. This leads to several interdependences: power price is also influenced by allowance prices and by RES-E production; green certificate prices are also influenced by power prices and green mandatory quota; allowance prices are also influenced by RES-E production and total emissions quota.

5. CONCLUSIONS

RES-E is expected to be economically competitive with conventional energy sources in the medium to long term. In this transitional phase, higher production costs lead to the need for subsidies or market based incentive schemes. In this paper we analyzed different support schemes; advantages and drawbacks of subsidies, and market-oriented incentive schemes are summarized in Tab. 1.

Table 1. Comparing incentive mechanisms

	Advantages	Drawbacks
Financial and Fiscal Subsidies	✓ Easy to implement ✓ Usually put in place to stimulate demand as a second tier incentive	✓ No incentive to operate RETs as efficiently as possible ✓ No RES-E output guarantee
Feed-in Tariffs	✓ Very effective regarding RES-E development ✓ Few regulatory and administrative costs ✓ Stable basic condition and high planning reliability ✓ Bankable project financing	✓ Non cost-efficient ✓ Non market-oriented approach
Tendering Procedures	✓ Competition between RES-E ✓ Awarded bids work as a feed-in tariffs scheme	✓ Complexity of tendering procedures
Green Pricing	✓ Consumers voluntarily promote RES-E	✓ Success depends on the willingness-to-pay on the part of consumers
Tradable Green Certificates	✓ Flexible and market-oriented ✓ Economic efficient	✓ Volatile certificate prices ✓ Higher insecurity for investors ✓ Weak bankable project financing
Tradable Pollution Permits	✓ Effective regarding the «polluter pays principle» ✓ Marked-oriented approach ✓ Innovation and investment in cleaner technology promotion	✓ Indirect form of RES-E promotion ✓ Possible windfall profits and market power abuse

The development of RES-E is a central aim of the European Commission's energy policy. Renewable energy has an important role to play in reducing carbon dioxide (CO_2) emissions, in enhancing sustainability by increasing the share of renewable sources in the energy balance, and also in improving the security of energy supply by reducing dependence on imported energy sources. But any support scheme should be integrated into a liberalized electricity market, reflecting different RES potentials, enhancing competition between generators, encouraging operation performance and technological efficiency and offering objective information to end users. Our analysis shows that tradable green certificates and tradable pollution permits are the incentive schemes that, respectively directly and indirectly, fulfil liberalized and competitive electricity market rules.

Giancarlo Pireddu
Department of Economics,
University of Milano Bicocca, Milan, Italy

REFERENCES

CIP 6/1992, Comitato Interministeriale dei Prezzi, "Prezzi dell'energia elettrica relativi a cessione, vettoriamento e produzione per conto dell'Enel, parametri relativi allo scambio e condizioni tecniche generali per l'ammissibilità a fonte rinnovabile", *Provvedimento No. 6/1992.*

Commission of the European Communities, *Green Paper. A European strategy for sustainable, competitive and secure energy*, COM (2006) 105 final.

Directive 2001/77/EC on the promotion of electricity produced from renewable energy sources in the internal electricity market.

Directive 2003/87/EC establishing a scheme for greenhouse gas emission allowance trading within the Community and amending Council Directive 96/61/EC.

Joskow P.L., 2003. The difficult transition to competitive electricity markets in the US, *Center for Energy and Environmental Policy Research*, MIT, 03-008 WP.

Kahn E., 1988. *Electric Utility: Planning and Regulation*, Washington DC: American Council for an Energy-Efficient Economy.

Morthorst P.E., Skytte K. and Fristrup P., 2003. Special Issue. Green certificates and emission trading, *Energy Policy*, Vol. 31, No. 1.

Pireddu G., 1996. Pollution, externalities and optimal provision of public goods: A computable general equilibrium approach, in Fossati A. (ed.), *Economic modelling under the applied general equilibrium approach*, Aldershot, Avebury.

Pireddu G., 2002. *Economia dell'ambiente. Un'introduzione in equilibrio generale*, Milano, Apogeo.

B. K. BUCHNER

CDM – A POLICY TO FOSTER SUSTAINABLE DEVELOPMENT?

Abstract. The close links between climate change and sustainable development suggest that decision making related to climate change is a crucial aspect of making decisions about the future's sustainability. Objective of this chapter is to analyze how sustainable development objectives can be promoted by appropriately designing policies to reduce the risk of climate change. One potential climate strategy will be investigated, the Clean Development Mechanism (CDM), introduced to assist industrialized countries in meeting their emission targets in a flexible, cost-effective way while enabling developing countries to achieve sustainable pathways. The potential of this policy in inducing countries to take action on climate change while improving also social and economic aspects of the different countries will be analyzed. By investigating the implementation of the CDM in the real world, we will verify whether this policy instrument actually holds what it promises in terms of setting incentives for a more sustainable path.

1. INTRODUCTION

The concept of sustainable development was from its very beginning meant to be relevant for a comprehensive philosophy including – apart from environmental aspects – a variety of issues. In fact, the pioneering work of the World Council on Environment and Development (WCED, 1987), refers to sustainable development as "development that meets the needs of the present without compromising the ability of future generations to meet their own needs." In this manner, the Brundtland Commission placed sustainability on international political and scientific agendas.

This analysis will not contribute to the already vast literature on the strengths and weaknesses of sustainability definitions, but will look at the necessity of inducing sustainable development from another point of view. In particular, by focusing on the economic dimension of sustainability, we want to explore which changes in the policy settings are sustainable in the sense of being able to support economic welfare in the long-run without creating burdens on social, economic and environmental resources. Due to the broad context of sustainability and its numerous questions, this study will investigate one dimension needed to resolve the difficulties, focusing thereby on only one explicit area. In particular, we chose the part related to incentives arising from policy instruments, while the search for incentive strategies will be carried out for the area of climate change.

The choice of climate change as field of application is a consequence of the increased attention that has been paid to this environmental problem. Climate change has evolved as one of the most serious threats to the sustainability of the world socio-economic system, representing one of the most important symptoms of "unsustainability" (IPCC, 2001b). As emphasized by the IPCC Third Assessment Report, climate change could jeopardize economic activities, social welfare and equity in an unprecedented manner. Due to its public good characteristic, its global diffusion and its close linkages to the use of energy, an effective climate change control symbolizes a crucial component of a sustainable future. In addition, the need to tackle the problem of climate change on a global scale, comprising as many countries as possible without having a supra-national authority available, suggests that this area could constitute an example for other fields of how to approach sustainability. Despite the close interactions between climate change and sustainability, they have traditionally been pursued as largely separate discourses, denying thus that the impacts of climate change may substantially affect efforts to enable the transition to sustainable development.

Objective of this study, therefore, is to analyze how sustainable development objectives can be promoted by appropriately designing policies to reduce the risk of climate change. The close links between climate change and sustainable development suggest that decision making related to climate change is a crucial aspect of making decisions about the future's sustainability. Therefore, after having given a short overview on the development of climate policy in section 2, this analysis will investigate one potential strategy aimed at improving climate–change control. We will focus on the Clean Development Mechanism (CDM), a policy instrument that has been introduced as part of the flexible mechanisms by the Kyoto Protocol in order to assist industrialized countries in meeting their emission targets in a flexible, cost-effective way while enabling developing countries to achieve sustainable pathways. Section 3 will introduce the CDM in detail, discussing its concept and advantages as well as disadvantages. The potential of this policy in inducing countries to take action on climate change while improving also social and economic aspects of the different countries will be analyzed. By investigating the implementation of the CDM in the real world, the central issue of section 4 is to verify whether this policy instrument actually holds what it promises in terms of setting incentives for a more sustainable path. Section 5 will conclude on the potential of the CDM.

Throughout this study, climate policy is taken as a proxy and step towards sustainable development. Since sustainability and climate change can be viewed as twin goals, this study will contribute to the overall goal of inducing a transition towards sustainable development by highlighting the role of the CDM as a policy instrument aimed at fostering sustainable development.

2. AN OVERVIEW ON DEVELOPMENTS IN CLIMATE POLICY

Over the last decades, climate change has evolved as one of the most serious challenges to the world's sustainability. The interplay of the global and potentially very serious implications of climate change and the uncertainties still prevailing with respect to its emergence and its consequences illustrate the problem of complexity which is inherent in climate change. Aggravated by the necessity to tackle climate change on a global scale in order to implement an efficient approach, requiring thus international cooperation among very different countries, also the aspects of efficiency and intra-national, international and intergenerational equity reinforce the challenge of climate change.

The international political response to climate change was introduced within the United Nations Framework Convention on Climate Change (UNFCCC), which was adopted in 1992 at the "Earth Summit" in Rio de Janeiro, Brazil. The convention has been signed by 155 countries who in this way agreed to prevent "dangerous" warming from greenhouse gases[1]. The UNFCCC sets the initial target of reducing emissions from industrialized countries to 1990 levels by the year 2000, although in a non-binding fashion, and recognizes for the first time officially the need for industrialized countries to somehow control their emissions of greenhouse gases.

In 1997, at the Third Conference of the Parties to the UNFCCC, the Kyoto Protocol has been agreed to, setting for the first time binding emissions reduction targets for industrialized countries. By 2012 the world wide GHG emissions should decline by an average 5.2% below their 1990 levels. Article 3 of the UNFCCC requires the Parties to engage in the protection of the climate system with "common but differentiated responsibilities", and consequently the burden of reducing emissions has been distributed in a differentiated way[2]. For instance, the United States have accepted an emission reduction target of -7%, the European Union of -8%, and Japan of -6%. Economies in transition, including Russia and several states from the former Soviet Union, have been conceded to catch up with their economic development after some serious recessions in the early 1990s. Consequently, the choice of the base year 1990 allows these countries to dispose of notably large amounts of emission credits due to their currently lower emission levels compared to 1990. Developing countries, including the large economies of India and China, were exempted from binding reduction targets in the first commitment period since their per capita emissions are much lower than the industrialized countries' emissions.

To achieve these targets, the Kyoto Protocol foresees two broad strategies: mitigation and adaptation, with a clear focus on the former. A variety of policies and measures is indicated that should be employed in order to achieve the targets, amongst others the promotion of energy efficiency, technological innovation, renewable energy and forest protection. The Kyoto Protocol foresees that the national measures are to be integrated with international measures aimed at achieving the targets in a flexible, cost-effective way. In particular, the protocol establishes a series of flexibility measures that enable countries to meet their targets by cooperating on emission reductions across country borders and by establishing carbon sinks such as certain forestry and land-use activities to soak up emissions.

The Kyoto Protocol represents remarkable progress in the history of international environmental agreements and its implementation could lay the basis for effective global action. However, the Kyoto Protocol determined only the targets, methods and timetables for global action, while the definition of specific rules and operational details was postponed to later meetings. The negotiations on the rules which were needed to prepare the Kyoto Protocol's entry into force turned out to be very difficult, amongst others due to the US withdrawal from the protocol. After long and complicated negotiations it was Russia's ratification of the Kyoto Protocol on November 4, 2004 that opened the way for the Protocol's coming into force on February 16, 2005. The emissions targets taken on for the 2008-2012 period by more than 30 developed countries (including the EU, Russia, Japan, Canada, New Zealand, Norway and Switzerland) have thus become legally binding.

The Kyoto Protocol's late coming into force has created a challenging situation: only few countries have already started to implement emission reduction measures, and therefore emissions in most countries have increased considerably, rendering the achievement of the Kyoto targets by 2012 demanding. The above-mentioned features of climate change – the public good characteristic and the structural differences among countries – induce strong difficulties in achieving a broad participation in international climate change control[3], which however is required to tackle global environmental problems in the most effective way. As a consequence, the Kyoto Protocol's flexible mechanisms are likely to play a key role in helping countries to achieve their reduction commitments. The main reason for their prominence is the fact that they take two interrelated objectives into account that are considered as crucial in inducing countries to take action (IPCC TAR, 2001b):

- a cost-effective reduction of emissions;
- ways of providing the incentives for countries to sign the agreement.

The design of the CDM has particularly taken these considerations into account by emphasizing the broader goal of sustainable development.

3. A FOCUS ON THE CLEAN DEVELOPMENT MECHANISM

The CDM has attracted a lot of attention as it is the only flexible mechanism introduced by the Kyoto Protocol that includes the participation of developing countries. The core objective of the CDM was to bring developing countries to the climate change debate, creating the opportunity for them to promote cost-effective GHG mitigation along with sustainable development, while industrialized countries were given the opportunity to comply with their reduction targets at the lowest cost. This section provides a general description of the details and key elements that characterize the CDM, giving particular attention to the incentives that this policy instrument provides both to developed and developing countries, and how the component of sustainable development is incorporated.

3.1. A Description of the CDM

The acknowledged basis for the CDM is a Brazilian proposal for a "Clean Development Fund", suggesting a financial penalty system that would impose a fine to industrialized countries if they failed to reach the targets. The revenues from the fine would be recycled back to non-Annex I Parties to support GHG emissions mitigation projects and adaptation measures in the developing countries most affected by climate change. Industrialized countries opposed to this proposal but combined it with the experience gained through the AIJ pilot phase[4]. The result was the establishment of the CDM in the Kyoto negotiations, which would devote a 2% levy to adaptation, but otherwise function as a market-based measure.

The CDM is defined in Article 12 of the Kyoto Protocol, granting Annex I parties the right to generate or purchase emissions reduction credits from projects undertaken within non-Annex I countries. In exchange, developing country parties will have access to resources and technology to assist in the development of their economies in a sustainable manner[5]. Parties not included in Annex I will thus benefit from project activities resulting in emission reductions, while parties included in Annex I may use the emission reductions accruing from such project activities to contribute to compliance. The rationale behind the CDM allows Annex I countries to buy credits (the so-called Certified Emission Reductions, CERs) from Non-Annex I countries, which is cheaper for the first group and profitable for the second group as well as positive for the environment on a global scale.

The institutional structure of the CDM is dominated by the "Executive Board (EB)" that is responsible for supervision of all steps related to CDM. The rules governing the CDM were finalized in 2003 and are available in the "Modalities and procedures for a clean development mechanism (CDM M&P)"[6] in the Marrakech Accords[7], the decisions of the CDM EB and subsequent decisions of the Conference of the Parties. According to these rules, projects must meet certain requirements in order to qualify as CDM. These requirements include (Cf. IETA, 2005):

- compliance with the normal project approval process and sustainability development criteria;
- the project validation and registration process (incl. additionality requirements);
- monitoring requirements;
- verification and certification requirements; and
- rules governing the issuance of CERs.

The Kyoto Protocol envisages a prompt start to the CDM, allowing CERs to accrue from projects from the year 2000 onwards. This has created a lot of interest, in particular after the entry-into-force of the Kyoto Protocol. However, as indicated by the list of requirements, the process that leads to the implementation of a CDM project and the issuances of the related CERs is very complex and has consequently created barriers for the quick start of this mechanism. A particular requirement is that emission reductions resulting from each project activity are to be externally verified and certified on the basis of the following criteria:

- voluntary participation approved by each Party involved;
- real, measurable, and long-term benefits related to the mitigation of climate change;
- reductions in emissions that are additional to any that would occur in the absence of the certified project activity ("additionality" requirement).

The two main issues that have been held responsible for the initially slow process of CDM are related to the setting of baselines and to the issue of additionality. The first and crucial step involved in estimating the emission credits generated by a CDM project concerns the setting of the baseline. The baseline represents a so-called "business as usual" reference scenario of what would have happened in the absence of the proposed CDM project, depicting thus a counter-factual situation. Requirements for the baseline are simplicity, conservatism and the possibility to be adjustable to changing conditions. The difference between the emissions prevailing in the baseline scenario and in the scenario of the implemented CDM project activity enables the calculation of the emission reductions (i.e., the CERs claimed). Closely related to the construction of a baseline is the second key issue of CDM projects, the question of whether a project is additional. According to Article 12, paragraph 5 (3), of the Kyoto Protocol, "a CDM project activity is additional if anthropogenic emissions of greenhouse gases by sources are reduced below those that would have occurred in the absence of the registered CDM project activity." Therefore, again a baseline scenario and a project activity scenario need to be compared in order to determine whether additional emission reductions are induced by project activities.

The various requirements that need to be met in order to implement a CDM project as well as the complicated institutional process that includes the discussion and consequent approval or disapproval of each methodology has impeded a fast success of the CDM. Given the difficulties, is it useful to support this instrument?

3.2. Implications of the CDM

Let us now provide an overview on the CDM's most important positive and negative effects in order to understand better which types of incentives it induces.

The advantages of CDM:

- *Incentive for new investments in developing countries*
  CDM project activity accelerates investment in developing countries, induces a number of positive impacts on the economic, social and environmental development of the developing country as well as positive economic impulses for the investing country.
- *Incentive for innovative technologies*
  Given its objective and consequent requirements, CDM is able to set positive incentives for investments in innovative, environmental-friendly technologies that would otherwise appear too expensive to attract attention. As a consequence, industrialized countries have the opportunity to capitalize in technology upgrades which the CDM can deliver.
- *Incentive for sustainability considerations*
  The need to account for sustainable development goals in the design of CDM projects gives more room for sustainability considerations and requires thoughtful analysis of how to come closer to a sustainable path. In this way, the more sustainable and economic use of resources is encouraged through CDM.

The problems of CDM:

- *Complexity*
  As indicated above, a CDM project activity needs to meet a number of requirements and to go through a long process to attain approval. As a consequence, CDM triggers high transaction costs and requires intense negotiations efforts, for instance to get approval of a certain project methodology or to prove the additionality of a project activity.

- *Risks*
  CDM is still in an early stage of implementation, and therefore is struggling with a number of risks, including the danger that the political framework of climate policy changes or becomes (even) less severe. In addition, also risks related to the success of the performance of the proposed technologies, to the danger of getting the project implemented in a slower way than it was expected, entailing thereby higher costs as well as licensing/regulatory risks (e.g., appraisal, certification, monitoring, verification, registration of the CDM project) and in general financial risks related to the success of the project activity (e.g., currency changes, or price changes) are threatening the attraction of CDM.
- *Market uncertainties (price differences)*
  A final difficulty related to CDM project activity is the uncertainty that the carbon market faces currently. Given the lack of a long-term political framework, as well as the multiple carbon markets that have emerged throughout the world, the price level of credits obtained through CDM projects is uncertain. In addition, the differences between the allowance-based carbon markets, as for instance the EU Emissions Trading Scheme, and the project-based markets, as CDM or JI, create difficulties as certain project types do not appear to be attractive any more (see also next sections).

3.3. Incentives for Sustainable Development

As described above, the special feature of CDM is its twin objective: to assist Annex I countries in achieving their emission reduction targets in a cost efficient manner, while also supporting sustainable development initiatives within developing countries. The novelty of CDM is that an international market mechanism grants for the first time equal importance to GHG mitigation and development concerns by setting them both as primary objectives of the instrument (Figueres, 2006). The main advantage of CDM for developing countries consists thus of attracting foreign investments in order to enhance sustainable development and reduce emissions.

This design has been the consequence of an increasing support for the argument that GHG mitigation and sustainable development goals can be simultaneously pursued. Yet, experiences from recent CDM project activity observe a certain trade-off between the two objectives, questioning therefore whether the current design of the CDM is actually able to fulfill both objectives[8]. The main issue is related to the fact that profits are measured in terms of reduced emissions, implying that CDM projects with high reduction potential will commonly be the first option. However, these types of project may not be the most preferable from a sustainable development perspective.

In general, there is a concern that both developed and developing countries focus more on the financial abatement component of CDM, having lost sight of the sustainable development component of the CDM. Currently, the host country has the power to decide whether a CDM project activity assists it in achieving sustainable development. Developing countries can therefore define the sustainability requirements for CDM projects developed within their borders according to their own perceptions and they have the right to accept or reject CDM projects. This has led to a situation in which the reduction of GHG emissions is often equalized with sustainable development, making the objective of foreign investments the principal goal of CDM project activity. There is a concern that developing countries compete against each other in order to attract as many CDM projects as possible, triggering a so-called "race to the bottom" in sustainable development standards. As pointed out by Chiavari (2006), this fear is caused by the fact that most individual host countries do not have a strong position in the carbon market due to their underlying social, economic and political circumstances, and could try to improve their investment setting through lower environmental requirements.

An additional difficulty is posed by the vague concept of sustainable development. Given to its complexities, and differences according to different national and local circumstances, a generally accepted definition of sustainable development is missing. Yet, the sustainability considerations of CDM need to achieve more attention again in order to live up to the initial expectations of this instrument. A more pragmatic approach can help, with a stronger emphasis on immediate development objectives such as poverty reduction, local environmental health benefits, employment generation and economic growth prospects (UNEP, 2003). Currently, certain countries have already started to implement this strategy by giving more weight to sustainability assessments of CDM projects through the establishment of guidelines, checklists, criteria and specific standards that identify sustainable development criteria and project priority areas for CDM project. Again, the responsibility of host countries to find a more successful approach to ensure sustainability considerations represents a significant burden for them.

The next section will depict the current trends of CDM project activities, thereby exploring whether the current concerns materialize in reality or whether the sustainable development goal has more weight than expected.

4. CDM IN THE REAL WORLD

After this qualitative analysis, this section provides an overview on the current situation of CDM applications in the real world, trying to identify the most important areas in which CDM has recently succeeded. Through this background information on the current situation of the CDM market, a further indication on the contribution to sustainable development induced by CDM projects is given.

4.1. An Overview of CDM Projects

During the last years, a steady growth of the market for project-based emission reductions took place as a consequence of a higher regulatory certainty. Both the Kyoto Protocol and the EU Emissions Trading Scheme have entered into force, the latter with a possibility to use a certain amount of CDM and JI credits in its market.

From 2003 to 2004, the market volume has increased by 38%, and in 2005, 374 million tCO_2e were transacted at a value of US \$2.7 billion with an average price climbing over US $7.23. Compared to 2004, these numbers reflect an almost fourfold increase in volumes from project-based transactions and over five times growth of the financial value traded. Given the strong price signal from the European carbon market, also the CER prices have increased, and in the first three months of 2006, prices for project-based emission reductions rose significantly with an average reported price of US $11.45 per tCO_2e. Overall, the last years have seen an increase both of volumes exchanged and of number of projects under development (with a notably large supply of unilateral CDM projects). The spiralling trend for CDM activities continues, as is evidenced by the fact that in the first three months alone of 2006, 79 million tons CO_2e have been transacted on the market, corresponding to a value of nearly US $0.9 billion (Capoor and Ambrosi, 2006).

During the last year, also the participation of developing countries became more evident. Developing countries began to participate meaningfully in the carbon market and enabled real emission reductions. The market share of CDM credits from developing countries was about 49.2% of overall volumes transacted globally, reflecting however only about 23.2% of the overall value of contracts signed in 2005. In the first three months of 2006, the CDM market share of the overall carbon market volume was about 27.2%.

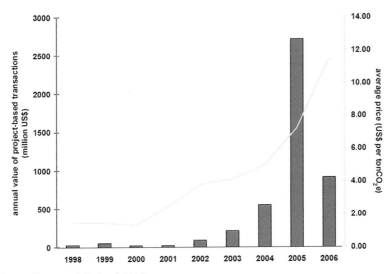

Source: Capoor and Ambrosi, 2006

Figure 1. Annual volumes (million tCO_2e) of project-based emission reductions transactions and annual average price in US$ per tCO_2.

CDM has thus tremendously increased in importance over the last years. Fig. 1 depicts the strong boost in the project-based market, showing both the increases in transaction volumes and prices, whereby CDM holds the main part of the market.

Most market transactions were in the hands of European and Japanese private entities on the buy side and Asian countries on the sell side. In particular, buyers based in Europe and in Japan together accounted for more than 90% of the market in 2005, both having increased their shares with respect to the previous year. On the sell side, Asia has become the main player, having 73% of contracted volume of project-based transactions signed between January 2005 and March 2006 (Capoor and Ambrosi, 2006). China dominated the market, accounting for 66% of the global volume in 2005, while the dominant player of 2004 – India (48% in 2004) – declined its share to 3%. Looking at the number of transactions per country for 2005-06, Asia had 32% (with China at 11%), while Latin America holds a share of 26-28% of transactions, with countries other than Brazil entering the market.

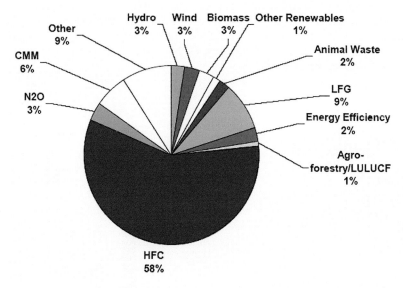

Source: Capoor and Ambrosi, 2006

Figure 2. Technology share of emission reduction projects for the period of January 2005 to March 2006 (as a share of volume contracted).

Let us now close this overview on CDM activities by analyzing the most important element related to sustainability considerations, the type of project that is implemented. The largest part of CDM projects – 58% in terms of volumes transacted in 2005 (compared to 36% in 2004) – has used the technology of HFC destruction. These so-called "synthetic" or "industrial" gases represent the extreme "low hanging fruits" in climate mitigation, as HFC projects are well-known for very low reduction costs while having the advantage of being implemented very quickly. Similarly, also projects that reduce nitrous oxide (N_2O) are considered very efficient and are slowly entering the market, whereas the coal mine methane and landfill gas assets are improving their importance on the market (as is shown by Fig. 2). World Bank data (see Fig. 2) shows that fuel switch, energy-efficiency, biomass and other renewables together represent a share of 10% by volume and 51% by number of projects – and 71% if landfill gas and animal waste management are included. With respect to the previous year, the share of biomass transactions has decreased, while that of energy efficiency transactions remained constant[9].

This section has shown how CDM has improved its role on the carbon market over the last years. Let us now verify whether the component aimed at improving sustainable development in the host countries has had a crucial role during the boom of project activities.

4.2. What Is the Role of Sustainable Development in Current CDM Projects?

As discussed above, double benefits of CDM project activity are envisaged, given that increased investment flows are induced under the requirement that these investments foster sustainable development goals, as for example:

- technological and financial transfer;
- local environmental benefits;
- poverty alleviation;
- equity consideration;
- sustainable energy consumption.

However, have these theoretical sustainable development impacts really materialized in reality? Tab. 1 summarizes the findings on last years' successful project activities. Both the number of projects and the related transaction volumes are indicated in total numbers and in percentage share of the total. All projects currently in the pipeline are included, comprising already registered projects, projects in the validation stage and projects currently under review as well as under request to be registered. In total, at the end of March 2006 654 CDM projects were in discussion, accounting for 836 $MtCO_2eq$ of CERs to be issued until 2012. This number resembles the significant weight of CDM project activities in the global carbon market. Yet, Tab. 1 shows also that reality is characterized by a high weight of N_2O and HFC-23 CDM projects as well as a low weight of small-scale projects. As indicated in the previous section, these types of projects are actually very cost-effective but could contribute to the grounding of sustainable development goals. Let us briefly discuss the related main argumentation.

HFC-23 is a by-product from HCFC-22 production, which is both a powerful greenhouse gas and an ozone depleting substance. It is mainly controlled by the Montreal Protocol, being scheduled for complete phase out by 2040. Equipping HCFC-22 plants with HFC-23 destruction technology would greatly aid efforts in mitigating climate change as the annual emissions of such plants are typically in the order of several million tons CO_2eq. Yet, several experts have argued that it would be advisable not to endorse registration of HFC-23 decomposition projects as they could have adverse effects on the implementation of the Montreal Protocol. The main reason is that HFCF-22 production costs are lowered considerably and likely to get even negative, thanks to the economic revenue from the CERs. This may create a perverse incentive to developing country markets to become more addicted to low priced HCFC-22, and early phase out is less likely to happen.

Table 1. Overview on CDM project types

Type	Number		CERs/year (tCO$_2$eq)		Accumulated 2012 CERs (tCO$_2$eq)	
HFC & N$_2$O reduction	11	2%	62,781	52%	396,560	47%
CH$_4$ reduction & Cement & Coal mine/bed	150	23%	27,351	23%	205,510	25%
Renewables	373	57%	21,206	18%	160,055	19%
Energy efficiency	90	14%	7,798	6%	61,701	7%
Fuel switch	28	4%	1,334	1%	11,820	1%
Afforestation & Reforestation	2	0%	72	0%	619	0%
Total	654				836 MtCO$_2$eq	

Source: own calculations based on UNEP data up to March 2006

Looking at recent developments on the carbon market, current CDM efforts tend thus, even if unintentionally, to give more weight to helping industrialized countries meet their targets than to fostering sustainability. Still, the recent project activities are also characterized by positive signs that indicate that CDM can be a valuable tool to move towards a sustainable path. In particular, recent examples emphasize that the host country's focus on the economic dimension and on technology transfer often helps to foster contemporarily also sustainable development.

One of the clearest examples in this context is China's approach to CDM that demonstrates that – if appropriately designed – CDM actually can promote sustainable development. This is particularly important as the previous section has shown that China has become the dominating player at the sell side of the carbon market. The next section will briefly outline China's CDM strategy.

4.3. The Example of China's Approach towards CDM

China has ratified the Kyoto Protocol, but faces – as a developing country – no obligations at the moment. During the last decade, China has seen a strong economic growth, and correspondingly a strong growth of emissions. Yet, China also disposes of a wide array of low-cost abatement opportunities, being characterized by low abatement costs. Given its circumstances, China was from the beginning of the CDM activities expected to represent one of the largest future hosts of CDM project offers. After initial delays, two recent positive trends reflect that China is now taking a proactive and sustainable policy towards the CDM:

- China's strong efforts on CDM capacity building projects at national, local and enterprise levels aimed at gaining more insight into the CDM and increasing its capacity to initiate and undertake CDM project.
- China's strong emphasis on clear institutional structures and implementations strategies aimed at streamlining CDM procedures and at sounding and clearing governance of responsibility and functions.

The latter part of China's strategy is crucial for the achievement of sustainability goals. In fact, the temporary law named "Interim Measures for Operation and Management of Clean Development Mechanism Projects" indicates a number of priority areas for CDM investment, enforcing thereby the transfer of environmentally friendly technologies to China. Instead, a differentiated CER tax is put upon CDM projects based on HFC-23 (65% tax rate) or N_2O (30% tax rate) technologies. Tax revenues are to be used for investments in sustainability purposes.

China assesses thus the social and environmental dimension of sustainable development when choosing CDM projects. This strategy shows that an appropriate design of a host country's CDM approach can help to put more weight on the sustainability component of this instrument. CDM can become a way of fostering sustainability.

5. CONCLUSIONS ON THE ROLE OF CDM AND THE LINK BETWEEN SUSTAINABLE DEVELOPMENT AND CLIMATE CHANGE

Being one of the most pressing dangers to the world's sustainability, climate change is inseparably linked to sustainability issues. In particular, as emphasized by the IPCC TAR (2001b), the climate change issue is part of the larger challenge of sustainable development because of dual linkages between these two fields. On one hand, the countries' ability to approach the goals of sustainability is seriously affected by the impacts of climate change and instability as well as by the activities implemented through the climate policy and the associated socio-economic development. On the other hand, the pursuit of the objectives focussed on sustainable development will have strong consequences for the climate policies in affecting both their opportunities and their overall success. As a consequence it seems obvious that "climate policies can be more effective when consistently embedded within broader strategies designed to make national and regional development paths more sustainable." (IPCC, 2001a, p. 3)

The particular features of climate change – i.e. the public good characteristic, the structural differences among countries that make it difficult to share the burden of emission reductions and the absence of a supra-national authority that is able to enforce environmental policies on a global scale – explain the difficulties in finding a broad participation in international climate policy. At the same time, the involved complexities and the potentially strong impacts of climate change emphasize that the climate-change issue is an essential part of the larger challenge of sustainable development and that urgent action is needed. These insights suggest that the future approach to climate policy should be based on two elements:

- First, a successful approach needs to search for strategies focussed on incentives to increase the participation in order to effectively cope with the threat of climate change.
- Second, the broader context of sustainable development has to be taken into account, acknowledging for the strong linkages between climate change and sustainability issues.

As a consequence, the strategy of aligning climate-change control with the overall objective of sustainable development seems to be a promising way of improving the long-term performance of climate policy. In this way, the adverse effects of climate change on the Earth's sustainability can be minimized and an approach compatible with the long-term requirements of a sustainable future can be initiated. The control of climate change could thus set the example of combining the objectives of a relatively narrow field, i.e. climate change, with those of the broader sustainability challenge. This strategy could provide important insights into how to approach sustainability; the climate issue might therefore serve as an instrument for making a crucial step towards the overall goal of sustainable development.

This article has analyzed one possible way of embedding climate policies in sustainability issues by looking at the CDM. The objective was to verify whether this mechanism sets incentives for emission reductions as well as for sustainability considerations. We have demonstrated that this policy instrument – if properly designed – can help to promote both cost-effective emission reductions and sustainability goals.

This insight is important as CDM has globally become a crucial instrument to induce emission reductions for compliance purposes, as shown by this study. In addition, recent developments indicate that more is to be expected from this policy instrument. Indeed, at the last climate conference in Montreal, December 2005, several decisions have been adopted that provide more certainty to the future of CDM. In particular, steps to strengthen and streamline the CDM have been adopted and the need to ensure CDM's continuity beyond 2012 has been recognized[10]. Yet, true implications will become visible during the year 2006.

Concluding, the longer-term challenge consists in appropriately designing policy instruments that support the transition to more sustainable structures. Undeniably, most mitigation efforts in developing countries such as China have common drivers in form of economic growth, energy security, and local environmental protection. The use of appropriate design of market mechanisms enforcing innovative technologies can therefore be a successful way of enabling both environmental and broader sustainability benefits.

Barbara K. Buchner
Energy Efficiency and Environment Division,
International Energy Agency, Paris, France

NOTES

[1] More states have signed it since then. For a detailed and up-dated list of signatories see http://unfccc.int/resource/conv/ratlist.pdf.

[2] The specific targets are defined in the Annex B to the Kyoto Protocol which represents an up-dated version of the industrialized countries defined in the Annex I to the protocol, including in addition to the OECD countries also the economies in transition to a market economy.

[3] Both theory and practice confirm this perspective (Carraro and Siniscalco, 1993; climate negotiations related to the Kyoto Protocol, 1997-2004).

[4] The AIJ (Activity Implemented Jointly) pilot phase was an emission reduction mechanism with the primary purpose of implementing activities (i.e. projects) that would be financed by industrialized countries to achieve GHG reductions. Between 1995 and 2000 several developed countries supported the goals of AIJ and the mechanism was evaluated in Kyoto in 1997, having produced poor results.

[5] Article 12.2 of the Kyoto Protocol states that "The purpose of the clean development mechanism shall be to assist Parties not included in Annex I in achieving sustainable development and in contributing to the ultimate objective of the Convention, and to assist Parties included in Annex I in achieving compliance with their quantified emission limitation and reduction commitments under Article 3.".

[6] Decision 17/CP.7 available at http://cdm.unfccc.int/Reference/Documents/cdmmp/English/mpeng.pdf

[7] The Marrakech Accords are available for download at http://www.unfccc.int Document FCCC/CP/2001/13/ and addenda.

[8] For an excellent discussion of this argument see Chiavari (2006).

[9] For a detailed discussion see Capoor and Ambrosi (2006), "State and Trends of the Carbon Market 2006", International Emissions Trading Association and Carbon Finance (World Bank).

[10] For a detailed discussion of the results obtained in Montreal see for example: International Institute for Sustainable Development (IISD) (2006), Summary Of The Eleventh Conference Of The Parties To The UN Framework Convention On Climate Change And First Conference Of The Parties Serving As The Meeting Of The Parties To The Kyoto Protocol: 28 November – 10 December 2005. Earth Negotiations Bulletin, online at http://www.iisd.ca/climate/cop11/

REFERENCES

Capoor K. and Ambrosi P., 2006. State and Trends of the Carbon Market 2006, *International Emissions Trading Association and Carbon Finance* (World Bank).

Carraro C. and Siniscalco D., 1993. Strategies for the International Protection of the Environment, *Journal of Public Economics*, 52, pp. 309-328.

Chiavari J., 2006. *The CDM Additionality Criterion: Legal Evolution, Current Limitations And Recommendations For A More Constructive Interpretation Of The Term*. PhD thesis, University Ca' Foscari, Venice, Italy.

Figueres C., 2006. Sectoral CDM: Opening the CDM to the Unrealized Goal of Sustainable Development, *International Journal of Sustainable Development Law & Policies (JSDLP)*, 2(1).

International Institute for Sustainable Development (IISD), 2006, Summary Of The Eleventh Conference Of The Parties To The Un Framework Convention On Climate Change And First Conference Of The Parties Serving As The Meeting Of The Parties To The Kyoto Protocol: 28 November – 10 December 2005. *Earth Negotiations Bulletin*, online at http://www.iisd.ca/climate/cop11/

Intergovernmental Panel on Climate Change (IPCC), 2001a. *Climate Change 2001: Synthesis Report*, Cambridge: Cambridge University Press.

Intergovernmental Panel on Climate Change (IPCC), 2001b. *Climate Change 2001: Mitigation*, Cambridge: Cambridge University Press.

IETA, 2005. IETA's *Guidance note* through the CDM Project Approval Process Vol. 1.5, May 2005. Online at www.ieta.org.

UNEP, 2003. *CDM Information and Guidebook*, Unep Riso Centre.

WCED (World Commission on Environment and Development), 1987. *Our Common Future*. Oxford, Oxford University Press.

VI. SUSTAINABLE INDUSTRIAL DEVELOPMENT

M. R. CHERTOW

INDUSTRIAL ECOLOGY IN A DEVELOPING CONTEXT

Abstract. Industrial ecology has emerged in recent years as a new multi-disciplinary field at the nexus of environmental science, engineering, business, and policy. Reflecting a systems view, industrial ecology sees industry embedded in the natural systems that surround it. This chapter offers explanations of industrial ecology concepts as well as practical examples and short case descriptions. It examines principles of industrial ecology; describes its core elements including design for environment, lifecycle analysis, material flow analysis, and industrial symbiosis; reviews policy approaches, discusses the relevance of industrial ecology in a developing world context, and discusses the parallel relationship of industrial ecology to the notion of the circular economy as it is developing in China.

1. OVERVIEW

The great contribution of modern ecology has been to recognize, in a formal, scientific way, the interconnectedness of natural phenomena. Despite the advances of the modern age that make it seem possible for humans to sustain an urban existence apart from nature, we ultimately recognize that life on earth is held together by sources of energy created by the sun, by a finite cycle of water, by the oxygen in the air around us. Joining together the words "industrial" and "ecology" is a bold acknowledgement of this interdependent condition.

Industrial ecology is a systems science. It is industrial in its resolute focus on each phase of the production processes of goods and services. It is ecological because it borrows from nature the notion of cycling – that the industrial system should emulate the natural one by conserving and reusing resources as completely as possible in production and consumption. While some writers of a philosophical bent conceive of the word "industry" quite broadly to cover the range of human activities, examining industrial ecology in a developing context requires a narrower focus, given that the challenges to sustainable development brought on by high urban population density and accelerated industrialization are so immediate in many developing countries.

China, as a major manufacturing centre for the world, exhibits this problem writ large. On the one hand, economic globalization has brought hundreds of millions of Chinese citizens out of poverty. On the other hand, swiftly depleting stocks of natural resources and the inability of nature to absorb the waste and pollution of intensive production raise questions about the system's ultimate sustainability. While China has garnered the world's envy by its rapidly rising GDP, what is the value of a 10 percent annual growth rate if, as long-time China researcher Vaclav

Smil (2004) has observed, the costs of China's ecosystem decline and environmental pollution also equals at least 10 percent of GDP annually?

Industrial ecology does not focus on remediation of past environmental ills, or on end-of-pipe controls of contemporary sources of pollution, though both are significant environmental concerns. Its purpose is to avoid environmental damage in the first instance through systems analysis, through product, process, and facility design, and through technological innovation. This chapter offers explanations of industrial ecology concepts as well as practical examples and short case descriptions. It examines principles of industrial ecology, describes its core elements, discusses the relevance of industrial ecology in a developing world context, and discusses the parallel relationship of industrial ecology to the notion of the circular economy as it is developing in China.

2. PRINCIPAL THEMES OF INDUSTRIAL ECOLOGY

Industrial ecology has emerged in recent years as a new multi-disciplinary field at the nexus of environmental science, engineering, business, and policy. Although its orientation is practical, it does not reflect a simple view: rather it depends upon a systems perspective. The importance of a systems view can be seen at the level of a factory emitting pollutants to the air. If we address the air pollution by installing scrubbers that capture the particles and create a sludge, then we have only succeeded in replacing one type of pollution with another. If the sludge is improperly handled, it may run off into streams and become a water pollution problem. Rather than divide pollution into many separate problems of air, water, and land, industrial ecology, as part of the quest for sustainability, brings forth a more comprehensive integration of environment and economy.

Another way to envision the systems approach embedded in industrial ecology is to think of a landscape revealing itself as concentric circles, starting with the soil and then expanding outward to the shrubs, then the larger trees, and then the entire landscape of forest and clouds and sun. In industrial terms, think of altering an automotive component, for example, which may reduce pollution but, at the next level, does not change the basic size of the vehicle. Similarly, a change in overall vehicular design does not substantially alter the problem of highway congestion and availability of road networks. At the broadest level, road networks involve land use and lifestyle choices as well as societal needs and wants (Fig. 1). In the words of ecologist James Kay, "sustainability issues can only be understood in terms of systems embedded in systems which are also embedded in systems" (Kay, 2002).

The Automotive System

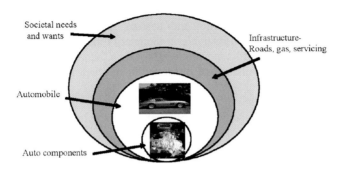

*Figure 1. Concentric circles of an industrial system
(Source: Thomas Graedel, Yale University).*

2.1. Industrial ecology embeds different scales and levels of activity

The hallmark of industrial ecology is its principal concern with the flow of materials and energy through systems at different spatial scales, from products to factories up to national and global levels. As shown in Fig. 2, industrial ecology allows focus at the facility level, across firms and other organizations, or, more broadly, at the regional and global level. Fig. 2 highlights that the goal of industrial ecology is sustainability, a still abstract term to which industrial ecology adds some tangibility.

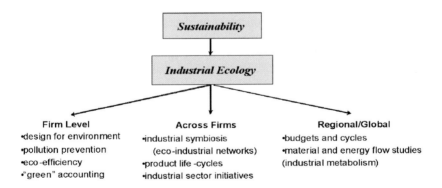

Figure 2. Industrial ecology operates at three levels.

Although there is no single declaration that "explains" industrial ecology, it has proven to be "an effective framework for applying many existing methods and tools, as well as for developing new ones" (Lowe and Warren, 1996). The three scales at which industrial ecology operates are described below:

Facility or Firm. Within a firm, tools such as "green" or "full-cost" accounting (Bennett and James, 1998), pollution prevention, and eco-efficiency (Huppes and Ishikawa, 2005) have proven to be useful ways of drawing together economic and environmental considerations into one system, by making improvements in environmental performance measurable in monetary terms. Design for environment implies recognition of what happens not only within the walls of the firm, but also what happens to firm outputs subsequent to manufacturing. It is described in greater depth in section 3 below.

Across Firms/Organizations. Crossing organizational boundaries implies cooperation among firms and organizations through resource and information sharing within a single industry sector or across different sectors. Thinking of supply chains, firms have recognized that their products cross many company boundaries during their life cycles from design and manufacture to distribution to use to final disposal. Life cycle perspectives and industrial symbiosis are further discussed in section 3.

Regionally and globally. Tracking flows of material and energy across regions, economies, and the globe illuminates what happens to the constituents of industrial and commercial products. The analogy has been made from *human* metabolism – the sum of all the processes by which a particular substance is handled in the living body – to *industrial* metabolism – the sum of the processes through which energy and materials move through industrial and consumer systems from extraction to final disposal, wherever that may be (Ayres, 1989).

2.2. Focus on material and energy across the product lifecycle

Central to industrial ecology is the notion of "embedded energy and materials". To create a typical product, resources are used for extraction of materials, transportation, primary and secondary manufacturing, and distribution. The total quantity of energy and materials used is the amount embedded in a product, process, facility, or service and, more broadly, in a region or economy. If a material or product is thrown away, so is the energy used to make it. Aluminum cans provide a dramatic example: if they are recycled rather than discarded, 95 percent of the energy that went into making them is retained.

Industrial ecology is committed to preserving embedded energy and materials as much as possible and accounting for them across their lifecycles. The most basic questions an industrial ecologist asks are: where does material come from and where does it go? How does it move from place to place? What happens when the original purpose of a product or facility is spent? Which materials stay behind and which dissipate into the environment? How much of the embedded energy and materials can be recirculated or recovered?

A classic story of industrial ecology comes from two researchers in the 1990s who plotted the sources and uses of close to 6 million metric tons of lead consumed annually around the globe (Thomas and Spiro, 1994). Their study constituted a "budget" for lead showing that its largest use was for lead acid batteries, and most of that lead is eventually recycled. In many other uses such as ammunition, pigments, solder, and fishing weights, however, lead use was "dissipative" – meaning it was released into the environment with little recovery. This type of analysis has clear implications for public policy: from a regulatory perspective, it is more important to confront the largest uses that are released directly into the environment than the uses where most lead is recycled. Historically, lead additives to gasoline were a large, dissipative use and while this is no longer significant in western countries, it is still part of the regional lead cycle in some developing countries (Graedel and Allenby, 1995).

2.3. Elevating technology and industry as part of the transition to more sustainable industrial systems

Although many environmentalists tend to be technological pessimists, focusing on the multitude of negative environmental impacts of our technological society, industrial ecologists recognize that technology can be effectively channeled toward environmental benefit. According to the first textbook in the field (Graedel and Allenby, 1995), industrial ecology has a "Master Equation" which reveals a new logic for technological optimism. This equation reads as follows:

$$Environmental\ impact = Population \times \frac{GDP}{Person} \times \frac{Environmental\ impact}{unit\ of\ per\ capita\ GDP} \quad (1)$$

In this conceptual equation, environmental impact is identified as a function of three terms: population; GDP per person, which is a means of measuring affluence; and a "technology term." Graedel and Allenby define this third term, qualitatively, as the degree of environmental impact per unit of per capita gross domestic product – or a measure of how much each unit of production or consumption pollutes. They describe this term as

> an expression of the degree to which technology is available to permit development without serious environmental consequences and the degree to which that available technology is deployed (Graedel and Allenby, 1995, 7)

The form of the equation is based on an earlier mathematical identity from the 1970s known as the "IPAT equation." As originally conceived, the IPAT equation states that environmental impact (I) is the product of 1) population (P); 2) affluence (A); and 3) technology (T), or I=PAT. Generally credited to ecologist Paul Ehrlich, the IPAT formulation arose from a dispute in the early 1970s among the most prominent environmental thinkers of the day about the sources of environmental impact. Ehrlich and John Holdren identified population size and growth as the most urgent IPAT factor, whereas ecologist Barry Commoner argued that post-WWII production technologies were the dominant reason for environmental degradation (Chertow, 2001).

The viewpoint articulated in a study by the World Resources Institute (Heaton et al., 1991) and inherent in the Master Equation of industrial ecology reversed this original 1970s logic. In the transition from IPAT to the Master Equation there is a recognition that increases in population are expected to continue for several decades more, and that increases in affluence actually have the potential to improve quality of life for billions of people around the world. So, if population increases by 50 percent between now and 2050, as projected by UN estimates, and if affluence only doubles over the same time period, then environmental impact would increase threefold. Such demographic realities put the burden of sustainability largely on the third term, the technology term, as an essential counterweight to increases in population and GDP/person, implicitly requiring environmentally effective technological choices that will reduce pollution per unit of economic impact.

Appreciation by industrial ecologists of the role that technology can play in solving environmental problems carries through to an acknowledgement of the important role that private sector firms play both in allocating resources and in implementing technological innovation. In this regard, industrial ecology positions firms as key players in environmental protection, not merely as the villains they were thought to be in the early days of the environmental movement (Powers and Chertow, 1997).

2.4. Requiring new models of cooperation and collaboration

Just as industrial ecologists recognize the important roles private actors play in environmental performance, it is also important that these same actors realize, in turn, that the future envisioned by industrial ecologists requires some shifts in mindset. More collaborative approaches for addressing environmental issues among firms, industries, sectors and often non-governmental organizations, government and academia are proving necessary to bring societal change. In a marketplace traditionally characterized by competition, inter-firm cooperation has recently demonstrated significant potential for maintaining or enhancing market advantages along supply chains, across lifecycle stages, or through resource exchanges of water, energy, and materials.

Many scholars have argued that "business as usual" will not achieve the goal of sustainability. They contend, instead, that a deeper structural change in human and social dynamics is needed "in which individuals, firms, governments and other institutions act responsibly in taking care of the future" for humans, other species and nature itself (Ehrenfeld, 2000). Continuing to achieve small annual reductions in pollution, for example, is necessary but not sufficient for achieving the ambition of industrial ecology. In the long run, the transformation to sustainability embeds a paradigmatic shift to models of cooperation and resource sharing anticipated by industrial ecology.

3. ELEMENTS OF INDUSTRIAL ECOLOGY

Five elements of industrial ecology are described in this section embracing useful approaches, tools and practices. These are: design for environment, life cycle analysis and assessment, material flow analysis, industrial symbiosis, and policy approaches suggested by industrial ecological principles.

3.1. Design for environment

Design for environment (DFE) is an approach to design in which the environmental characteristics of a product, process, or facility are internalized and optimized from the earliest stages. It has been estimated that some 70 percent of the cost of a product is determined during the design stage. Building in environmental considerations before capital equipment is purchased and distribution channels are developed, is arguably the least expensive time to make proactive decisions which can, in turn, influence the entire life cycle chain.

In industrial product design there are multiple claims in addition to environmental concerns. These are sometimes called "Design for X," where X represents many useful design attributes including ease of assembly and consideration of how design affects reliability, safety, or serviceability. A central tenet of DFE is that DFE actions should not compromise other design attributes of a product such as performance, reliability, aesthetics, maintainability, cost, or time to market (Graedel and Allenby, 1995).

The World Business Council for Sustainable Development (WBCSD, 2000) has identified seven elements that businesses can use to improve eco-efficiency, which provides a useful introduction to DFE ideas at a practical level as follows:

- Reduce material requirements (total mass)
- Reduce energy intensity
- Reduce dispersion of toxic substances
- Enhance recyclability
- Maximize use of renewable resources (avoid depletion of finite resources)
- Extend product durability/product life
- Increase service intensity

DFE is associated with a life cycle perspective in the industrial ecology literature because it builds in longer term considerations beyond production to use, reuse, and disposal recognizing that every engineering decision is also an environmental one.

3.2. Life cycle assessment

As described by Graedel and Allenby (1995), industrial ecology

> is a systems view in which one seeks to optimize the total materials cycle from virgin material, to finished material, to component, to product, to obsolete product, and to ultimate disposal.

The breadth of focus from a life cycle perspective is not limited to what happens within one facility or factory, but considers the entire set of environmental impacts that occurs at each stage of industrial development and use across entities. Such thinking creates new awareness about precisely which life cycle stages most effect the environment for different products or services. Many consumer electronics, for example, are manufactured efficiently but are difficult to dispose at the end of their useful lives. Motor vehicles generate some 90 percent of their environmental impacts not in the factory but in the use stage – e.g. when cars are being driven or airplanes flown well after the vehicle has left the manufacturing site.

As a formal methodology, Life Cycle Assessment (LCA) is an analytical tool for the systematic evaluation of the environmental aspects of a product or service system through all stages of its life cycle. Based on procedures of the Society of Environmental Toxicology and Chemistry (SETAC), the formal structure of LCA contains three stages as well as on-going interpretation. These stages are:

1) *Goal and scope definition* to define the boundaries of what is being studied – for example all stages of a new telephone design.
2) *Inventory analysis* to total up the types and quantities of energy and materials used in the system being studied as well as resulting environmental releases.
3) *Impact analysis* to group and quantify the resources used and emissions generated into environmental and toxicological impact categories which are then to be weighted for importance.

Formal LCA, then, can offer a quantitative comparison between alternative product or process designs such as whether it would be environmentally preferable to use cloth or disposable diapers. Because such analysis can also be complex and expensive, industrial ecologists have also worked on streamlined life cycle methodologies or considered life cycle management more broadly but less formally to provide general guideposts in thinking beyond a product or process. In 2002 SETAC and the United Nations Environment Programme created The Lifecycle Initiative inspired by the notion of managing production and consumption impacts as part of creating a lifecycle economy (UNEP, 2006).

3.3. Material flow analysis

Material flow analysis methods are used to map and quantify the flow of materials through a network of actors, be they in a single facility or group of facilities, a defined region, or along a product supply chain. Within a defined region, the actors of interest are approached individually to identify and quantify all of their energy, water and material inflows and outflows, as well as various attributes of the facility. Generic data, compiled from a broad range of sources, can be used to create expected material flow data templates for different industries. At other scales, a single material or substance can be tracked nationally or globally or many materials can be tracked more locally. Tools and software have been developed for material flow analysis (MFA) and substance flow analysis (SFA) that formalize tracking practices (Brunner and Rechberger, 2003).

3.4. Industrial symbiosis

The concept of industrial symbiosis (IS) is broadly based on the idea of exchange, where one facility's waste (energy, water, or materials) becomes another facility's feedstock. Thus, inherent to industrial symbiosis is a cooperative approach to competitive advantage among traditionally unrelated firms. The keys to industrial symbiosis are collaboration and the synergistic possibilities offered by geographic proximity (Chertow, 2000).

The term "industrial symbiosis" was coined in the small municipality of Kalundborg, Denmark, where a well-developed network of dense firm interactions was encountered. The primary partners in Kalundborg, including an oil refinery, power station, gypsum board facility, and a pharmaceutical company, share ground water, surface water, and wastewater, steam, and fuel, and also exchange a variety of by-products that become feedstocks in other processes (Fig. 3). High levels of environmental and economic efficiency have been achieved which has led to many other less tangible benefits involving personnel, equipment, and information sharing.

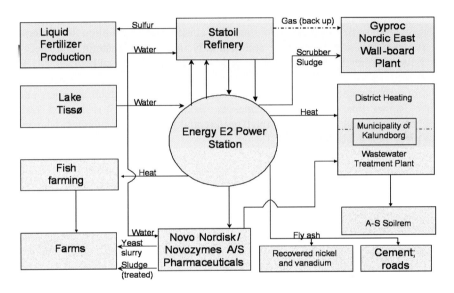

Figure 3. The industrial ecosystem of Kalundborg, Denmark.

Currently, industrial symbiosis takes many shapes and forms around the world. Less successful have been planned "eco-industrial parks" (Gibbs et al., 2002), while more successful have been industrial ecosystems based on principles of self-organization as in Kalundborg (Jacobsen and Anderberg, 2005; Chertow, 2006). Near Perth, Australia, is a mineral processing region with extensive exchange of energy, water, and numerous materials (Altham and van Berkel, 2004). Jurong Island in Singapore is a large petrochemical complex with coordinated provision of numerous services from fire suppression to hazardous material treatment to pipeline delivery of gas and water. Japan has designated many "eco-town" projects, as in Kawasaki, where industrial partners in that city are pursuing waste and material reuse across firms. A state-owned sugar refining company in China went beyond sugar refining into related industries that use materials from its two key by-product streams: molasses (the sugar refining residue) and bagasse (the fibrous waste product) from sugar cane production (Zhu and Côté, 2004).

3.5. Policy approaches suggested by industrial ecological principles

In this definition of industrial ecology, Robert White, former president of the U.S. National Academy of Engineering, reflects upon how to use the insights gained from the study of industrial ecology for policy:

> Industrial ecology is the study of the flows of materials and energy in industrial and consumer activities, of the effects of these flows on the environment, and of the influences of economic, political, regulatory, and social factors on the flow, use, and transformation of resources.

The opening for policy is through the explicit mention of "economic, political, regulatory, and social factors" that must be considered in addition to systems analysis techniques and quantitative and qualitative analysis of material and energy flows. Several policy ideas rooted in industrial ecology are described below.

Greening the Supply Chain refers to buyer companies requiring a defined level of environmental responsibility in the core business practices of their suppliers and vendors. Rather than regulation, businesses seek "compliance" with their own internal environmental standards through preferential treatment of suppliers who are able to meet their goals (Fig. 4).

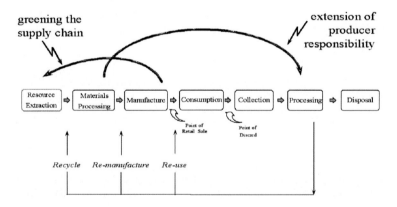

Figure 4. Greening the supply chain and extended producer responsibility illustrated on a product lifecycle diagram (Source: Reid Lifset, Yale University).

Extended Producer Responsibility (EPR) is based on the concept of manufacturers assuming responsibility for their products after they have been used for their original purposes. In industrial ecology terms, EPR embeds a life cycle approach by addressing the environmental impacts of products beyond the factory to the end-of-life stages (Fig. 4). EPR began in Germany with a focus on packaging waste and quickly spread through Europe with European Union directives on packaging waste, end of life vehicles, and waste electronics and electrical equipment (WEEE). China, as key producer of electronics, must respond to EU developments that could affect its export markets, deals with growth of its domestic electronics market, and addresses the environmental challenges of electronic waste imports (Tong *et al.*, 2005). Japan now requires return of several appliances through a reverse logistic system and more such laws are expected to be adopted in Asia.

Environmental Certification is a process of singling out particular products or classes of products as environmentally superior according to some predetermined criteria. If a product is successfully certified it generally receives an "ecolabel" – such as Germany's Blue Angel or that of the Korea Eco-labelling Program – awarded by a third party entity to inform consumers that a product or service credibly meets specific environmental standards (Global Ecolabelling Network, 2006). The certification movement is experiencing rapid growth from forests and forest products to coffee, apparel and eco-tourism as a voluntary means of regulation (Cashore *et al.*, 2004). While it was not begun by industrial ecologists, certification embodies a lifecycle perspective as it looks across the value chain to assess whether sustainable practices were used in growing and harvesting for food and forest products as well as production of a broad and growing assemblage of goods and services.

Products to Services expresses a means of changing the focus of consumption from ownership of goods themselves to the desirable services they provide. The implication is, for example, that it is less important to own a car than to get the transport service it offers through alternatives such as car sharing. Similarly, it is less important to own an actual television than to be able to enjoy the programming when we want it. The environmental implication is that alternative ownership schemes such as leasing or renting can provide incentives for producers to think long-term about issues such as greater product durability, less intensive use of material and energy, or favorable economies of scale in servicing and end-of-life reclamation.

4. INDUSTRIAL ECOLOGY IN CHINA AND THE DEVELOPING WORLD

The first history of industrial ecology clearly shows its roots in developed countries, based on ideas beginning in the late 1960s, especially in Northern Europe, Japan, the US and Canada (Erkman, 1997). In this regard formative ideas of industrial ecology track the development of the broader environmental movement as a response to booming post-World War II industrial growth and its attendant pollution. New science, technology, law, and management practices were needed to deal with the intensity of air and water pollution and, with much trial and error, environmental methods were developed. More recently, much of the developed world has been de-industrializing in favor of China, India, Mexico, Indonesia and other developing countries. Indeed, the first book to focus on industrial ecology in developing countries, published in India, casts aside the notion that only rich countries can afford it stating:

> Industrial ecology is a very relevant and an urgently needed strategy for developing countries. As a matter of fact, inefficient use of resources and getting rid of the waste with end-of-pipe 'pollution control' technologies are truly unaffordable luxuries for developing countries!...The magnitude and urgency of problems need a preventive approach like Industrial Ecology, not only to avoid irreversible degradation of environment and exhaustion of resources, but also to allow continued access to basic resources like water needed for all economic activities. It is thus crucial for developing countries to have a tool like Industrial Ecology for anticipation, early detection, and prevention of environmental problems (Erkman and Ramaswamy, 2003, viii).

4.1. Applicability of industrial ecology in developing countries

There are many important reasons for uptake of industrial ecology ideas in developing countries. The most obvious reason is that the developing world is where most of the world's population and industrial production now reside. With rising incomes and quality of life, consumption, too, will soon be dominated by the developing world. Arguably, the world's environmental future may well be determined by what happens in the rapidly growing urban communities of Asia where the most prodigious expansion of industrial activity in the world is occurring. Recent entry into the World Trade Organization (WTO) by China and accession by other developing countries reduces many trade restrictions, but often forces countries to upgrade systems in order to meet more stringent international environmental standards. Therefore, industrial ecology principles of resource productivity and eco-efficient industry are desirable in the developing world to help manage growth. Industrial ecology, through its strong focus on optimizing materials and energy flows, offers the promise of expanded development but with a more sustainable trajectory.

Another important recognition is that industrial ecology, with its emphasis on design for environment and life cycle perspectives, is forward-looking and proactive. A 1999 estimate suggested that at least 80 percent of the industrial stock that will be in place in China and Indonesia in the next quarter century had not yet been built (Rock *et al.*, 1999). This could constitute an unprecedented opportunity to shape infrastructure and development in ways that were unavailable to industrialized countries in the past through the use of industrial ecology principles that are both technical and conceptual in nature.

4.2. Using industrial ecology tools in the developing world

The developing world also poses some challenges to the way industrial ecology has taken shape. The focus on technology described in section 2.3 requires significant capital investment, which is often unavailable in developing countries. In addition, rather than dominance by fewer, larger firms, the structure of the economy in many developing countries includes thousands or even millions of small or informal sector businesses making coordination difficult (Erkman and Ramaswamy, 2003). Industrial ecology has not been used extensively with planned economies as in East

Asia. Therefore, many of the tools of industrial ecology would need some adjustment to these conditions.

It seems likely that such adaptations are possible. The emphasis on five year plans in China, for example, provides excellent opportunities for the incorporation of more environmental practices at industrial park sites and also provides transitional time for adoption of policies that would foster gains in energy intensity as well as in materials recovery. The passage of the Cleaner Production Promotion Law in China in 2002 has laid solid groundwork for industrial ecology not only because it defines and sets targets for clean industry, but also because it clearly stipulates implementation responsibilities (Mol and Liu, 2005).

Erkman and Ramaswamy (2003) have specifically worked to adapt western-based industrial ecology tools to improve public planning processes. Material flow analysis, for example, was usefully modified for developing world circumstances as illustrated by a story that comes from a highly polluted part of India where untreated smoke from coal burning was spoiling the air. Although prevailing opinion was that the origin of the smoke was predominantly from the many coal-fired power plants in the region, material flow analysis showed that at least as large a contribution was found to be from the burning of coal in individual homes and cottage industries. From a policy perspective, then, actions would have to be found that would emphasize minimizing the use of coal in these smaller, more disbursed entities.

4.3. China and the circular economy

In 2002, China's 16th National Congress pledged to realize an overall "well-off society" (*xiao kang*) by the year 2020. The stated goal is to quadruple the country's GDP while simultaneously advancing enhanced social equality and environmental protection. Given the enormous challenge of sustaining rapid economic growth while reversing environmental degradation, the Chinese are pursuing

> a new path to industrialization featuring high technology, good economic returns, low resource-consumption, low environment pollution and full use of human resources (Hu, 2006).

A key policy advanced for this new development path and incorporated in the 11th Five Year Plan is that of the "Circular Economy."

The Circular Economy concept is developing in China as a strategy for reducing both the demand for natural resources as well as the environmental damage this demand creates. As envisioned thus far, the Circular Economy concept calls for high efficiency in resource flows as a way of sustaining improvement in quality of life within natural and economic constraints. With respect to the product lifecycle, the circular economy is both an upstream strategy related to resource productivity as well as a downstream strategy related to end-of-life management and improved use of by-products. So far, projects have been planned at the individual plant level, across firms as in eco-industrial parks, and more broadly, at the level of the eco-city or eco-province (Yuan *et al.,* 2006). There is already a lot of interest in industrial ecology in China with a Center for Industrial Ecology at Tsinghua University, a broad network of professors who have come together to advance it, numerous translated educational materials, and the availability, in Chinese, of all abstracts in the Yale-owned *Journal of Industrial Ecology*. The fit between industrial ecology as described in this article and the goals of the circular economy is strong indeed: industrial ecology provides an intellectual foundation for the emerging ideas of the circular economy.

5. CONCLUSION: CIRCULAR ECOLOGIES AND CIRCULAR ECONOMIES

In searching for greater economic prosperity communities, anywhere in the world may take on industrial practices that leave them worse off in the long run when damage to local ecosystems is realized. We see this at a global level where the aggregate amount of carbon dioxide deposited in the atmosphere from the burning of fossil fuels is enough to begin to change our climate. The lessons of industrial ecology are targeted at improving the environmental performance of our technological society by grappling with the consequences of production and consumption. For countries that are still designing, still developing, this forward-looking field raises enormous opportunities to improve productivity of resource use while reducing environmental impacts. Just as we are dependent on natural systems to manage resources effectively, filter pollution, and minimize waste, we must strive for economic systems, too, that emulate the model of circular ecologies with those of circular economies using the systems principles of industrial ecology.

Marian R. Chertow
Industrial Environmental Management Program,
Yale School of Forestry and Environmental Studies, Connecticut, USA

REFERENCES

Altham J. and van Berkel R., 2004. Industrial symbiosis for regional sustainability: An update on Australian initiatives. *11th International Sustainable Development Research Conference*, Manchester, U.K. March, pp. 29-30.
Ayres R.U., 1989. Industrial metabolism. In J.H. Ausubel and H.E. Sladovich (eds.), *Technology and Environment*. Washington, D.C.: National Academy Press.
Bennett M. and James P., 1998. *The Green Bottom Line: Environmental. Accounting for Management*, Greenleaf Publishing, Sheffield England, pp. 164-183.
Brunner P. and Rechberger H., 2003. *Practical Handbook of Material Flow Analysis*, Baton Rouge, Louisiana: CRC Lewis Publisher.
Cashore B., Auld G. and Newsom D., 2004. *Governing Through Markets: Forest Certification and the Emergence of Non-State Authority*. New Haven, Connecticut: Yale University Press.
Chertow M.R., 2000. Industrial symbiosis: Literature and taxonomy. *Annual Review of Energy and Environment*, 25, pp. 313-337.
Chertow M.R., 2001. The IPAT equation and its variants: Changing views of technology and environmental impact. *Journal of Industrial Ecology*, 4, 4, pp. 13-29.
Chertow M.R., 2007. "Uncovering" industrial symbiosis. *Journal of Industrial Ecology*, 11(1), pp. 11-30.
Ehrenfeld J., 2000. Industrial ecology: Paradigm shift or normal science? *American Behavioral Scientist*, 44, 2, pp. 229-244.
Erkman S., 1997. Industrial ecology: An historical overview. *Journal of Cleaner Production* 5(1-2), pp. 1-10.

Erkman S. and Rameswamy R., 2003. *Applied Industrial Ecology: A New Platform for Planning Sustainable Societies.* Bangalore, India: Aicra Publishers.
Gibbs D., Deutz P. and Procter A., 2002. "Sustainability and the local economy: The role of eco-industrial parks." Presented to Ecosites and Eco-Centres in Europe, Brussels, Belgium, June 19.
Global Ecolabelling Network, 2006. Retrieved May 6, 2006, from http://www.gen.gr.jp/members.html
Graedel T.E. and Allenby B.E., 1995. *Industrial Ecology.* Englewood Cliffs, NJ: Prentice Hall.
Heaton G., Repetto R. and Sobin R., 1991. *Transforming technology: An agenda for environmentally sustainable growth in the 21st century.* Washington, D.C.: World Resources Institute.
Hu J., 2006. Speech by Chinese President Hu Jintao at Yale University. New Haven, Connecticut, April 21.
Huppes G. and Ishikawa M., 2005. A framework for quantified eco-efficiency analysis. *Journal of Industrial Ecology,* 9(4), pp. 25-41.
Jacobsen N.B. and Anderberg S., 2005. Understanding the evolution of industrial symbiotic networks – The case of Kalundborg. In J.C. J. M. van den Bergh, M. Janssen (eds.), *Economics of Industrial Ecology: Materials, Structural Change, and Spatial Scales.* Cambridge, MA: MIT Press.
Kay J., 2002. On complexity theory, exergy, and industrial ecology. In C. Kibert, J. Sendzimir, G. Guy (eds.), *Construction Ecology: Nature as a Basis for Green Buildings,* Spon Press, pp. 72-107.
Lifset R. and Graedel T., 2002. Industrial ecology: Goals and definitions. In R.U. Ayres and L.W. Ayres, (eds.), *A Handbook of Industrial Ecology.* Edward Elgar, Cheltenham, UK.
Lowe E. and Warren J., 1996. *The Source of Value: An Executive Briefing and Sourcebook on Industrial Ecology.* Richland, Washington: Pacific Northwest Laboratory, 3.2.
Mol A. and Liu Y., 2005. Institutionalising cleaner production in China: The Cleaner Production Promotion Law, *International Journal of Environment and Sustainable Development,* 4(3), pp. 227-245.
Powers C. and Chertow M.R., 1997. Industrial ecology: Overcoming policy fragmentation. In M. Chertow and D. Esty, (eds.) *Thinking Ecologically: The Next Generation of Environmental Policy,* New Haven, Connecticut: Yale University Press.
Rock M., Angel D. and Feridhanusetyawan T., 1999. Industrial ecology and clean development in East Asia. *Journal of Industrial Ecology,* 3(4), pp. 29-42.
Smil V., 2004. *China's Past, China's Future: Energy, Food, Environment.* New York and London: Routledge Curzon.
Thomas V. and Spiro T., 1994. "Emissions and exposure to metals: Cadmium and lead." In R. Socolow, C. Andrews, F. Berkhout, V. Thomas, (eds.) *Industrial Ecology and Global Change,* Cambridge, UK: Cambridge University Press.
Tong X., Lifset R., and Lindhqvist T., 2005. Extended producer responsibility in China: Where is "best practice"? *Journal of Industrial Ecology,* 8(4), pp. 6-9.
UNEP, 2006. The Life Cycle Initiative homepage. Retrieved May 8, 2006 from http://jp1.estis.net/builder/includes/page.asp?site=lcinit&page_id=15CFD910-956F-457D-BD0D-3EF35AB93D60
WBCSD, 2000. *Eco-efficiency: Creating More Value with Less Impact.* 32pp. Available at http://www.wbcsd.org/web/publications/eco_efficiency_creating_more_value.pdf
Yuan Z., Bi J. and Moriguchi Y., 2006. The circular economy: A new development strategy in China. *Journal of Industrial Ecology,* 10(1-2), pp. 4-8.
Zhu Q. and Côté R., 2004. Integrating green supply chain management into an embryonic eco-industrial development: A case study of the Guitang Group. *Journal of Cleaner Production* 12, pp. 1025-1035.

G. CHIELLINO

STRATEGIC ENVIRONMENTAL ASSESSMENT (SEA): INTEGRATION AND SYNERGY WITH EMAS MANAGEMENT SYSTEMS ON THE TERRITORY

Case Study: SEA applied to regulatory territorial plans

Abstract. This article aims at presenting an analysis of the common factors between the Environmental Management System EMAS that addresses territorial authorities, and the Strategic Environmental Assessment (SEA), included in the European Directive 2001/42/EC. The work highlights the methodological analogies and the possibility to achieve sustainable development objectives following the synergetic implementation of the two procedures.

1. INTRODUCTION

Even though the EMAS regulation is widely accredited as being the best instrument regarding environmental and social performance, it is often disregarded by many local authorities and organizations that, even recognizing its achievements, choose not to invest in resources that if implemented would lead to a management system (EMAS).

The Strategic Environmental Assessment (SEA) with its associative properties (in accordance with the European Directive 2001/42/EC and the Regional Law 11/2004 in Veneto Region), is now applied in most Italian Regions, through town councils and local authorities. These institutions have drawn up sector plans and programmes (European Directive 2001/42/EC) such as the Regulatory Town Plan, even though they are not fully regarded for their additional value in terms of guaranteeing sustainability when using the resources to evaluate the local territory and to improve the environment quality.

To help to overcome this transition from voluntary to compulsory legislation, the SEA applied to the Regulatory Territorial Plans could, in addition, develop its aspects regarding the implementation of a continuous improvement process, becoming a valid support for further EMAS registrations by local authorities. The EMAS management systems already implemented by Public Administrations could, conversely, be acknowledged as valuable supports when applying the Strategic Environmental Assessment.

2. REPORT

2.1. Similarities between the strategic environmental assessment (SEA) and the EMAS management system

The SEA derives from the European Union's need for a structured procedure to promote sustainable development in the preparation and adoption of plans and programmes. The procedure follows the Directive 2001/42 of the European Parliament and Government bodies concerning the evaluation of the effects of plans and programmes on the environment.

There are numerous similarities between the regulations that define the SEA (European Directive 2001/42/CE) and those of Eco-Management and Audit Scheme – EMAS (European Regulations 2001-761). A synergy between the two is not only easily understandable, but also beneficial. A Town Council or local authority that intends to apply SEA in the process of planning or programming, can benefit from the studies and processes already implemented aiming at obtaining a further added value through an Environmental Management System (EMS) of the EMAS registered body. In the same way, a local authority that chooses to implement EMAS would see its work significantly supported once the SEA has been carried out. The common aspects between the two procedures are numerous, as it can be seen in Fig. 1.

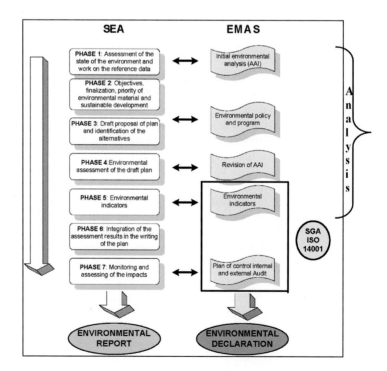

Figure 1. Integration and synergy between EMAS and SEA procedures.

1) The SEA is a systematic process meant to evaluate the environmental effects of the actions proposed by the plans, programmes, policies, interventions and initiatives. It aims at guaranteeing that any effect is taken into consideration in the most appropriate way, starting from the initial decision-making process, equally considering the economic and social factors (Chiellino et al., 2004). The assessment process of the plan/programme (P/P) is realized at two distinctive stages.

2) The first is the realization of basic research analysis in order to establish, for reference purposes, an initial picture of the environmental situation. This deals with a guide that outlines the 'invariants' within the environmental system and a compendium on the state of territorial resources, summarizing the complex cross-information gathered by experts (Brunetta and Peano, 2003). At this phase it is possible to define the indicators that give a mock idea of the state of the environment and allow, over time, the compatibility and incompatibility of specific actions on the environment to be monitored.

3) The second stage outlines and utilizes the methodology that helps the decision-making process, enabling more concrete decisions to be made (Plan/Programme of actions) according to an open flow diagram. This diagram can be closely compared with the continuous improvement cycle proposed by Deming, this follows the logical and retroactive sequence of PLAN-DO-CHECK-ACT, which allows improved actions to be carried out in accordance with the defined environmental, economic and social indicators.

The Deming Cycle realized in environmental management systems in accordance with EMAS can be found in the implementation phases of the SEA as shown in Fig. 2.

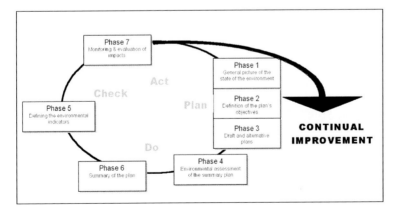

Figure 2. Possible comparisons between the Deming Cycle and SEA phases.

In this model the methodological and procedural sequence for the SEA is close to that of EMAS, as follows:

- PLAN: analysis of environmental facts, definition of objectives and plan/programme strategies, analysis of the risks/impacts, and identification of alternatives for the plans/programmes. The PLAN phase of EMAS makes provisions for analysis of both the direct and indirect impacts of the local authorities' activities. Therefore, the same phase in both procedures can be understood starting from the analysis of environmental impacts realized at different levels but integrated: territorial for the SEA and local for EMAS.
- DO: implementation phase, representation and management of plans/programmes that define the outlined regulations (e.g. a Regulatory Plan with its own technical regulations), comparisons between the procedure and operative instructions established in the environmental management system according to EMAS. This results in the management and monitoring of the main environmental impacts which derive from direct and indirect activities of local authorities and result in the PLAN phase.
- CHECK: verifying and monitoring the established objectives in the PLAN phase, and the efficiency of the actions. This concept is applicable to both the operative procedure and Improvement Programmes in the EMAS accredited local authorities, and to the actions expected to realize the plans/programmes subjected to SEA. For both, the actions' monitoring and control is based on the specific indicators previously chosen in the analysis phase (PLAN). The indicators are valid tools to follow the development of the planned decisions both in terms of an EMAS accredited Control and Improvement Plan, and in a Territorial Government Plan (e.g. Indicators chosen for the technical Regulations drawing).
- ACT: Critical revision of all the processes carried out, any retroaction that can be changed, modifications of plans/programmes for SEA or Environmental Policies for EMAS. For both procedures this phase represents the adoption of the system and the course to a better level in which the cycle is repeated and the improvements continue towards new objectives.

These concepts are widened in the case study illustrated in section 3. From here on, the SEA applied to the Town Regulatory Plan, named Territorial Country Plan (Piano Assetto Territorio, PAT) in the Veneto Region Town Planning Legislation, is evaluated.

2.2. Plan: the analysis

In the "PLAN" phase EMAS proposes to establish environmental policies and plan management systems. This is made possible due to the data collected in the Initial Environmental Analysis (IEA), from which it is also possible to establish the initial position of the local authority/organisation with respect to the environmental conditions leading to registration. The information gathered from the Initial Environmental Analysis constitutes the basis of the entire Environmental Management System (EMS). It is not difficult to distinguish in this preliminary study many of the aspects considered in the SEA during the initial analysis of the environmental characteristics that are summarised in the Environmental State Report for the preparation of the PAT.

In both EMAS and SEA, the planning phase premises the collection and elaboration of data and quantitative indicators that give indications of the territory in relation to the predominant context.

- *air* (principle sources of atmospheric pollution, evident pollutants and their diffusion...)
- *water* (quality of surface and underground water, waste water drainage, presence and state of sewage networks and water treatment plants...)
- *biodiversity* (Sites of Community Importance, Special Protection Zone-SPZ, faunistic aspects, vegetation, landscapes and ecosystems...)
- *waste* (quantity and type of waste, presence of disposal sites, recycling plants...)
- *noise* (sources of acoustic pollution, acoustic areas...)
- *electromagnetism* (sources of electromagnetism, sensitive areas, power and distribution of mobile phone stations, electric lines...)
- *soil* (geological aspects, geomorphological and pedological, soil use, polluted areas...)
- *population* (residential population statistics, social-economic aspects...)

The next step is the recognition of all the specified environmental aspects, those with the greatest significance which are called significant environmental impacts. They constitute the basis for defining the improvement objectives to be followed through local authority policies.

Starting from an accurate analysis of the territory subjected to the SEA, it is possible to identify and describe through indicators the most critical aspects that, from a territorial planning point of view, must be taken into particular consideration in the strategic decision-making phase. These critical aspects, which prove the negative state of environmental resources or elements, define the improvement objectives as well as the programming choices, by integrating town planning with environment.

Just like IEA represents an instrument to identify the environmental policies and programmes, analogically the initial territorial analysis in the SEA allows well thought choices to be made, which lead to greater environmental sustainability from the programming options.

2.3. Do: implementation

The "DO" phase brings to light the main differences between the two regulations that converge to make practical and operative choices. In fact, what EMAS intends to achieve is an Environmental Management System, or more precisely a rationalization of programme processes and elaboration, with objectives and targets aimed at continuous environmental improvements; the SEA, on the other hand, is a tool created by PAT related to the development (not only Town Planning) of the interested town. Both implementation phases produce a document that defines the obligations and responsibilities as well as the indicators for the best management of environmental impacts at town (EMAS) and territorial levels (SEA and PAT).

2.4. Check: indicators and monitoring

The overall picture is formed even further in the monitoring phase "CHECK". In both the EMAS and SEA cases, the local authority tries to identify the effects of both managerial and strategic decision taken on the surrounding territory. The methodology itself that realizes the monitoring is similar, particularly when deciding on the environmental indicators. The EMAS indicators describe the direct environmental aspects tied to the activities of the local authority (e.g. waste water drainage in town offices, traffic flow problems etc.) as represented in Fig. 3.

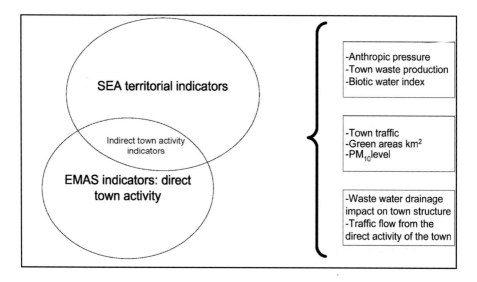

Figure 3. Integration of the indicators for EMAS and SEA on different territorial scales.

2.5. Act: the answer

At the end, both processes are given to the Town Administration, in the "ACT" phase, to confront any possible negative feedback that arises from the presence of the EMAS Environmental Programme and/or the plan in action (PAT), in order to establish new improving objectives that take into account the state of the indicators concerning the territory and its citizens. This is carried out through various instruments that allow the revision of the actions: e.g. changes to PAT, the insertion of new costs in the Town Budget, the involvement of the population through Forums etc.

2.6. Communication and participation

Another strong common point between SEA and EMAS is "participation", i.e. the importance of communicating with the public – specifying by this term the "stakeholders" of the territory. The SEA assumes that the systematic consultation with the population allows to define phases and to write the Plan/Programme. In its environmental programme EMAS includes meetings with those interested, to illustrate the Environmental Declaration in the EMAS registration phase.

Similarly to the EMAS, the SEA presents a strong vocation for written communication. The draft of the SEA details in the Environmental Report, for all stakeholders, presents a strong likeness to the Environmental Report expected from EMAS. It aims at divulgating on a wider scale the policies, programmes and decisions effected and more generally the Town's environmental performance. Both the Environmental Report and the Environmental Declaration can be the starting points for the preparation of the Agenda 21 document, undersigned by Italy and which contains the commitments of a local authority from the economic, social and environmental points of view. This document also aims at involving all institutional and economic actors to promote clarity and participation.

Finally, it should not be forgotten that both SEA and EMAS are fundamentally inspired by the principles of sustainable development, also through:

- the great control of the decisional processes and the strategic decisions, with consequential reductions of environmental risks;
- an increased efficiency in terms of reducing the environmental impacts, containing pollution in all its forms, reducing the consumption of resources, etc.;
- improved relationships with the local community;
- greater attention paid to environmental problems;
- more knowledgeable and thought out research aimed at resolving the questions related to the environment and territory.

3. CASE STUDY: THE SEA OF THE PAT IN BELLUNO, ITALY

By analyzing the Case Study of the town of Feltre (Belluno, Italy) it is possible to highlight some of the similarities between the SEA and EMAS management systems.

Recently, the town of Feltre came face to face with the need to update its Town Regulatory Plan, dating back to the 60's, in order to meet the new Town Planning requirements. Significant over-dimensioning, excessive number of roads, excessive use of agricultural land with vast areas left underused are just some of the most relevant factors demonstrating the inadequacy of the old town planning. In fact, the old town plan was inspired by optimistic hypothesis of demographic growth and economic development without taking into account the sensibility of the territory, part of which is made up of a highly valued natural reserve area in the National Park of the Dolomiti Bellunesi.

Therefore the Town Council decided to work on a new Regulatory Town Plan capable of taking into consideration the environmental requirements and current social and economic problems. Right from the beginning of the work, the realization of the Town Council Administration Plan activated the SEA procedure for the PAT 1 (Territorial Country Plan), as foreseen by the Regional Law 11/2004 of the Veneto Region, intending to take full advantage of its validity rather than considering it a mere compulsory amendment to the law (The town of Feltre, Territorial Country Plan; Town planning laws in the Veneto Region).

In the initial SEA phase process (see Tab. 1) applied to the PAT for Feltre, the collection and analysis of environmental and territorial data were necessary to give a clear image of the territorial situation as a starting point for the Report on the State of the Environment (RSA). The environmental components taken into consideration when writing the RSA follow many of those commonly considered in the IEA of an Environmental Management System (Sistema di Gestione Ambientale, SGA) in EMAS (Tab. 1).

Table 1. *Fundamental environmental components analyzed in the State of the Environment Report for the Feltre Town Council (Feltre Town Council, Strategic Environmental Assessment for the PAT)*

ENVIRONMENTAL COMPONENTS	BASIC ENVIRONMENTAL DATA
Air	Air quality level, presence of monitoring networks, punctual and diffused emissions, sources of atmospheric pollution.
Water	Surface and underground water quality, waste water drainage, presence of sewage network, % of connections, water consumption, water treatment plants.
Biodiversity	Important Community Sites (SIC) and Special Protection Zone (SPZ), natural oasis, ecological corridors, list and distribution of animals and plants density.
Waste	Quantity and type of waste, presence of ecological areas, method and % of recycled waste.
Noise	Level of noise pollution, noise zones, presence of sensitive areas.
Electromagnetism	Source of pollution, sensitive areas, power and distribution of mobile phone stations, electric lines.
Soil	Geological aspects, hydrogeological, geomorphological, pedological, use of the soil, polluted areas.
Social-economic aspects	Structure and demographic dynamics, economic, housing, historical-cultural value…

In the Report on the State of the Environment, problems and territorial potentialities emerged: weak points in terms of use of resources and pollution due to the main anthropic-productive pressures, and positive points regarding the environment and natural resources to be conserved and developed. The resulting picture gave useful elements to the policy makers, as well as citizens involved through Forums for the implementation of policies, for the prevention and environmental improvement to be carried out with the new PAT, in analogy with the intentions and results of the IEA in EMAS.

During the preparation of the Report, further proposals of indicators for each environmental aspect were considered. These indicators, in the initial analysis phase, represent the current aspects of the state of the environment. They are then selected on the basis of the planning choices and the need for monitoring, in order to meet the final phase of the SEA for the selection of indicators necessary for the complete assessment of the PAT and its actions achievable in the Plan of Interventions (Town planning laws in the Veneto Region). These indicators present many consistencies and offer interesting starting points to choose the indicators destined for an EMAS environmental management system which the Town Council of Feltre has recently decided to implement.

The territorial evidences and criticalities emerging from the RSA were understood during the draft output for the PAT, realizing the operative basis to define the objectives and actions of the plan, just as the analysis of the significant environmental aspects of a local authority that implements the environmental policy in the EMAS system.

The environmental assessment phase in the draft is configured according to the Deming Cycle and its PLAN-DO-CHECK-ACT process that redefines and improves the strategies and actions to meet the objectives outlined. The same cycle process can be seen in the revision of the initial environmental analysis and in the environmental policies/programmes in the drafting of a management system according to EMAS, such as the one being implemented in Feltre. The assessment phase is concluded with the Environmental Report according to the foreseen statute legislation. The Environmental Report represents the main document to be communicated to the public (European Directive 2001/42/EC). It has many similarities with the Recommendation of the Environmental Declaration contained in the Attachment 3 of the EMAS 2 regulations (European Regulations 2001-761).

As far as the technical work description is concerned, the communication problem that the environmental report causes in the representation of the territory was solved by using Geographical Information System (GIS) that allowed the environmental indicators chosen to evaluate the planning actions, in the town topographic cartography, to be studied. In SEA the communication terms were on a broader scale compared to EMAS and graphics on territorial dimension were used in addition to the bi-dimensional ones.

Fig. 4 shows an example of the topographic cartography related to the town recycling services used in the evaluation for the PAT of Feltre.

The graph represents with a gray scale the highest and lowest distribution of recycling bins in the central housing zone of Feltre.

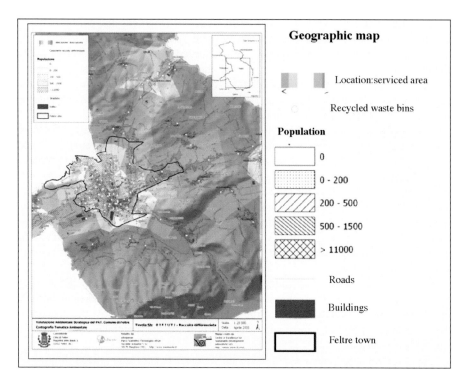

Figure 4. Map of recycled waste in Feltre. Made in cooperation with eAmbiente-CESD.

The map shows the distribution of area served by separate waste collection and the location of the bins. The example of recycled waste can be useful just to show how near the visualisation of an indicator is to both the decisions of PAT and the environment. These indicators establish not only the environmental improvement programmes for EMAS, but also the planning choices for the town territory (e.g. a new waste recycling centre).

The results of the environmental assessment process in the proofing of the PAT, are currently in course, and will be integrated in the drafting phase finalized and approved by the PAT. This follows the definition of foreseen measures with regards to the monitoring and evaluation of the actions' implementation impacts expected by the PAT (e.g. the resulting impacts of new housing and/or production buildings), in a similar methodological way as for the control plan and the internal and external audit plan for the SGA of the town. The feedback from the monitoring activity includes regular checking to see if the objectives are met and which the possible improvements are.

4. CONCLUSION

The SEA represents the first significant step towards territorial planning that takes into account the aspects of environmental sustainability. In order for the SEA to really result as being an efficient, supporting tool for effective territorial requirements and limitations, an exchange of information and continuous research to develop instruments and applicable methods for the planning process is necessary. Such a proposal is important and needs stimulus and synergy with other planning and management systems; first and foremost EMAS, with which the SEA shares many aspects, some of which include the type of environmental data used, the methodological approach, the objectives followed, public relations and the desired sustainability described in the previous paragraphs.

The reciprocal advantages that derive from the two systems are also in their availability for operative management control, outlining a flow chart that defines roles and jobs within the local authority/Town Council, that takes responsibility for the management of environmental aspects connected to their direct and indirect activities (planning, traffic management, tourism, etc.).

Whether a town council finds it necessary to implement SEA or voluntarily chooses an EMAS management system, both cases have available an important quantity of data and environmental analysis already organized and finalized for a plan or management which would be extremely attentive to the environmental, social and economic requirements. SEA and EMAS can represent two synergetic and complementary instruments for environment protection and sustainable territorial development.

Gabriella Chiellino
E-Ambiente, Italy

REFERENCES

Brunetta G. and Peano A., 2003. Strategic Environmental Assessment methodological aspects, procedures and criticism. *Books on Environment and Security*. Ed. Il Sole 24 Ore.

Chiellino G., Bassi S. and Giacomella S., 2004. Environmental territorial analysis as a voluntary management instrument and statute for sustainable development. Case Study: a) Parco Scientifico Tecnologico VEGA (Venezia), b) Comune di Cavallino. *Ecomondo – Seminar proceedings*. Rimini 3-6 November 2004.

EMAS, *Statistics of EMAS registrations in Italy*, available on the website www.minambiente.it.

European Directive 2001/42/CE from the European Parliament and Council, concerning the environmental assessment of Plans and Programmes, Art. 3.

European Directive 2001/42/CE, Attachment I, European Parliament and Council.

European Regulations 2001-761 (EMAS 2)

Feltre Town Council, Strategic Environmental Assessment for the PAT, *Report on the current state of the environment*, eAmbiente s.a.s., 2005.

The town of Feltre, Territorial Country Plan, Preliminary Document (ex art 3 L.R. 11/2004)

Town planning laws in the Veneto Region, LR 11/2004 "Regulations for governing the territory".

VII. SUSTAINABLE URBAN DEVELOPMENT

A. COSTA

GENERAL ASPECTS OF SUSTAINABLE URBAN DEVELOPMENT (SUD)

Abstract. Features of stable ecological structures are explored and compared with the characteristics of urban structures. Having a high efficiency in recycling material flows, while exploiting a renewable energy resource (sunlight), such ecosystems represent a good example of natural solution towards self-sustainability. Part of their dynamics could be taken as a reference for the environmental sustainability of anthropic systems. By identifying the standpoint from which the anthropic system differentiates from the natural one, this analysis allows the introduction of a set of ecological metaphors useful to interpret urban environmental inefficiencies and used to suggest possible pathways to cope with such issues.

After having introduced a number of urban issues that have the most relevant influence on the sustainability of cities, a refined and enhanced metabolic model (named new urban metabolic model), that takes explicitly into account the role of institutional activities, is deepened to produce a tool for the description and the analysis of environmental dynamics and performances in cities.

1. INTRODUCTION

Nowadays, more than 50% of the world population lives in cities. Such threshold will be soon achieved by the Asian continent itself, while the current figure for Europe reaches an impressive 80%. Due to such large diffusion, urban settlements have a great impact on their environment, being simultaneously the place of highest consumption of energy and transformation of material resources; the most relevant source of contamination/pollution and waste; the most relevant contributor to regional and global environmental threats.

Moreover, cities are characterized by a high degree of complexity, because of the broad interaction of actors, processes and dynamics that take place mainly within the urban settlement but also between the city and its surroundings.

This complexity, which is reflected on the main challenges that the city has to face to carry on with its development without compromising local and global environment, needs to be properly handled to achieve local and global sustainability.

A significant way to cope with such an issue is to couple technological knowledge with the study of natural dynamics that ensure self-sustainability of ecosystems, in order to find further inspiration for the definition of good policies and practices.

2. URBAN ECOLOGY

With a high efficiency in recycling material flows, while exploiting energy obtained from a renewable resource (sunlight), ecological systems represent a good example of natural solution towards self-sustainability. Part of their dynamics could be taken as a term of reference for the environmental sustainability of anthropic systems, best represented by urban settlements (Costa, 2004). The knowledge that rises from an ecosystem approach can be used to find new solutions to improve environmental performances of cities.

2.1. The city as an ecosystem

In ecological terms, cities can be seen as a peculiar kind of ecosystem. Such analogy is based on the fact that a city contains, in a limited space, a heterogeneous community of living species, heavily dominated by mankind. Humans have the ability to transform to a high degree their supporting system – represented not necessarily only by their spatial surroundings – to meet their needs. These transformations are based on the interchange of matter, energy and information with the supporting system. The supporting system represents at a time a reservoir of resources that are necessary to maintain urban processes, and a "sump" where disposing exhausted materials (waste and heat).

Other typical features of urban ecosystems are the almost complete dependency on exosomatic energy (energy that circulates outside organisms) produced elsewhere (which makes the city a *heterotrophic* ecosystem) and a higher degree of complexity deriving from cultural artefacts, both material and immaterial (Rueda, 1995). It is the cultural factor that governs the existence and the evolution of the city, while it is the continuous flow of matter and energy that feeds the complexity of urban structures.

Such predominance of the human dimension over the physical and spatial layout of urban areas indicates that the environmental analysis of the city system cannot ignore the integration of scientific and socioeconomic variables. Otherwise, the interpretation would be incomplete.

The term *urban ecology*, thus, means the study of dynamics, interrelations and processes that occur among those functional sectors that form urban artificial ecosystems.

2.2. Flows in cities

The functioning of ecosystems is strongly related to flows of energy and matter circulating throughout them. The study of such stream dynamics in urban areas, and of the mechanisms lain behind them, represents a useful exercise to evaluate possibilities of optimization – therefore of better sustainability – of urban processes.

An interesting approach to such an issue, which comes from a biological transposition, is the study of *urban metabolism* (Wolman, 1965; Newman, 1999). Through it, city processes are split into two phases:

- *anabolism*, or constructive activity: the complex of processes of matter and energy transformation entering in the city system to satisfy demands and needs expressed by urban activities;
- *catabolism*, or destructive activity: the complex of processes related to the production and removal of waste deriving from urban activities.

In its catabolic phase, the city pushes away towards its supporting system the entropy produced, that is represented by both waste and energy with lower information contents (e.g. infrared radiation).

Basically, the city system is seen as a black box, which increases its own internal order at the outer environment's expenses. Major environmental problems – with their high economic costs – expressed by urban systems relate to the growth of entering streams and to the management of outputs' increase. According to the II Principle of Thermodynamics, the higher the entropy produced by the black box, the lower the level of the city's sustainability on the surrounding environment (Prigogine, 1954).

Therefore, it is possible to define the goal of urban sustainability as the combination of the increase in urban living standards, with the decrease in the exploitation of both natural resources and production of waste that put the city in the position to better cope with the carrying capacity of local, regional, and global ecosystems. A more articulated concept of urban metabolism rises from such consideration, which is based on an input/output approach: *input* standing for matter, energy and information, while *output* standing for information/knowledge, goods/services, heat and waste.

In terms of differences between urban and natural ecosystems, it can be observed that the former bases its efficiency on high rates of circular metabolism, while the latter is strongly characterized by linearity. This fact gives an explanation for the relatively low efficiency of city ecosystems and suggests that a move toward *circular economy* could easily improve environmental performances of urban settlements. The closure of energy cycles needs to undergo a thorough change in energy policies, through a shift from traditional fossil sources towards renewable sources, also exploiting technologies to recover energy from residual materials.

On the other hand, the closure of material cycles should be focused on an appropriate inter-linkage among different productive processes, so that the use of resources is maximized, and by-products of a productive process feed other productive processes. A further step in such approach is represented by the so-called *3R philosophy*, that encourages – where feasible – the *recycling, reduction* and *reuse* of goods, extending their usual life cycle. The implementation of a circular economy approach is highly beneficial to contain anthropic environmental impacts.

2.3. Urban diversity

Much of the stability of an ecosystem is based on the functional diversification of its living community. This occurs because an ecosystem rich in species has a more complex organization and a higher number of feedback mechanisms capable to stabilize its global function. Diversity is also responsible for the specific spatial distribution and occupation/transformation of an environment.

Transposing such approach to the urban environment, *cultural diversity* among people could be considered a measure of the resilience of a city. The city is found on the interaction among its actors: citizens, activities, and institutions. Each individual possesses a pattern of *attributes* that determine its functional specialization (Rueda,

1995). Specialization affects paths of energy, matter and information within the urban system, through which the individual interrelates with attributes of other actors, either physical or juridical. A city has a range of categories based on the convergence of single actors, each one finding its own *locus* of existence, a sort of *ecological niche*.

The web of mutual interrelations is regulated by institutional mechanisms (e.g. regulations, laws, other codified actions) and fostered by the presence of networks (e.g. transport, energy, communication, cultural interactions) and their related infrastructures.

Each urban range of diversity implies a different impact on the environment. An increase in the density of diversity, if favoured through the selection of activities functionally mixed and compatible with the carrying capacity of the supporting system, allows for a more efficient and sustainable territorial use.

2.4. Feedback mechanisms in cities

As in any other cybernetic system, the presence and the quality of suitable feedback mechanisms are of primary importance in order to guarantee along time the existence of a city (Odum, 1996). Due to the city's nature, urban self-regulative mechanisms are man-driven. The intervention of such artificial controls allows for the management of both energy and matter flows, permeating urban ecosystems.

Anthropic feedback mechanisms can be subdivided into two different groups: aware mechanisms and unaware mechanisms. The first class is composed by urban institutions (public, private, organizations, councils, etc.), regulations that have a direct or indirect effect on urban dynamics, market incentives for the enhancement of urban quality of life, and all those services and technologies available for maximizing the use of resources while minimizing the impacts on the environment. The second class is composed by cultural diversity and functional specialization of urban actors, which just by interacting can express a certain degree of control on territorial dynamics and on the use of resources. Moreover, anthropic feedback mechanisms follow a codified hierarchy, where someone controls the action of others. At the apex of the hierarchy stand those institutions – like urban or supra-urban administrations – appointed to rule on less complex institutions.

The definition of a set of anthropic feedback mechanisms that imply increasing levels of institutional organization and technological back-up should allow for the activation of an urban *ecological government* (mainly on flows and on cultural diversity), suitable to keep the city pressure below its carrying capacity, while ensuring at the same time a high standard of quality of life.

3. URBAN ENVIRONMENTAL ISSUES

Leaving the ecosystem approach to sustainability, this paragraph proposes an overview on the most relevant dynamics that have environmental implications for cities.

With respect to rural conditions, urban appeal lies on the perspective of finding a better set of opportunities and a better level of quality of life. In this sense, urban satisfaction can be represented by the following four factors:

- quality, variety and accessibility of services provided to citizens and firms;
- living and working opportunities;
- the existence of a healthy and recreational environment;
- individual income and welfare.

Classical patterns of industrial development have privileged economic growth, neglecting the correct relevancy to quality of life and environmental integrity. Such biased situation needs to be rebalanced through the application of the principles of sustainable development.

To avoid negative consequences of urbanization, as those experienced by western countries during the last 40-50 years, more effort should be put by those countries in which cities are growing at a very fast pace, a phenomenon that nowadays characterizes – e.g. – many Asian states.

3.1. Different patterns of urbanization

Focusing on patterns of urbanization, two different Models of Territorial Development (MTD) can be observed, each one of them exerting a different pressure on the environment. Such empirical classification is mainly determined by the spatial density of population (Newman and Kenworthy, 1999). The two categories are:

- the *compact city* (more than 45 dwellings per hectare), observed in Europe;
- the *spread city* (10-12 dwellings per hectare), more usual in USA.

Due to their spatial dimensions, Asian, African, and American *megacities* (urban systems with more than 10 million inhabitants) combine together features belonging to both MTDs, mostly emphasizing management drawbacks than opportunities.

Although having a lower rate of *naturality* (reduced presence of urban parks, green corridors, green buffers, etc.), the highly transformed environment of a compact city allows for a shared effort in service and infrastructural provision, as well as a shared consumption of resources.

On the contrary, the spread city consents a better dilution of contamination and higher rates of naturality. But low density implies higher consumption of resources *per capita* and lower efficiency in service and infrastructural provision.

Whether the compact city is more sustainable than the spread city is a dispute that dragged multidisciplinary opinions into it, during the last decades. With respect to those issues that have a more direct connection to the economic dimension of environmental management, it can be easily seen that the compact pattern is more beneficial. Preserving rural space from urbanization, reducing transit needs by fostering proximity between residency and employment, promoting a compatible mix of activities in a single area are further factors of basic importance for the environmental performance of an urban settlement. They can be better achieved in a dense environment than in a diffuse one. In the end, it is the critical mass of the compact city that offers the best guarantee of closure for urban metabolic cycles.

3.2. Most relevant environmental urban issues

This paragraph is devoted to depict the main issues that concur in determine the environmental performance of an urban settlement. For each one of them a brief description of its features is given, together with the suggestion of good practices to be implemented for sustainable management.

What should be clearly borne in mind is that such issues are strongly intertwined. It is only through an integrated approach that takes into account the criticalities of all issues that a policy of sustainable urban development will achieve effective and fruitful outcomes.

3.2.1. Transport
It is widely acknowledged that mobility of people and freight is the most impacting environmental urban issue. Financial commitment for infrastructural provision, energy consumption, environmental pollution, congestion and land fragmentation, deriving from coping with the need for mobility, impose a strong pressure on the environment.

There is a strong interdependency between the transport network of a city and its MTD. In fact, spread cities need a higher rate of transport infrastructure *per capita*.

Indistinctly favouring road mode and individual means of transport as common practice, constitutes the breeding ground for the rise of diseconomies. The right to use a private vehicle is not questioned, but the negative evidence so far described should encourage the reliance on collective means of transport. Individual means have a lower energetic, economic and environmental efficiency. Collective means perform better, but have limited door-to-door efficacy.

This very last consideration can be extended to other more environmental-friendly networks, like railways and waterways.

Actually, most common policies found their success on intermodality, relying on park-and-ride systems to connect different transport networks, and trying to find a better balance between collective and individual transportation. But, to be more effective, this approach needs to be integrated with further initiatives, co-ordinating transport policies with spatial development policies. Enhancing proximity between home and working place has the double benefit of reducing the commuting flow, while making the use of collective transport more viable. To reduce traffic congestion, transport of people and freights can be decoupled, not necessarily through the use of different networks, but promoting the mobility of each category at its most appropriate time-course (i.e. people during day and freights on night time).

With respect to technological issues, the use of intelligent transport systems and the shift towards less impacting means of transport – through the implementation of the use of new generations of engines and fuels – should also be endorsed. Replacing oil with gas and fostering the use of energy from renewable sources is a first, clear step towards the goal of sustainable mobility.

3.2.2. Energy consumption

Because of the high concentration of human activities, the city is a hot-spot for energy consumption. Environmental concerns rising from energy consumption are due to the current energy base of our socioeconomic system, which is almost entirely dependent on the exploitation of non renewable sources (e.g. oil, coal, nuclear). The use of fossil fuels embeds two key negative aspects:

- the emission of green-house gases as by-product of their transformation, that is responsible for global warming;
- geopolitical implications, related to the uneven distribution of fields and to their progressive depletion, reflected on constantly increasing energy price.

To be successful, sustainable energy policies should take into account at a time two main approaches: the maximization of use, through the containment of waste, and the endorsement towards a shift of energy base, enhancing the reliance on renewable sources (e.g. sun, wind, hydro, geothermal).

Recovering energy from waste, co-generation and measures of energy-efficiency, are some examples of maximization in energy use. They are all lined up with the 3R philosophy described in par. 2.2.

The shift towards renewable energies does not only have to bridge the difficult gaps represented by cultural attitude and technological change. It also implies a massive transformation in the energy supply economy sector. Financially speaking, updating the energy base requires a great effort by both producers and customers. The use of public-driven market incentives for the development of renewable-based utilities is a tool that can be beneficial to the growth of such new economic sector.

3.2.3. Green spaces

On top of their aesthetic and recreational function, urban green spaces provide the city with a set of free services, better known as *ecological services* (Costanza *et al.*, 1997). Microclimatic and noise control, filtering pollution, reducing soil erosion are all examples of such services. Unfortunately, urban development has frequently compromised the integrity of local natural capital, being this phenomenon a direct consequence of anthropic perturbation, polluting charges and fragmentation of green areas.

Restoration of green spaces through their correct planning and design allows the city to take advantage from the provision of ecological services.

Technological and scientific know-how applied to green space management, can successfully enhance natural effectiveness of ecological services. Choosing native species for reafforestation (they have longer chances of survival and lower maintenance costs) and selecting the most appropriate set of plants for wetland filtering represent remarkable models of correct application of current knowledge.

Engineered wetlands for water treatment, green buffers alongside transport infrastructures, and green belts for the containment of water run-off are all examples of how green presence can be a significative means for urban sustainability.

3.2.4. Built environment

Built environment is the most characteristic visual feature of the city. Building and infrastructural continuity is the clearest sign of territorial transformation under the scope of human coexistence. Unfortunately, its development causes a series of complex and interrelated environmental impacts, in terms of energy and material consumption, soil sealing, green-house gases emission, microclimatic alteration, etc.

Traditionally, design has not taken into the due account the environmental performances of infrastructural and building construction and management activities. Most technological advancements in the construction sector were focused on the containment of building costs, and not lead by any environmental awareness.

Given the current growing concern about environmental issues – and the increasing energy and material costs – a new approach in such sector has been taking place. Although retrofitting present structures with environmental technologies can improve the situation, the most appropriate response to environmental issues raised by the built environment is to tackle them from the earliest stages of planning and design. This new tack is based on the principle that reducing environmental impacts of built environment, both during its construction and in day-by-day management phases, allows for the containment of negative externalities, being, thus, in the mid and long term more economically efficient. Such new approach, named *Environmental Design and Construction* (EDC), focuses on embedding in urban buildings and infrastructures solutions of energy-efficiency, waste reduction while delivering enhanced liveability. It is applied to the design, choice of materials and construction yard activity phases. In terms of design, it can take advantage from passive solutions (shape & orientation of buildings to maximize climatic conditions) as well as from technological endowment (e.g. photo-voltaic panels for energy production, high-efficiency cooling/heating systems, water recycling devices), the outcome being low-emission structures. When inspiration is taken from environmentally successful natural features (e.g. termite-mound to exploit natural ventilation), such approach can also be referred to as *bioarchitecture*.

Promoting the use of recycled and environmental-friendly construction materials is a further step towards the optimization of the use of resources.

To make yard activities more cost-effective and less polluting, a series of *Environmental Management Systems* (EMS) have been developed. EMS are a set of procedures and codes of practices that can help assessing the impact of construction activities, thus finding the best solution to minimize it. EMS are often linked to international quality control standard systems, like ISO or the European EMAS.

Another relevant factor for built environment sustainability is the pre-emptive assessment of interventions' impact. Environmental Impact Assessment (EIA) and Strategic Environmental Assessment (SEA) – each one for its respective scale of application – are two effective tools to evaluate the potential impact of a development operation before it is materially implemented, thus allowing for corrective or mitigating measures, if the case.

EIA and SEA have a great value also at larger spatial scale. They can be applied to territorial development in order to address correctly the planning processes of an entire urban area towards its sustainability.

3.2.5. Water

The city is a high consumer of water, a resource scarce worldwide. Natural water cycle in urban environment has been deeply modified for two main reasons: because a considerable amount of resource has been diverted from natural paths to fulfil anthropic needs; because surface sealing exerted by built structures forces water flow through new paths, frequently leading to unexpected floods. Water run-off, often enhanced by water-proof surfaces, collects the contamination proceeding from the large number of non-point sources and sends it to the final receiving bodies (rivers, lakes, aquifers). This determines a potential hazard for the integrity of the receiving bodies and for those ecosystems positioned downstream. The overall result of all such dynamics is a new water cycle that can be very hard to manage.

On top of technological solutions for water treatment and water use reduction, policies of sustainable water management should build on a cultural shift towards an environmentally aware use of such resource. Schemes of combined waste water and storm water recycling, besides having a concrete effect on containing water consumption, have a sound cultural value, for they help raising the awareness on the importance of preserving such scarce resource.

3.2.6. Waste

The wide range of urban processes entails the production of a large and composite amount of waste, which requires a high financial commitment for its management and final disposal.

Apart from hazardous waste and toxic waste that need specific treatments for their disposal, most part of collected urban waste is either stored in landfill or incinerated. Both disposal methods have strong environmental implications: the former being a potential cause of ground contamination, while the latter spreading pollution in the atmosphere.

The application of principles of 3R philosophy is a first approach to the containment of waste production that, while improving the efficiency of material flows in the city, reduces the need for more space and financial resources to be devoted to disposal facilities and operations. However, what is becoming every day clearer is that waste itself is to be considered a new resource: collecting biogases from landfill, co-generating heat from waste incineration, composting organic waste, are some examples of how the *culture of waste as a resource* can be also an economically viable sector.

Once again, it should be reminded that the best outcome in terms of sustainability is achieved when all different solutions are exploited at the same time.

3.2.7. Contamination

Almost any transformation of matter and energy produces a correspondent amount of polluting substances which are responsible for the contamination of the environment. Decreasing environmental quality and high costs of decontamination are the most relevant impacts of pollution.

Contamination load can be split with respect to the environmental compartment that receives it: atmosphere, soil and water bodies. Atmospheric and water contamination is generally managed either through waste water and emission treatment right before their release into the environment, or through their dilution to levels of concentration considered not harmful for the ecosystem. Soil contamination, on the other hand, calls for more articulated interventions of land remediation.

Commonly, contamination control system is based on the definition of concentration thresholds of polluting substances on top of which a legal action is taken to stop their release into the environment.

Leaving such end-of-pipe solitary approach to contamination control is a praxis that must be enforced to obtain more effective results on pollution management. Fostering the development of clean productive processes is a pre-emptive policy to reduce the polluting load proceeding from the industrial sector.

For what refers to land remediation, favouring when possible on-site treatment from off-site techniques is in general a better practice, because it avoids the risk of transferring the contamination elsewhere, while saving handling costs.

What is also significant is that so far, due to lacks in knowledge or finance, the combined effects of different contaminants have been seldom assessed. To develop a more successful management policy, a stronger effort should be put in studying synergic effects of multi pollutant contamination.

3.2.8. Abandoned areas

Abandoned areas are urban parcels of land formerly devoted to productive activities – that for economic and/or environmental reasons have been delocalised elsewhere – which are no longer hosting urban processes. Having lost their original functionality, they can be recovered for new development, a dynamic definable as *territorial recycling*. In most cases, even urban space can be seen as a scarce resource and the presence of such areas can provide the city with new places where to settle new activities. Such areas are generally highly suitable for redevelopment, owing already a favourable infrastructural endowment and a good level of connectivity with the urban system.

However, unfortunately, updating abandoned areas is an issue strongly connected to environmental contamination. Due to their industrial past, these areas usually need to undergo expensive remediation intervention to remove contamination before allowing their redevelopment. Thus, financial availability is often the main constraint for resolving environmental contamination. Public-private partnerships and project-financing mechanisms are two possible approaches to gather the needed amount of money to carry-out successful remediation and redevelopment operations.

Once reclaimed, abandoned areas are ready to host new urban functions which shall be chosen with respect to local environmental compatibility.

3.2.9. Sense of belonging to a community

Promoting the sense of belonging to a local community is a factor of great importance for urban sustainability. Man establishes a wide set of relationships with the place in which he acts (lives, works), a fact that enhances individual and social attention for local quality of life (Rees, 1997) A community shares concerns about local issues and shares the responsibility for the conditions and fate of a common space. In an urban context, social equity, referred to access to opportunities and availability of services, allows for the development of a virtuous community that helps territorial management and enhances the control over it. This is beneficial in case of emergencies for early warning, but also to promote a fruitful debate on local issues.

A policy suitable to strengthen such sense of belonging is to encourage individual participation to local management. Participative approach can be a significant feedback mechanism of the urban ecosystem, because through their daily experience its inhabitants can help policy makers giving correct relevancy to local criticalities.

Active participation of a community is facilitated by its cultural diversity that allows for the comparison of different opinions on the best way to cope with a contingent issue.

Fostering this kind of governance through the establishment of discussion panels on local issues and through the provision of spaces for social aggregation is a good policy to enhance community control on urban environment.

4. THE METABOLIC MODEL

To handle the evident complexity and dynamism of urban environment, principles, dynamics and practices described in the previous paragraphs are here condensed through a metabolic approach to built conceptual metabolic models that give a different insight on the way cities work and develop. A metabolic model is a schematic and synthetic representation of urban flows and processes. Such diagram is used to organize the set of information collected about a city, in order to:

- identify whether there is enough information about a specific topic and to recognize where there is a need for tackling criticalities;
- have a simultaneous outlook on active urban dynamics;
- develop performance indicators to assess urban policies' effectiveness;
- define standardized monitoring procedures to allow the comparison of performances among different cities.

If customized to focus on environmental aspects of city development, metabolic models provide a powerful tool to frame urban features, help the evaluation of their environmental conditions and, eventually, address policies and management strategies toward urban sustainable development.

4.1. Evolution of metabolic models and their main features

Fig. 1 shows the metabolic model originally proposed by A. Wolman (Wolman, 1965) in the mid 60's. This diagram takes into account amounts of input and output, allowing the definition of efficiency indicators (e.g. the ratio between emissions and fuel gives a measure of the pollution raising from energy consumption). However, its strongest limit is that it gives no insight on urban dynamics, considering the city a mere black box.

Such a weakness has been addressed by the model evolution occurred at the end of the 90's. Enhanced metabolic models encompass a specific and valuable remark related to urban systems: as observed by P. Newman (Newman, 1999), there is more to a city than a mechanism to process resources and produce waste. The city is a place where human opportunities are created and for which the population has hopes and expectations.

Through such development, the model's conceptual scheme becomes more articulated, with a more analytic list of elements forming the input/output sequence. Two important considerations emerge from the enhanced diagram. Firstly, it allows for a set of internal dynamics that take place in the city and that have control over the input's repartition and transformation of human settlements, and, consequently, over their output's quantity and fate. Secondly, it defines two different areas of output, acknowledging that urban dynamics do not only determine the production of waste, but they are in place mostly for the development of services, goods and quality of life.

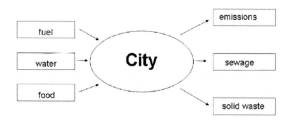

Figure 1. The original metabolic model from Wolman.

Giving a deeper look at the enhanced diagram, it can be observed that:

- it accepts performance indicators and quantitative goals (standards and benchmarks), to assess local environmental impact;
- it represents a broad spectrum of urban key sectors (housing, health, transport, employment, consumption, leisure, economic activities, etc.) and puts forward a systematic vision of the city;
- it allows representations of causal relationships within the system, through DPSIR sequence (Driving force/Pressure/State/Impact/Response);
- it allows effective representations of those dynamic processes associated with sustainability, following the endowment/processes/output sequence.

The enhanced metabolic model has been successfully applied as a reference scheme for urban environmental issues discussed in the *Australian State of the Environment of 2001*. Fig. 2 describes the used metabolic diagram (Newton et al., 1998).

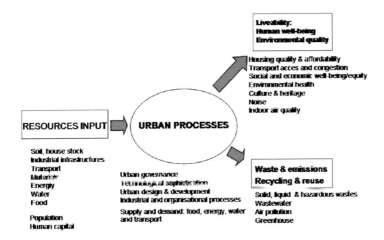

Figure 2. Newton's Extended Urban Metabolic Model (EUMM).

4.2. A latest development: The New Urban Metabolic Model (NUMM)

With the intention of better integrating sustainability principles and the role of control mechanisms (feedbacks) within the metabolic scheme, a further development of the metabolic model, called New Urban Metabolic Model (NUMM), has been proposed (Costa, 2004).

Last decades' international debate on sustainability acknowledges the importance of institutions as a new pillar of sustainable development, adding this independent factor to the old conceptualization based on three components (ecology, economy and society). In a heavily anthropic environment, most effective feedbacks for urban dynamics and processes are those established by humans. Institutions (administrative, juridical, social, economical, financial, cultural and so on) represent those urban actors that provide the system with rules. The term *institutions* is referred to those organizations that govern collective practices. They can be mainly subdivided into two groups: managers of the *res publica* and managers of resources.

The institutional role has been internalized into the metabolic diagram, ensuring that sustainability principles are admitted as basic principles of institutional activities.

Such behaviour is put into practice especially through the development of the following activities:

- evaluation of urban development proposals and strategic planning;
- urban management day-to-day activities;
- environmental command & control activities.

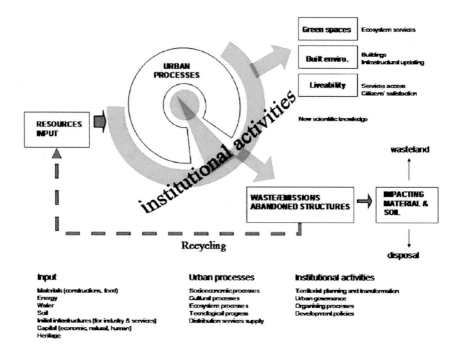

Figure 3. The New Urban Metabolic Model.

Institutions, thus, act as a "filter" that mediates relationships among different components of the urban ecosystem. They govern paths and flows in the city.

A further essential factor encompassed by NUMM is that urban environmental policies can gain from citizens' participation to decisional processes. Such vision entails *governance* based on the synthesis of local groups' needs and expectations.

The second relevant novelty is that NUMM is a metabolic scheme that explicitly takes into account the 3R philosophy, through the cyclization of flows.

Combining the remarks referred to 3R philosophy with those on institutions enables the definition of a new metabolic scheme that has been conceived to openly evaluate how different sectoral environmental policies take part in the definition of a project for a sustainable urban environment. NUMM diagram is presented in Fig. 3.

The section that refers to inputs embodies material, energetic and water resources, but also land availability and infrastructural equipment (e.g. transport, communication, manufacturing and environmental services). Particular attention has to be devoted to different kinds of capital (natural, economic and human). The flow of inputs must be conveyed through the intervention of different institutions, organizing and regulating urban processes through institutional interaction.

Such dynamic feeds two areas of output: one referred to material and intangible urban production, a second one referred to the production of waste/emissions and physical elements (soil and buildings), no longer used, and that need to be either disposed or reclaimed.

The first of these areas of output focuses on three different kinds of urban products: green areas, built environment and all the elements that concur to urban quality of life. In addition, new scientific knowledge – that is central to technological progress – must be taken into account as an additional outcome.

The second area of output focuses on further interesting elements. Thanks to sound environmental policies carried out at institutional level, a fraction of the catabolic flow can be transformed into a positive cycle, hence feeding once again the flow of inputs. The more considerable is this fraction of recycle, the more relevant is the degree of closure of the system. At the same time, as a matter of fact, the need for waste disposal, within or outside the city system, is reduced. Again, the effectiveness of closure of the cycle resides largely at governance/institutional level. This model is, thus, strongly characterized by a focus on an explicit circularity and on the analysis of the institutional feedbacks within urban processes.

5. CONCLUSIONS

Being the city really a dynamic and complex ecosystem, it requires an integrated approach to produce successful politics for its sustainable development.

In such sense, NUMM can be a useful tool to describe urban dynamics and can be applied to support the decision making process carried out by actors/institutions devoted to the development and management of urban areas. As a matter of fact, it is of remarkable relevance to the focus of this analysis that enhanced metabolic models embed the four key principles of urban sustainability:

- the reduction of the resources' input through the maximization of their use;
- the increase of urban liveability;
- the reduction of waste and emissions produced by anthropic activities;
- a higher efficiency in the management of urban processes.

The ultimate outcome of the application of such approach determines the following list of rules that can be generally considered valid for every urban system:

- promote the culture of waste as resource (3R philosophy), broadening this approach to land policies, giving priority to regeneration of abandoned areas instead of encouraging urban expansion;
- promoting practices of energy efficiency, focusing on energy reuse and on exploitation of renewable sources;
- restore natural capital equilibrium, in order to regenerate ecosystem services;
- minimize the emissions produced by anthropic activities;
- minimize the environmental impact of transport systems;
- foment the sense of belonging to a community;
- enhance social equity and access to urban opportunities;

- develop policies of spatial compatible multi-functionality, with particular reference to coexistence of different activities (like housing and business);
- promote efficient urban services;
- compacting the density of urban sprawl;
- develop a governance model open to active participation of citizens in the debate on urban development and transformation.

To conclude referring to the ecological view of the city, it can be stated that the task of local policy-makers must consist in managing urban ecosystem aiming at transforming the linear flow resources/goods/waste into circular and self regulated flow, typical of a stable ecosystem. Such control process should not affect negatively the quality of life; on the contrary it should increase the liveability of the city itself.

Alessandro Costa
Fondazione Eni Enrico Mattei, Italy

Sino-Italian Cooperation Program for Environmental Protection, China

REFERENCES

Costa A., 2004. *Developing the Urban Metabolism Approach into a New Urban Metabolic Model. The Sustainable City III*, Wit Press, pp. 31-40.

Costanza R., d'Arge R., de Groot R., Farber S., Grasso M., Hannon B., Limburg K., Naeem S., O'Neill R.V., Paruelo J., Raskin R.G., Sutton P. and van den Belt M., 1997. *The Value of the World's Ecosystem Services and Natural Capital. Nature 387*, London, pp. 253-260.

Newman P., 1999. Sustainability and Cities: Extending the Metabolism Model. *Landscape and Urban Planning 44*. Elsevier Science, USA.

Newman P. and Kenworthy J., 1999. *Sustainability and Cities. Overcoming Automobile Dependence*. Island Press, Washington D.C.

Newton P., Flood J., Berry M., Kuldeep B., Brown S., Cabelli A., Gomboso J., Higgins J., Richardson T. and Ritchie V., 1998. *Environmental Indicators for National State of environment Reporting – Human Settlements. Australia State of the Environment (Environmental Indicators Reports)*, Department of the Environment, Canberra.

Odum H.T., 1996. *Environmental Accounting, Emergy and Environmental Decision Making*. John Wiley & Sons, USA.

Prigogine I., 1954. *Introduction to Thermodynamics of Irreversible Processes*. C.C. Thomas, Springfield.

Rees W.E., 1997. Urban Ecosystems. The Human Dimension. *Urban Ecosystems 1*, Chapman & Hall, pp. 63-75.

Rueda S., 1995. *Ecologia Urbana. Barcelona I la Seva Regió Metropolitana com a Referents*. Beta Editorial, Barcelona.

Wolman A., 1965. The Metabolism of Cities. *Scientific American 213*.

M. SAVINO

A NEW PERSPECTIVE FOR THE FUTURE: ENVIRONMENTAL PROTECTION AND SUSTAINABILITY IN URBAN PLANNING

The Italian Case

Abstract. Since the origins, planning has been conceived as a technique, then as a science for the transformation of natural resources (first of all the soil) to respond to economic and social needs of a small community, first, and then for a fast-growing and more complex society. According to this deep-rooted approach, environment has been actually considered as a *means* for development to be used unconditionally. In Italy, since the 1970s the economic development involved a strong process of urbanization, by the sprawl of residential areas far away from big cities' areas as well as the sprawl of factories from big industrial plants concentration. In the last years, various regional governments (that in Italy are in charge of planning and environment protection) have been trying to introduce sustainable principles and a framework to conceive, draw up and carry out plans and projects for a concrete sustainable land use and an effective environmental protection.

1. INTRODUCTION

Planning activities have been traditionally voted to solve settlement problems for production activities and residential uses, briefly to assure the best conditions for the development of local communities.

Since its origins, planning has been conceived as a technique – and only afterwards as a science – for the transformation of natural resources for the development of human activities in order to respond to the economic and social needs of a small community, first, and then for an increasingly fast-growing and more complex society. According to this deep-rooted approach that still runs in our contemporary and globalized society, natural resources have been actually considered as a *means* of development to be used unconditionally and every development pattern produced in these last fifty years never paid any attention to the environmental effects or to the irreversible impacts on natural environment.

Since the 1990s, due to some well-known international agreements and not only for the environmental emergencies spread all over the world or for some environmental accidents, institutions, people and economic actors agree on the fact that natural resources are not reproducible; degradation and pollution of environment are a serious danger for our society (and maybe a hazard for economy itself) and, for the first time, they are realizing that our development pattern, urban organizations, housing habits and lifestyles are wrong and must be changed. Therefore, the introduction of sustainability principles is progressively improving land use patterns and urban development. In the last ten years it has been possible to record a very intensive and vigorous effort to improve planning goals, techniques and visions, in a sustainable key, looking for new approaches to urban areas development, combining economic growth, qualitative improvement of urban life and protection of natural resources.

Planning seems to have a decisive and strategic function in this "revolution" of social awareness, for it could explore alternatives – better than imposing strict controls or restrictions – to build a different environment for community development and to assert the principles of sustainability.

In many countries and in many local experiences, planners are achieving new projects and new interventions for regulating land use (with new approaches for soil exploitation control, by the containment of urbanization and build-up areas), organizing spatial locations (rationalizing industrial plants location, or with brownfields redevelopment), changing land mobility patterns (strengthening and supporting public transport networks; enhancing integration among different transport modalities). These efforts take place with some complications due to the lack of a specific legislative framework but with a very strong popular support and participation. However, it is not an easy task because often (and frequently in Mediterranean urban areas) the urbanization and the city expansion are not the effect but the cause and tool of urban economic growth.

In Italy also, sustainable development principles have been introduced thanks to an increasing political and social awareness, as well as for the juridical prescriptions imposed by the EU or by international agreements. We can now witness an interesting debate on how to change policies, practices and interventions in a sustainable key. It is a very exciting and difficult moment (we cannot deny it), because we have to face a legislative system which is not so flexible, a mainstream of thinking which is not completely persuaded of sustainability.

It may be that in Italy this process is carried out with some difficulties – due to a very peculiar economic and political situation which implies for instance more disadvantages in the less prosperous areas in the south of the country – but we have to recognize that in many urban areas and in some regions – in line with other European countries – significant results have been achieved and interesting experiences have been performed, thus indicating the diffusion of the new perspective of sustainable development.

2. THE TRADITIONAL ROLE OF PLANNING FOR THE DEVELOPMENT OF SOCIETY

Although planning, as a set of techniques and objectives, theoretical explanations, different methodologies, and ways of thinking, finds its contemporary roots in the XIX century, around the time of the Industrial Revolution, as a means for society to adjust to new productive typologies, it actually dates back to ancient times when construction techniques were inextricably linked to religious, military, economic and symbolic elements.

The modernization of the city, its capacity to embrace more and more inhabitants and the organization of an efficient network for the exchange of goods and labour force have become, with time, decisive factors in the development of the production system. Urban planning is therefore the tool to achieve efficiency and functionality with regard to both the social system and its surrounding territory.

With time, planning has changed aims and theoretical references, bending towards the construction and organization of the urban system and the improvement of the quality of life and always guaranteeing the provision of services and infrastructures: all this has always been achieved considering man and its needs as the focal point of theoretical elaborations and concrete solutions. This approach, in line with the tradition that has characterized civilization throughout history, implies a vision of nature as an ensemble of resources at man's disposal, for his needs. Only in very few cases has he been able to use such resources without fatally compromising his habitat.

In this respect, planning has represented a fundamental means to point out how to use at best one of the main natural resources, the soil, not only for settlement or productive purposes, but also for leisure activities, which are necessarily linked to the economic conditions and the specific traits of social and residential structures.

Some specific topics of the discipline have been fine-tuned to this purpose: from the analysis of a community's past, present and future needs to demographic and economic growth projections; from the evaluation of planning measures necessary to meet these needs to the identification of the most suitable plan or project tool for this purpose, as well as the extent of the action and the different regulation forms concerning soil exploitation for the preservation of the territory's specific structure. Last but not least, the modalities of exercising (or denying) property rights, the management and the control of the process regarding the transformation of the territory. Within this system, the main attention has always been paid to the city while raw spaces, areas soon to be transformed into urban soil or areas functional to the city, have received only a reflected interest for their potential development.

However, from the 1970s onwards, a new vision and a new consciousness with regard to the preservation of natural resources seem to have emerged with vigour, finally being institutionalized within the framework of the 1992 *Rio de Janeiro Agreement* for the introduction of sustainable development. This progressive change is not only channelled towards the preservation of the natural heritage. More frequently, planning has been re-oriented towards a new perception of the territory, understood no longer just as a space where all can be done without any form of restriction or limitation, but rather as one where natural resources management becomes of paramount importance in order not to compromise their very existence.

This newly found will has not been met with the same interest by institutional action and private practice, although there are different approaches and results in various countries in Europe and in other industrialized nations. Extensive delays and gaps have been registered between the formulation of new legislation and regulations in this respect and their implementation on an individual, family and business levels; this could be explained by the economic costs and the mental efforts involved to bring about this attitude change and to start living and working in accordance with the principles of sustainability, although these are widely recognized as being highly honourable and just objectives.

This approach also found some resistance, on a theoretical and practical level, in those regions where the development is lagging behind. The introduction of the principles of sustainable development is perceived as a social and economic limitation to their progress and their advancement. It has also been understood not as an alternative model for a long-lasting development (one which would be able to contain flaws and adjust to new circumstances) but as an impediment to, or even the denial of, development itself.

The uncertainties about these forms of intervention, but above all the strength of the development pattern consolidated in the past fifty years, have guaranteed an overall improvement in the quality of life of city dwellers but this has not come without a price: the built-up environment has increased the consumption of natural resources and rural areas; it has developed impervious surfaces altering the natural flow of water within a watershed; the developed built infrastructures, which interrupt feeding, dispersal and breeding patterns, have deleted wilderness and have affected the population and the diversity of species across a wide area; meanwhile a sustained sprawl of houses and economic activities have completely changed the features and mechanism of the environment.

However, different factors are now contributing to a substantial change of this scenario: people's increased awareness and understanding of the importance of the principles of sustainability; the perception of a progressive deterioration of life conditions (changes in natural conditions, built environments and natural spaces, air pollution, diminished water reserves, shrinking of green spaces, etc.); a more rigorous institutional approach, in particular by the European Union. We must not forget the process of a slow but steadily increasing mobilization of society, notwithstanding contemporary society's focus on individuality; local communities witness a newly-found strength in defending their issues, where local and global are inextricably linked. Moreover, the deep transformation of some urban development processes determines a different approach for planning (such as the decrease in manufacture production and a boom of the third sector; the change of demographic growth in terms of its redistribution over a wider territory and the reduction of its density, which is a phenomenon increasingly observed first in metropolitan areas and mostly in small to medium cities; the increasing frequency in urban fabric of derelict and vacant lands, outcomes of the deindustrialization process or an unsystematic urban development; environmental and built-up deterioration frequently joined to social unease in urban districts. All these problems seem to be the result of an inadequate, or anyway flawed planning technique).

It is clear, therefore, that planning is now at a crucial point and on the verge of a drastic change, in a context of ideological resistance, technical difficulties but also growing social sensitivities, together with a new approach by the institutions which are interested in the reformulation of objectives and instruments at their disposal in order to be better prepared to face the question of society's development.

3. THE NEW ROLE OF PLANNING IN A SUSTAINABLE PERSPECTIVE

3.1. The framework of sustainable goals

In many countries of our continent, therefore, the reflection on sustainability and on planning in the respect of such principle has increasingly been taking shape, and not without contradictions, during the 1980s and 1990s and often in the absence of a cross-cutting and cross-disciplinary coordination. It is thanks to the European Union and its promulgation of documents and recommendations (more than directives), that the progressive coordination and clarification of the planning objectives through sustainability have been possible, and this has positively reflected on national governments and their planning of territorial intervention.

The steps taken by the EU have been numerous, although they have often been marked by an excess of rhetoric (in the declarations) or an oversimplification of the analysis of the recent territorial change, other than by mistakes in the evaluation of the processes that determine such transformation. It also needs to be noted that the efforts to create a homogenous and concise vision of reality is counterproductive, especially with regard to the diversity which characterizes our continent (and one only needs thinking of Italy itself for an example), as it happened with the *European Spatial Development Perspective – ESDP*, approved in Potsdam in May 1999.

However, it is from this document – but above all from the results of other Community activities and from the strategies put into place at the time of the Structural Funds 2000-2006 programming – that the most recent community policy lines for the city and the territory have stemmed as an indication of sustainable planning pursued by the various institution at all government levels.

In particular, in line with the decisions taken in Lisbon and structured in the 2004 document entitled *Towards a thematic strategy for urban environment – COM(2004)60*, the following fundamental issues are emphasized:

- the central role of the city as a strategic place and a structural key for social and economic development for the promotion of entrepreneurship, local employment and community development; for the improvement of the quality of life, for the elimination of social unease, for the strengthening of security and the promotion of economic, social and cultural integration; the city is therefore identified as a complex organization where it is necessary to promote and to assure a sustainable development;
- the priority themes considered essential to the long-term sustainability of towns and cities, such as sustainable urban management, sustainable urban transport, sustainable construction and sustainable urban design;
- new contents and goals for planning, as well as for avoiding the consumption and waste of renewable resources; keeping pollution within the limits of regeneration capabilities of natural resources; controlling and limiting consumption of non-renewable resources (soil is one of these); keeping air, soil, water quality in sustainable conditions; keeping and enhancing biomass and biodiversity; avoiding social exclusion.

At this stage the direction seems to be clearer, but, above all, the framework within which planning needs to redefine not only its specific objectives but also, and especially, its tools for intervention, its role and the legitimacy of its action is clearly determined.

3.2. Sustainable planning

First of all, it needs to be asked whether a technique or a science built to "consume" resources (and not always with the efficacy one could have expected in the ability to control land uses) could still be useful to plan their safeguard and to stimulate (rather than impose) their sustainable use.

Various examples of initiatives, plans and projects in cities of different dimension, sporadic at first but more and more widespread, show that with a coherent normative system (now present in most of the EU Regions and States) and above all in the presence of a collective will for sustainable action, planning has the capacity to adapt its tools, even traditional ones, to the principle and spirit of sustainability, but even more it always manages to put in place new solutions to emerging issues.

It seems evident, also in the light of an intense debate in recent years over a supposed crisis of the discipline, that planning has elaborated methodologies which still present a specific validity despite such a marked change as the one introduced with the policies on sustainability.

Rather than exclusively providing new solutions for the organization and the shape of urban centres, the planning process can still guarantee to contribute (also through a comprehensive approach which has been highly criticized in recent years although fully implemented in action) to the following elements: the build-up of thorough frameworks of analysis (despite the inability to precisely know the needs and possible future emergencies that may arise within the society), the ability to produce forecasts (to be understood not in terms of accuracy of the projections but as a means to create future scenarios through the analysis of social and economic trends), the translation of such knowledge in effective plans and projects and the application of an evaluating process (technical and political) for the gathering of all necessary information for the decision-making process. It needs to be remembered that other than the comprehensive approach mentioned above, the increasingly direct participation of the people is a further tool that enriches the work of planners.

Within this system, we cannot talk about a crisis of the discipline but rather of a difficult moment for the institutions that have to modify their powers and roles, sometimes by taking a step back when faced with public mobilization, other times by reaffirming their authority especially when dealing with economic actors and particular and local interests which often declare to act in the name of the principles of sustainability but actually deny their very essence.

The discipline can still play its role in a sustainable perspective, with the objective of creating plans which satisfy economic and social needs with due respect to the preservation of natural resources and which contain planning solutions and adequate techniques, that can be summarized through the following points:

- *to know* ecological-environmental system features and the state of the local environment; conditions and trends of environmental elements (air, water, soil, vegetation, animals, etc.), human activities compatibility with environmental protection; environment weakness and fragility. In fact, planning (as it happens in Italy's planning activity) could represent a moment for collecting and coordinating all the general and specific prescriptions for environmental protection; could help to set a cognitive framework and list all the incompatible land uses with environmental conditions and assess environmental impacts;
- *to forecast* evolving/degenerative conditions and tendencies of local environment changes, features and trends of environmental elements, effects of different/alternative localization decisions;
- *to process* general goals for a sustainable development; specific goals and practices compatible with environment protection; alternative scenarios for the future of local communities; policies to protect and preserve natural resources; interaction and matching of different and particular policies for environmental protection;
- *to assess* environmental impacts of interventions, works and projects; goals and contents of plans and interventions; eco-compatibility of all the policies included or involved by planning; "embedded" or "cumulative" (often negative) effects.

For the decision-making processes, even when participation is a determining factor, planning still represents an opportunity for a comparison and a coordination of research necessary for the creation of shaped territorial and city scenarios, as a fundamental synthesis and synopsis for decision-making and the constitution of a framework useful for the management of urban and territorial transformation processes.

Planning can also directly support sustainability through its specific tasks in the territory's organization, its control function with regard to land uses and localization of industrial plants, residential areas, infrastructures and equipments, determining furthermore the limits of development and land transformation and mostly affecting development patterns, land works, soil consumption practices and urban design.

4. THE SPECIFIC SUSTAINABLE GOALS FOR URBAN PLANNING

Other than the role planning can have for the implementation of the principle of sustainability in the formulation of cohesion and development policies, the discipline is reviewing its own specific objectives, which implies a change in the traditional ones, at times subverting them, at times channelling them towards new ones.

In fact, the containment of urban congestion and the control of concentration processes for activities and people represent a turning point in planning theory which, being a reflection of an ideology of constant and limitless growth, has traditionally foreseen a progressive urban growth through the annexation of rural spaces following an extremely positive vision of urbanization. This meant satisfying the settlement demand of people and businesses through an intensive consumption of soil patterns. A clear consequence of this process is the growth of the city through the agglomeration of low-density outskirts and the emergence of a sprawl, once a peculiarity of North-American conurbations and North-European cities and now a continental widespread phenomenon.

In this sense, it is not by chance that planning has addressed the first measures and forms of urban design towards a limit to soil consumption, by trying to guide the real estate market requests and the residential demand of dwellers in the direction of urbanization models that have proper urban densities, space-saving spatial policies, infill development, sustainable cluster development, other than the support of a strategic reutilization of existing housing stock (as a deterrent to the occupation of free agricultural and natural soil) and a development reduction of impervious surfaces and improvements in water retention.

Planning should also be an opportunity to generally promote sustainable housing and sustainable construction techniques; there is also a growing need, for productive and urban purposes, of a revitalisation and regeneration of derelict and brownfields. These processes are becoming increasingly frequent in the Mediterranean area of the EU and have included these planning provisions at a later stage and a slower pace. The zoning technique for the organization of urban space, introduced by the Movimento Moderno and one of the main tools for planning, is today progressively replaced with the promotion of mixed uses in built-up areas as well as in developing areas, feasible to limit also vehicular and private travels. All this needs to be combined with an attentive transport planning and a different access and mobility policy able to optimize the use of lands in proximity of infrastructural junctions.

Despite these objectives being apparently simple, real estate owners and market forces oppose some resistance to their achievement: it is a struggle with uncertainty fostered not only by different real estate regimes but also by people's different perceptions about environmental impacts of single every-day practices.

Regarding these issues, the strength of planning has resulted more in the effort to build a sense of shared sustainable objectives amongst different social forces, where duties and advantages of individuals, families and communities are clear, rather than in its normative authority.

Only this seems to guarantee the success of measures and planning choices that are becoming more and more widespread especially on an urban scale, and include these first steps toward a sustainable planning:

- project for *infilling development*, which must occur in locations where some development has already taken place and infrastructures already exist. In urban areas, infill development is typically executed by converting old buildings and facilities into new uses (redevelopment) or by filling undeveloped space within these areas;
- indications for *cluster development*, mostly in newly developed areas; clustering development into concentrated areas can protect natural habitat. Cluster developments are built at gross densities comparable to conventional developments but leave more open space by reducing lot sizes;

- precise regulations for *economic activities development*, avoiding the sprawl of industrial plants, concentrating economic activities in few equipped areas, locating commercial sites near infrastructures and transport nodes, controlling industrial waste and emissions;
- preference for *brownfields redevelopment,* not only in big cities and metropolitan areas, but also in sprawl settlements where industrial waste lands are increasing due to technological innovation and to production replacement, while an incongruous demand for new urbanised industrial areas still keeps on growing. It is a shared need to re-use industrial waste lands even to avoid agricultural and natural soil consumption;
- promoting a *change of land mobility patterns*, strengthening and supporting public transport networks, enhancing multimodality opportunities, supporting transport nodes transformation into multi-purpose and facilities centres (by developing housing, commercial spaces, services, and job opportunities in close proximity to public transportation: a land benchmark, an urbanization point of agglomeration), promoting bicycle networks, increasing wide pedestrian areas, re-naturalising infrastructure systems.

This last point seems particularly important and interesting because, as many examples in Europe show, combined land-use and transport strategies for sustainable development are one of the best ways to implement sustainable development in our cities for the following reasons: the concentration of activities and facilities in limited areas reducing travel distances, the improvement of pedestrian areas and cycling accessibility, the location of homes and business near public transport nodes, the increase of settlement size and mixed uses areas, the encouragement of public transport utilization, the provision of critical mass for a viable public transport system, briefly by the design of high quality built-up environment, walk- and cycle-friendly development, green areas.

Literature abounds in different ways of building a city and governing territorial transformation in a sustainable key. The battle seems to be fought more on the political and social arena, rather than the technical one, where institutions are involved in a direct – and not always peaceful – confrontation with economic actors, private citizens and amongst institutions themselves. More than particular or public interests, the cleavage seems to concern two different ways of looking forward: on the one hand a short-term vision with a limited territorial scope, on the other a long-term vision with a seemingly endless horizon.

5. ITALIAN EXPERIENCES: DEVICES TO ASSERT A SUSTAINABLE FUTURE

5.1. *Some peculiarities*

In Italy as in the other EU Member States, efforts to introduce the principles of sustainability in various areas of public and social interventions are numerous and evident and have been under way for a while.

Planning is one of the strategic disciplines that contribute to the definition of these policy lines. First of all, the transversal character of sustainability but above all the needs for coherence and coordination required for the implementation of policies and actions that impact on different areas of interest of our administrative system, have always been kept separate. In the past years government action has clearly shown such unease not only as far as interventions were concerned (and understandably at the start-up stage), but also within the framework of policy lines and normative prescriptions.

Besides, as already pointed out, the Italian administrative system, in line with other countries presenting the same structures and bureaucratic traditions, favours the emergence of other complications due to the allocation of competencies, the creation of norms and regulations and the presence of various levels of territorial governance. In recent years this last aspect has registered a marked acceleration through the process of devolution and the widening of the scope of action on a regional level.

Undoubtedly, the principle of *subsidiarity* which is increasingly gaining ground in Italy in line with EU directives, is favouring the introduction of sustainability in various policies. Furthermore, Agenda 21 and participation are developing significant new approaches in policy-making.

In the planning field, the result of this power transfer to the regional level has allowed a more rapid innovation process coupled with a fragmentation of planning provisions, different from region to region, other than a variable speed for the absorption and the implementation of sustainability in the formulation of territorial policies. Some problems have not been overcome yet, as in the case of legal, economic and social disputes over soil exploitation, whose control in Italy is both the strength and weakness of planning.

It needs to be remembered that Italy also presents a strong regional segmentation in the social and economic development: the disparities are clear if one examines the level of production and wealth reached by some regions such as Lombardy and Veneto, making them two amongst the leaders in Europe, and regions with a slower growth rate. This is a factor that affects the modernization of policies even though it has not prevented the welcoming and the application of sustainability principles also in those Italian regions where development is lagging behind.

Sustainability as a new way of understanding territorial governance is now undoubtedly a process which has gathered considerable consensus and interest, other than having produced a growing awareness and participation of communities.

5.2. Levels of government

The different levels of government in Italy have also the downside of producing innovative policies at a different pace, both on a vertical level (in a hierarchical order, where the lowest level is nevertheless characterized by a remarkable dynamism) and a horizontal level (between different regions, provinces or the municipalities of the same region or province) with a peculiar articulation:

- a National level, where the government is responsible for the application of EU directives and indications, legislative framework construction, affirmation of general objectives, prescriptions for administrative procedures and management of public financing;
- a Regional and County level which has grown stronger and stronger in the last years, where governments have all the responsibility for Regional Planning Act, general objectives, regional Plans drawing, indications for specific administrative procedures and Counties and Municipalities Plans approbation, and also have the authority over other fields of intervention on land and environment[1]. The coordination of all the policies is a duty of this level of government other than the assessment of cumulative interventions and impacts at a general level;
- and, at the end, a Municipal level, the most strategic one for an effective and actual intervention to achieve sustainable development, with a more careful and direct land use regulation and with a sustainable re-designing of settlements. Municipalities are directly involved in the implementation of all European Union, national, regional/counties legislative indications, urban plans drawing and land use control prescriptions, infrastructures and public equipment building, local transport plans and management and local public financing management, private sector involvement in planning interventions, information and education about sustainable goals and tools, and empowerment of public participation.

Besides, municipalities also represent the level of governance closer to families and businesses and, although they are the weakest administrative level and often suffer from funding shortages, they are the ones that bear the burden of the practical implementation process of sustainability.

5.3. The first challenge of sustainable planning in Italy

Through an observation of the first actions promoted in Italy for a sustainable planning, it is interesting to note some specificities that emphasize the first areas where the new principles were implemented. These areas were already the object of some legislative initiatives, a couple of decades ago.

With a certain pride it could be said that some laws for the protection of the landscape and of the historical and architectural heritage (respectively the laws 1089 and 1497 issued in 1939), and the planning law itself (law 1150 approved in 1942) were at the time considerably modern measures, although their application has not always proceeded smoothly. Already in 1985 Italy prides itself on the drafting of a strongly principled law for the protection of the landscape and natural resources. In line with the historical tradition, natural resources (still considered as landscape components) have benefited from numerous laws and measures (at the national and regional level) which found a stronger legitimacy in the principle of sustainability, other than a redefinition of the objectives and tools for intervention.

More interestingly, other aspects have been at the centre of disciplinary innovation. For instance, the construction of a set of environmental information and the use of a trans-disciplinary approach as first steps for land use planning have been the more interesting innovation recorded mostly at regional or county level (according to national legislative frame); a set of effective information beyond the usual preliminary analysis (social, economic and physical investigations, a very consolidated tradition in Italy) which planners were used to arrange for decision making.

In these last years, planning has recorded an increase of studies and assessments, composition of indicators and their utilization to build a very considerable knowledge about natural processes and anthropization effects. However, frequent difficulties to convert analysis outcomes into decisions and operational prescriptions have emerged.

Regional and Counties master plans are usually the field of application of a new sustainable approach, chiefly with a deepening and detailing analysis of environmental impacts of land use and urban development. In fact, more frequently new Regional and Counties master plans are projected by a compilation of more environmental-focused information (according to Regional and County duties to assure environmental database and frame analysis even for municipalities' plans and projects), GIS and indicators, and environmental matrices aimed at knowing rural and urban zones features and potentialities, but also economic and residential activities location effects and low impact developing conditions.

In the Cremona County Master Plan, coordinated by M.C. Treu (2001), the quantity and complexity of the analysis are very interesting and appreciative, aimed at constructing a *Land and Environmental Sustainability Indicator* which measures ecological stability levels and assesses embedded effects of land use as criteria for development and localization decisions and prescriptions.

The complexity and costs of these analyses cannot be denied, mainly for municipalities which less frequently recur to likewise preliminary studies. However, in some cases, when applied this approach produces interesting developing indications.

As in the cases of Reggio Emilia's Master Plans, coordinated by G. Campos Venuti (1999), and Cesena's, coordinated by F. Oliva (2001), the introduction of a peculiar methodology such as the *Environmental Compatibility Assessment* for every development and land use indication (directly outcome from detailed investigations to check soil vulnerabilities and restraints, water detention, impervious surfaces increase, future waste and natural resources deprivation, built-up area density, etc.) drives to:

- a revision of building parameters;
- restrictions for build-up area density;
- new limitations and project indications for impervious surfaces, chiefly in developing residential zones;
- detailed indications for public and private green spaces;
- a new design for great and small infrastructures with a "re-naturalization" of environment and rural spaces;
- bio-architectural prescriptions for new buildings (not only for public housing).

This approach was originally limited to some sporadic cases; considering the Italian situation it was already a remarkable effort which private stakeholders tenaciously opposed. However, this approach is currently experiencing an increasing diffusion, thanks to the rising sensibility of public authorities, especially in the North of Italy.

More frequently, many municipal plans, although in the absence of elaborate information systems, attempt to apply new criteria in urban development for the containment of building-up and urbanization processes, as it occurs in the Po Valley metropolitan and urban areas, affected by sprawl and enduring soil consumption by new low-density development and small firms cluster areas. Once considered a solution for the betterment of the quality of life (opposed to the low quality of life in high-density areas and strong industrial presence, with the relative congestion and pollution and "high-density unease", etc.), sprawl as a development pattern in this last period is considered an environmental damaging factor and the cause of a rise of collective costs. For instance, in Italy, since the 1970s the economic development involved a strong process of urbanization, by the sprawl of residential areas far away from the big cities areas, as well as the sprawl of factories from the big industrial concentrations[2]. Also, in the last years the economic success of small and medium enterprises (the well-known "Veneto model") and the diffusion of residential trends for a single "countryside" house have increased the sprawling process and, as a consequence, have produced more soil exploitation, an increased usage of private means of transport, the worsening of air pollution and so on.

Today, this area has become one of the first where an attempt is being made to exercise control and reduce the impact; many plans try to avoid a leapfrog development and introduce restriction to new development, new land use control devices and incentives for high-density building practices, according to a revisited compact city pattern. Infill development strategies, on the contrary, are still not so widespread in plans.

At urban scale, one of the most interesting deals of the last years has been the reutilization of industrial derelict areas and brownfields. The first objective is to strike one of the causes of the progressive urban environmental deterioration, as it occurs in big industrial cities (affected by economic decline for production replacement or technological innovation or for local economic organization turned into more soft and less invasive activities). The second objective is to recover underutilized areas in order to turn them into new urban facilities or green and residential areas (where possible) and to root new economic activities as an answer to the still increasing demand for industrial development areas, quite always directed to rural spaces. The attempt in doing this is to:

- avoid sprawl of industrial plant and agricultural soil consumptions,
- contain economic activities in few and outfitted areas,
- concentrate commercial sites near infrastructures and transport nodes,
- monitor and checking industrial waste and emissions, otherwise very difficult.

From Genoa (the Ancient Port and some coastline industrial zones), Milan (where the phenomenon has astonishing dimensions) and Turin (as it happens at Castel di Lucento urban district) to Naples (the well-known Bagnoli district with 1,750,000 square meters of redeveloping surface), but also in small cities (such as Sesto San Giovanni, where the entire area of the municipality is characterized by the presence of derelict lands) industrial areas rehabilitation processes are proliferating in Italy. The attempt carried out by the Bologna County Master Plan issued in 2004 for the creation of limited, concentrated and controlled new industrial developments only in some municipalities and in some selected areas (for good location near infrastructures, low environmental impact, good accessibility, etc.) is very interesting. The Plan is also trying to transfer development rights and to reallocate financial, economic, social, and employment benefits among municipalities involved in development projects – the ones deprived by new development opportunities and others favoured by new locations.

The most innovative sustainable set of interventions (also for their technological contents) achieved in Italy concerns projects and plans for the change of mobility/land travel models, a sector where many Italian cities are putting increasing efforts and resources for the creation of integrated public transport systems. Such projects involve a more widespread presence of pedestrian areas and car bans as well as the creation of new infrastructures; it is worth noting that Italy's current economic junction and the reduction of public expenditure will lead to a redefinition of private mobility (strongly supported in the past to privilege national production in this sector) and a cut in investments for big road infrastructures.

Many municipalities (from North to South) are financing new mass public transports lines (with a strong preference for tramway, sometimes by underground railways improvement). However, efforts are aimed at organizing mostly integrated public transport system at regional or metropolitan scale:

- locating residential and industrial development opportunities only in cities which can provide access to housing, employment and entertainment centres by low environmental impact mobility; pointing up multimodality, integration of pedestrian and cycling mobility with public transport; limiting private vehicular transport (as it occurs in Bologna County Master Plan for the construction of an integrated public transport system project); or
- enhancing multimodality and transportation connections and trying to create "conspicuous infrastructures nodes" with services and public equipments and new development cluster just around a station (maybe on underutilized or vacant sites) not only parking surfaces, changing them into land benchmark (strategy performed by Veneto Regional and Metropolitan Rail System).

To conclude, we must remember the implementation at every level, of environmental assessment processes for every land use indication. In this case too, starting from EU directives, the development of evaluation processes has been given a boost, not only with regard to single interventions but also with reference to the different uses of the territory. These processes are trying to avoid a negative environmental impact or to maximize the containment of the damages produced by those interventions which resulted as necessary and which could not be located elsewhere. EIA results a process normally applied by private sector and public authorities at every level. Meanwhile, many Italian Regional governments have introduced SEA procedures in accordance to the European Directives.

Despite the obstacles and difficulties encountered along the way – as emphasized also by the international press and by the fact that the institutions are at times hesitating in the choice of the territorial policy to be pursued, causing many delays – the strengthening of these processes is now well under way throughout the country, as much as this is happening in other EU countries, meaning a really hard and exciting challenge for the future.

Michelangelo Savino
Department of Sciences for Engineering and Architecture,
Faculty of Engineering – University of Messina, Italy

NOTES

[1] E.g., for the Italian legislative system, Regions have political and administrative responsibilities for Environmental Protection Plans, Natural Resources, Historic and Cultural Heritage Protection Plans, Flood Risk Management Plans, Transport and Mobility Regional Plans, Water Protection and Management Plans, Mining Plans and Rules, Waste Management Plans, Water Protection and Management Plans.

[2] One reason of their increase also lies in the municipal practice to support local public budget by private financing contributions for residential or industrial development (according to some national planning acts), while municipalities must manage alone public works, services and infrastructures. A good reason for increasing developing areas or supporting urban extension.

[3] As it happens in Veneto, Emilia Romagna and Lombardia, regions affected by residential as well as industrial sprawl, for the diffusion of small and medium enterprise (SMEs) which show preferences for suburban or rural locations.

REFERENCES

Campos Venuti G., 1999. *Reggio Emilia Master Plan*, Reggio Emilia: Comune di Reggio Emilia.
Commission of the European Communities, 2004. *Towards a Thematic Strategy on the Urban Environment*. Brussels. COM(2004)60.
Commission of the European Communities, 2005. *Communication from the Commission to the Council and the European Parliament on Thematic strategy on the urban environment*, Brussels. COM(2005)718.
Breheny M.J. (ed.), 1992. *Sustainable development and urban form*. London: Pion.
Jenks M., Burton E. and Williams K. (eds.), 1996. *The Compact City: A Sustainable Urban Form?*, London: E & F.N. Spon.
Kunzmann K.R., 1998. *The European Spatial Planning Perspective: Much Ado About Nothing?* Urbanistica, No. 111, pp. 53-55.
Law 1089/1939, *Protection of Artistic and Historic Heritage*, now converted into Legislative decree 42/2004, *Code for Cultural, Historic and Artistic Heritage and Landscape*.
Law 1497/1939, *Protection of Natural Heritage*, now converted into Legislative decree 42/2004, *Code for Cultural, Historic and Artistic Heritage and Landscape*.
Law 1150/1942, *Planning Act*.
Law 142/1990, *Local governments rules*.

Oliva F., 2001. *Cesena Master Plan*, Cesena: Comune di Cesena.
Savino M., 2002. Esiste una pianificazione sostenibile? in Fregolent L., Indovina F. (eds.) (2002), *Un futuro amico. Sostenibilità ed equità*, Milan: FrancoAngeli, pp. 199-322.
Treu M.C., 2001. *Cremona County Master Plan*, Cremona: Provincia di Cremona.

M. TURVANI AND S. TONIN

BROWNFIELDS REMEDIATION AND REUSE: AN OPPORTUNITY FOR URBAN SUSTAINABLE DEVELOPMENT

Abstract. This paper focuses on brownfields remediation and reuse strategies as opportunities to favour environmental protection, improvement of economic and social conditions and enhancement of human health and safety. The reuse of brownfield sites can also promote and encourage urban sustainable development practices. Land management and soil protection are common priorities for all the modern Governments and the experience achieved by Western countries in this field may be of some utility for a sustainable urban regeneration policy in China. Finally, an overview of the main costs and benefits of brownfield redevelopment and cleanup projects and a brief introduction to the issue related to the economic valuation methodology used to quantify them will be investigated and discussed.

1. INTRODUCTION

Land management and soil protection are required to achieve goals of sustainability and therefore Governments worldwide have committed themselves to implement the necessary strategies. These strategies are largely intertwined because land is scarce and soil is a non renewable resource: land uses management and soil protection are the appropriate actions to be taken to ensure flexibility and a more rational way to use these scarce resources within a sustainable perspective of growth.

To protect the environment and to achieve the goals of improved land management and soil protection, the European legislation has been framed within the "polluter pays" principle: this implies both that the polluter may be required to invest in equipment and processes that reach environmental standards and, as in the new Directive on Environmental Liability, that responsibility for the cost of cleanup is placed on those parties who have contributed to creating the contamination problem. The system of environmental liabilities should both prevent future contamination problems and provide financing for any needed remediation.

In China the legal system is undergoing continuous and profound reforms (Cao et al., 1997): the Government has committed itself to move in the direction of incorporating sustainability principles into laws and regulations related to environmental and natural resource protection, and to develop legislation to fulfil China's international obligations in terms of environmental treaties and conventions[1].

In industrialized countries, soil remediation and protection stands out as a priority and programmes for redeveloping land are common: these programmes very often take place in urban areas and they imply large investments in cleanups which are the first step in the strategy of urban regeneration. Many major cities in China are facing the same challenge and experiencing the same problems of regeneration of urban cores and these problems may be exacerbated by the continuous and rapid process of urbanization and the conflict between alternative land uses. Land scarcity is a distinctive Chinese problem and the ongoing trend of land consumption is raising major concerns (Brown, 1995).

Land remediation and recycling poses a vast array of financial and governance problems: reusing previously developed land (so called brownfield site) is attractive because very often these areas are placed in high value location within the inner city, but several drawbacks are possible because of contamination problems negatively impacting on the social fabric hosting them, and posing hazards on human health and the ecosystem. In Europe and the USA there has been a politically driven emphasis on reclaiming these sites and a huge amount of human and financial resources has been devoted and is still required to address this social and environmental challenge. Public sector funding and management is necessary to promote both cleanups and processes of urban regeneration, framing them within a sustainable urban and land planning and achieving the recycling of urban land while protecting greenfields and rural land. Most interventions require the joint efforts of private and public agencies because of the scarcity of public financial resources and the existence of a variety of stakeholders and property owners involved.

Economists would recommend that when making decisions about these projects their costs should be compared to the benefits: in this essay we want to develop this recommendation by discussing in detail brownfield sites remediation and reuse costs and benefits, claiming that the use of economic valuation tools may actively support a sustainable development strategy in urban regeneration policies.

The remainder of the paper is organized as follows. In section 2 we introduce the issue of brownfield sites in terms of diffusion of the problems, and extent of the cleaning up costs in the Western experience. In section 3 we analyze the main costs and benefits of the remediation and regeneration of brownfield sites and we address the importance of developing benefit cost analysis to improve decision making and public participation in environmental and urban regeneration policy. In section 4 we discuss some implications of the Western experience in land remediation and reuse for China. Section 5 presents the conclusions.

2. BROWNFIELDS REMEDIATION AND REUSE

Changes in the economy, in the industrial composition and in technological innovations in the last few decades have resulted in the creation of large areas of underused and abandoned and possibly contaminated land in cities and suburbs all over the world. These areas are called brownfields:

> Real property, the expansion, redevelopment, or reuse of which may be complicated by the presence or potential presence of a hazardous substance, pollutant, or contaminant[2].

Economy evolves and this evolution leaves behind the legacy of a previous pattern of location and land uses: it is a recurrent phenomenon which is important to understand in order to be able to react with proper economic and social policies. Other factors that have contributed to brownfields creation as pointed out by Pellow (1998) include: i) a demographic shift away from the city to the suburbs and urban fringe areas; ii) expanded transportation networks that almost entirely by-pass the inner city and provide little incentive to develop there, contributing to suburban sprawl growth; iii) a regional shift of economic production and population centres that limits or restricts investment in the older industrial cities; iv) a global change in the technology of post-industrial economic production that renders much of the early 20th century development obsolete, particularly with respect to electronic communications; v) the rising global competition from trans-national corporations and their increasing drive to cut costs, maximize profits, and increase capital mobility.

The results on land use of these concurrent forces operating in the economy are impressive and some data may help to understand the scale of the problems involved. In the USA, the Office of Technology (1995) estimated 450,000 brownfield sites nationwide. A more recent report from the US Conference of Mayors provides a national tally of 600,000. The most contaminated brownfields are the 1,550 Superfund sites on the National Priorities List (USEPA, 2005).

In Europe, unfortunately, reliable data at the national level on the number of brownfield sites are not available for all countries. Almost 800,000 potential brownfield sites have been identified in Europe (Oliver et al., 2005). More than 300,000 potentially contaminated sites have been identified in Western Europe and the estimated number for the whole of Europe is much larger (Van-Camp, 2004). The diffusion of contaminated sites in the territory, and the extent of contamination in soil, in water, and groundwater constitute a menace to the population's health and to the ecosystem, and legal rules work both to limit this menace in the future and to address ongoing problems by identifying responsible parties and at least in part solving the problem of finding adequate financial resources for cleanups.

The burden for private and public finance is heavy: for example, Probst and Konisky (2001) estimate that in the low case, the total cost from fiscal year 2000 through fiscal year 2009 will be approximately $14 billions (or $15.6 billions, adjusted for inflation), and in the high case, the total cost is estimated at about $16.4 billions (or $18.3 billions, adjusted for inflation), approximately 8.6% more than in the base case.

European official reports (EEA, 2000; and Van-Camp et al., 2004), estimate that the total cleanup costs for the countries that have provided data are about 115 billion euros at current price, or 490 euros per capita.

It can be argued that land is a scarce and finite resource and that in theory a competitive market for land could be the solution to these problems. Obviously this is not the case: various institutional factors are affecting the working of the market for land: property rights regime and planning practices interact with demand and supply for different land uses and the rules and functioning of capital markets. In the case of brownfield sites a further element needs to be taken into account: the liability regimes for environmental and property damages. The liability systems have their own costs: brownfields may remain idle in part because of the threat of liability for brownfields developers under strict environmental laws (Urban Institute, 1997). In the US, the Comprehensive Environmental Response, Compensation and Liability Act (CERCLA), also know as Superfund law, is one of the most significant examples. Its severe liability regime with the characteristics of strict, retroactive and joint-and-several liability has made developers, potential buyers, and owners diffident to invest in brownfields remediation and redevelopment. But, experience and the awareness of the potential beneficial gains accruing from redeveloping brownfield sites, have encouraged US federal governments to activate a series of initiatives to promote redevelopment and reuse of these sites. Examples of these initiatives are a targeted tax deduction for brownfields redevelopment, a reduction of liability risk under CERCLA for lenders that become involved with brownfield sites, and the voluntary cleanup programs aiming at fostering the cleanup of brownfield sites. The new European legislation, which came into effect on April 30th 2007, (2004/35/CE), draws heavily on the experience of the contaminated sites statute in place in the United States – Superfund, which was established 25 years ago – while trying to avoid some of its drawbacks.

While liabilities may constrain the market for brownfield sites redevelopment, there are also various reasons for the market to be able to overcome these constraints. Brownfield sites are generally old industrial sites located adjacent to core areas and inner-suburban areas and are close to existing infrastructure, jobs, and other resources. Porter (1995) asserts that inner cities have some competitive advantages that are connected to brownfields redevelopment, such as strategic location, local market demand, integration with regional cluster, and human resources. Brownfield sites are generally located near inner cities where the existence of high rent areas, business centres, and nodes of transportation and communication favour brownfields redevelopment policies. Local market demand refers to untapped purchasing power in inner cities that can make up for lower per capita income. Inner city residents are particularly underserved by financial, retail, and personal services. Brownfields redevelopment could result in wider availability of local services in inner city neighbourhoods. Integration with regional economies refers to the opportunity to take advantages of the existing clusters of regional economic activity and the possibility to compete in downstream products and services. The presence of industry clusters may increase productivity, efficiency in the access to specialized inputs, services, employees, information, institutions and may enhance the ability to perceive innovation opportunities. Hence, brownfields redevelopment could result in new inner city business and local employment growth. Finally, the inner city may have underutilized labour pools and unexploited capacities for entrepreneurship. Brownfields redevelopment can increase the supply of moderate wage, low to moderate skilled jobs and small business opportunities in inner city neighbourhoods.

Nevertheless, brownfield sites redevelopment may incur a variety of problems: these properties are derelict or underused; they have in many cases real or perceived contamination problems, and the financial costs to remediate and reuse them may be impeded by the overall decline of the urban areas in which they are located. Large, idle brownfields may have a wider negative impact on the regional level (RESCUE, 2004) and for this reason their cleaning up and reuse has been considered a priority in many countries.

Whether enforcement-based or relying on collaboration between private entities, such as developers and investors, residents, and governments (as in many recent "brownfields" initiatives), addressing the problem of contaminated sites is judged to be an important component of sustainable urban regeneration. Remediation of contaminated brownfield sites of course needs to be combined with redevelopment policies to bring back these sites into their productive and socio-economic functions promoting urban revitalization while reducing development pressures of greenfields (Eisen, 1999).

RESCUE (2002) defines a sustainable regeneration of brownfield sites as the management, rehabilitation and return to beneficial use of the brownfields in such a manner as to ensure the attainment and continued satisfaction of human needs for present and future generations in environmentally sensitive, economically viable, institutionally robust and socially acceptable ways within the particular regional context. Within this framework, while contaminated sites remediation is costly, it also contributes to numerous benefits, offering economic and social advantages to the local and regional economy and to the community nearby. Benefits Cost Analysis is an appropriate economic valuation tool to understand the importance of remediation projects for urban regeneration policies.

3. IDENTIFYING THE COST AND BENEFITS OF BROWNFIELDS REMEDIATION AND REUSE

A comprehensive cost and benefit analysis of any brownfields intervention site must take into account various factors that can have positive or negative impacts on the value of the property, on the communities involved by the cleanup project of the site, and on the natural environment (Tonin, 2006). The costs of brownfields redevelopment and remediation can be divided into two broad categories: direct costs, linked to the remediation process itself, and indirect costs, i.e. costs related to

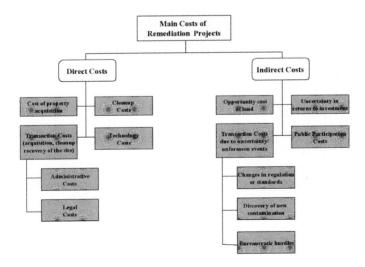

Figure 1. Main costs of remediation projects.

the effective management and implementation of brownfields redevelopment and remediation processes (see Fig. 1).

The direct costs depend on the type and severity of contamination (extent, mobility of contaminants), the characteristics of the site itself (location and historical conditions of the area), the choice of the best available cleanup technology, and finally the administrative and legal costs relative to the acquisition, cleaning and recovery of contaminated sites. Indirect costs are, for example, those due to the delay in using the land because the remediation process is slow, and those related to the uncertainty and the higher transaction cost of the project. Brownfield redevelopment projects pose higher levels of uncertainty to decision-makers than would occur with any other property investments, and especially greenfields. The higher risks refer to the site assessments needed to determine the type and extent of the pollutants, remediation planning, the execution of remediation plans, and the environmental damage liability claims associated with the past pollution of a site. Any one of these factors imply higher transaction costs that involve an array of measurement, information, bargaining, and contracting costs other than those associated with acquisition of unpolluted land. Other costs incurred due to contamination are related to the difficulties in accessing the necessary funding for development projects. Thus, although many contaminated sites have the potential for becoming profitable business ventures that generate new activities and new employment opportunities, public investment is often needed to catalyze private funds. This is because the cleanup interventions are very expensive, and financial capital is hard to find, and there is considerable delay between the initial investment and the time in which the site can be productively used again. According to OECD

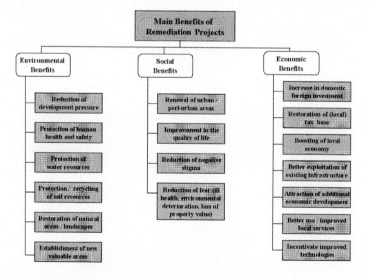

Figure 2. Main benefits of remediation projects.

(2000), the role of the public sector in these projects is to design the cleanup strategy, to pinpoint the appropriate areas for development, to initiate the remediation process, to provide funding and to encourage the participation of the private sector.

Economists recommend that the costs should be compared with the benefits of the project. Thus, focusing now on the positive side of the brownfields remediation and redevelopment process, the main benefits can be grouped into three broad categories: environmental benefits, social benefits and economic benefits (see Fig. 2).

The environmental benefits include the reduction of development pressure on greenfield sites, protection of public health and safety, protection of groundwater resources, protection and recycling of soil resources, restoration of former landscapes and establishment of new areas deemed to have ecological value. Restoring natural areas may also entail "non-use benefits", such as the option to conserve a natural resource to use it in the future, and the enjoyment of knowing that natural resources are preserved for future generations.

The social benefits include, among others, the renewal of urban cores, the improvement of the quality of life, the elimination or reduction of the negative social stigmas associated with the affected communities by revitalizing them, the reduction of the fear of ill health, environmental deterioration, and loss of property values in these communities. Another important positive effect included in this benefit category is the reduction of fear and anxiety related to the perceived health risks. The deaths, illness and injuries avoided by various interventions are generally considered and valued in monetary terms. But little attention has been devoted to the dread and uncertainty associated with these effects. Schelling (1968) was one of the first researchers to introduce the concepts of anxiety and fear as part of the consumer interest in reducing the risk of death so that it is worth for the consumer to reduce it. The explanation that fear, in contrast with death, illness or injury, is too intangible to be recognizable by risk regulators, cannot entirely be accepted. In fact, wide ranges of intangible benefits are now routinely recognized within environmental economics: the value of fear should thus be estimated as well (Adler, 2003).

There are many potential economic benefits of remediation and redevelopment of brownfield sites. For example, domestic and foreign investment can be attracted by the restoration of the tax base of the government, especially at the local level. Brownfields requalification can also increase employment opportunities (number of short-term and long-term jobs) thus boosting the local economy, and can increase the utilization of and reinvestment in existing municipal services. Finally, brownfields remediation can encourage the development of remediation/decontamination technology; and the exploitation of existing infrastructure systems. By returning these facilities to productive use, cities can reacquire their economic and social vitality and be the catalyst for additional economic development. Case studies have illustrated economic advantages of building in already developed areas (De Sousa, 2003), such as less expensive development costs, lower operating costs, lower infrastructure costs, and lower fiscal impact costs to local jurisdictions; however this is not necessarily true for all sites when the cost of environmental remediation is very high because of heavy pollution and the site location is not very attractive.

4. THE ECONOMIC VALUATION OF COSTS AND BENEFITS

In many situations, to estimate costs and benefits is not a trivial exercise: severe data limitation and methodological problems are well recognized and yet we contend that economic valuation tools are very important to support credible policies for a sustainable development.

The economic value of goods or services is generally captured by market price or, more precisely, by how much people are willing to pay to obtain them (WTP) or are willing to accept compensation for some given loss (WTA). There are however a number of goods, such as environmental quality, human health, and risk, which are not traded, and for which markets are absent. The monetary value of changes in society's well being due to a change in environmental goods is generally estimated through the total economic value (TEV) of the good to be appraised. Since an environmental resource provides a variety of services to society, the TEV can be disaggregated to consider the effects of any change on the well being derived from the existence of the good. The TEV is commonly divided into use values and non-use values. Use values include direct use values, which refer to the actual use of a resource; indirect values, which are society benefits from ecosystem functions; and option values, which refer to the values individuals are willing to pay for the option of using a resource in the future. Non-use values can take the form of existence value, which reflect the fact that people value resources for moral or altruistic reasons, unrelated to current or future use, and bequest values, which measure people's willingness to pay to make sure their heirs will be able to use a resource in the future.

The total economic value (TEV) of an environmental good is thus the sum of both use and non-use values:

TEV = use values + non-use values = direct use + indirect use + option + existence + bequest values.

Economists have resorted to a variety of techniques to place a value on these goods, which are generally termed non-market goods. In general, there are two main approaches for valuing non-market goods: revealed preference and stated preference methods (Freeman, 1993). With revealed preference techniques, preferences for non-market goods and income are revealed indirectly when individuals purchase other market goods that are in some way related to the non-market goods. Examples include the Travel Cost and the Hedonic Price Methods. By contrast, stated preference techniques ask people to report their willingness to pay in the course of an interview (Contingent Valuation) or infer preferences by asking people to choose among hypothetical alternatives (Conjoint Choice).

Some of the costs and benefits of brownfields remediation and redevelopment are not captured by market transactions: it is thus required to apply one of the non-market valuation techniques for the full assessment of the costs and benefits of these sites. This situation occurs more frequently to obtain sound estimates for the benefits of a given project: human health benefits and environmental quality improvements are more difficult to obtain because markets are not well functioning or do not exists. For example, in our own research in Italy, devoted to the study of the benefits of remediation policies, we used conjoint choice questions to investigate people's preferences for different brownfield cleanup policies that vary in terms of lives saved, duration of the benefits, number of years necessary to complete a specific cleanup program, population involved and cost of the cleanup (Turvani *et al.*, 2006).

Further difficulties may arise when we have to measure uncertain costs and risks, such as those related to environmental liability, information asymmetry, other transaction costs, or costs related to the perceived extent and severity of environmental contamination. Again, our own research experience shows the effectiveness of these evaluation techniques: we examined different market-based mechanisms and incentives intended to promote the environmental remediation and reuse of brownfields, such as reductions in regulatory burden, relief from liability for future cleanups, and subsidies for regeneration of brownfields. We used a questionnaire based on conjoint choice questions to assess the responses of real estate developers to different mixes of these incentives (Alberini *et al.*, 2005).

5. BROWNFIELD REMEDIATION AND REUSE: A CHALLENGE AND AN OPPORTUNITY FOR CHINA

In 1994, the Chinese Government published China's Agenda 21 – as a platform document for guiding the country's social and economic development. As far as land management is concerned, the Program reports:

> Implement the basic state policy of treasuring and rationally utilizing land and effectively protecting farmland, [...] Strengthen land resources survey, evaluation and monitoring [...] *Strengthen the management of land use for construction, control the scale of land use for such purposes*, [...] Improve land asset management, *deepening land-use restructuring by aggressively introducing a market-oriented approach* to land-use rights, improving the land pricing system and land tax system, and promoting efficient land use. Improve the land property system [...] Reform the land expropriation system. [...] Strengthen land legislation, improve laws and regulations and law enforcement.

More recently, China Council for international cooperation on environment and development (CCICED, 2006) reports:

> More scientific understanding and measurement on environmental resilience and carrying capacity in urban and surrounding areas are needed. *Avoiding urban sprawl, thus preventing cities from expanding in an excessive and uncontrolled manner which has negative environmental and social implications, is of utmost importance.*

Furthermore CCICED gave the following advice:

> *After appropriate environmental impact assessment, the use of remediated brownfield sites for housing should be given priority over greenfield sites, to avoid affecting natural and agricultural areas.*

The necessity to accelerate efforts to control the environmental impacts of cities and towns by setting in place essential laws, policies, knowledge and incentives within the Chinese strategy of growth is clearly acknowledged. The focus of the strategy is on circular economy, which is a major feature of Chinese way of welding further strong development and environmental protection by means of resource saving efforts, inter sectoral efficiency and recycling (Chen *et al.*, 2001). Land recycling and brownfield remediation and reuse become relevant for this approach because urban growth and development carry large increase in the consumption of land and demand for space for residential, industrial and commercial uses.

As far as official documents are taken into consideration we may trust that China is on the right track and it is going to avoid, at least to some relevant extent, what has been the fate of major industrial cities in the Western World, where a phase of urban decay, including the abandoning of factories and the generation of contaminated brownfields, was the consequence of the changing sectoral composition of the economy and of people changing lifestyles. Of course, the situation is quite different in the Western World, where urbanization is still taking place but at a much lower pace and where economic growth is much lower than in China. Differences notwithstanding, recycling urban land, i.e. finding new uses for previously used land and addressing issues of contamination due to previous hazardous activities taking places within urban areas, is a necessity for a more rational and sustainable land use and a base for sustainable development. Urban regeneration has proved to be a process able to trigger economic and social development in many cases, improving the quality of life in cities and economic development in Western Countries (Fox-Prezeworksi *et al.*, 1991) within a logic of sustainable development (Riddell, 2004).

China has shown great ability to learn from existing experience elsewhere in many fields: its ability to commit to sustainable development goals is demonstrated in some positive environmental improvements in cities thanks to ongoing policies and planning practices and, therefore we may trust that it will be able to capture the benefits of learning from Western success and defeats in brownfields remediation and reuse practices (Hanson, 2003).

Urbanization has been the distinctive feature of the rise of industrial societies throughout the world, but in China this process assumes a different character (Lin, 2002). A specific feature of urbanization in China has been the high number of floating population, due to the working of the *hukou* system[3] of home registration, which was introduced in the 50's to control the movements of population, to hold down the costs of housing and the benefits of urban workers which were mainly provided by state employment.

A recent research (Fujita and Thisse, 2003) shows that urban population in China is more dispersed in smaller cities with respect to more industrialized countries, while the overall urbanized population is little over 40% of the total, a percentage well below that one of other similar countries for size and income (Henderson, 2004). Given this situation a further urbanization of the population is expected by 2020 with a continuous reduction of the labour force engaged in agriculture. Urbanization over the coming two decades could bring as many as 300 to 400 million more people into cities, attracted into industrial and service sectors. The location of these activities will be a key determinant for the use of land in cities and in the greenfields surrounding them (Small, 2002).

Furthermore, in the past the development of town and village enterprises (TVE) has been a major strategy to spread industrial development into the countryside and raise the level of revenue of rural population. This choice on the one hand favoured the location of industry in areas beyond the east coast of China, helping economic growth substantially, but on the other hand consumed large parcels of land and contributed to the diffusion of hazardous industrial waste and environmental contamination in many towns and villages (Hanson, 2003).

As we previously discussed, the impact of industry on land and water resources has been considerable in all parts of the world. The burden of past industrialization has proved to be heavy for the environment and for social life in cities when traditional industries reduced employment and relocated in other areas. Liability associated with contaminated "brownfield" sites turned out to be an expensive and contentious issue to deal with, especially in North America and in both Eastern and Western Europe. The effects of liability regimes on the possibility to find new uses for redeveloping land were largely negative, affecting investment in existing industrial operations by local and foreign investors.

From the financial point of view yet, many brownfield sites are considered as the most valuable lands for new uses, ranging to less polluting industries, residential and light commercial land uses. Being able to redevelop them can lower the impact on suburban land sprawl that arises from the desire of industries to build on "greenfields" and the desire of residents to move from decaying city centers. This is the experience of the West and it has become an experience also for China.

Many large cities in China have experience of redevelopment of land (Chen, 2005; Li, 2003) and relocation of heavy industry (Chengri Ding, 2003) and of course redevelopment and land recycling is much needed in the future to provide new space for both citizens and industrial needs. With regards to these redevelopments, two important issues deserve attention: one is the problem of possible contamination, with implications for human health, ecosystem and property damages; the other relates to social justice and equity, given that the costs and benefits of redevelopment programmes affect citizens in a very uneven way. Vast areas within the urban cores have been redeveloped, displacing people from their neighborhoods to give room to other land uses which can pay higher rents. In the short run it is an issue of economic redistribution and fairness; in a longer term perspective, and learning from Western Countries errors, it is important to remember that urban quality in cities asks for the preservation of mixed land uses and it requires avoiding the creation of huge suburbs working as a dormitory to maintain the well functioning of the dynamic forces driving economic evolution, which find their roots in urban life and agglomeration (Fuijta et al., 2001; Jacobs, 1985), and to prevent the insurgence of costly and devastating social conflicts.

Contamination of land and underground water, due to polluting previous uses, poses similar problems to those encountered in Western Countries. The legacy of heavy industry and diffused polluting industries on the territory, in terms of soil and water contamination, may be enormous but it is also very important to look ahead and to avoid the continuous regeneration of such problems. Economy is an evolving system and even though land supply is fixed and scarce, land uses may change overtime. From this perspective it is important not only to have a full inventory of the existing problematic areas but it is necessary to develop hazardous waste management practices to avoid further problems that may arise in the years ahead.

> "The siting of toxic waste treatment centers, and the conditions and management of municipal land fills need careful attention. The culture of industrial development will need to shift so that both state-owned and private enterprises see themselves as stewards, prepared to invest in proper land care, and to leave sites in better condition than when they arrived. Special restoration funds set aside by companies for land management once industrial activities are completed is a workable mechanism that deserves attention in China (Hanson, 2003)".

Western Countries have developed complex liability systems, by assigning responsibility for the costs of cleanup on those parties who have contributed to creating the contamination problem. These liabilities should both prevent future contamination and provide a base for financing the necessary remediation: as we noticed, these systems, though important and just in principle, may carry very high cost of enforcement and may imply huge transaction costs, especially when the issue at hand is the cleaning up of past contamination (Stone McGuigan, 2000; Turvani and Trombetta, 2006).

In China, liability systems even though necessary to avoid and limit future contamination may encounter further problems of application due to the nature of land use rights granted to industrial firms rather than ownership, which is retained by the state. The possibility of enterprise restructuring may find strong limits because their land rights may be their only valuable asset and this value varies according to location. Old, poorly located industries may face the necessity to continue their polluting activities beyond any optimal lifespan since they do not have the financial means to afford the costs of restructuring and relocating plus the cost of remediation. 'It has been suggested by the World Bank that an administrative fund could be used to pool the revenues from sale of both high and lower value land use rights taken back from bankrupted enterprises in the region. The funds could then be used to meet social costs and rehabilitation fees of poorly-located industries wishing to declare bankruptcy (Hanson, 2003).

What has emerged from the Western experience is that the regimes of land ownership, title transfer and investment possibilities become very complex if uncertainty exists about who actually "owns" the liabilities associated with the land.

The ongoing system of land property rights and the way it is used by local governments to finance public urban expenditure and urban development by selling land use rights can be a major obstacle for the control of urban expansion and for the protection of greenfields surrounding cities and towns. The lack of well functioning property system may furthermore impede the development of an efficient land market and it has negative effects on the functioning of capital markets. The possibility to develop a workable system of liability to protect health, environment and property, not to say to enforce such systems, is necessarily constrained by the way property rights are now regulated and further reforms are needed (Chengri Ding, 2003). Furthermore many of the human health impacts associated with contaminated lands are not fully taken into account when counting the costs of land remediation while a public perspective on remediation costs and benefits should consider them. These costs may be quite relevant in financial terms and for the pursuing of a sustainable perspective in development which needs to place high value on human life, health and justice.

6. CONCLUSION

The European Union states that

> Protecting the environment is essential for the quality of life of current and future generations [...] Public participation is a central element in the common procedures applying across the EU for assessing the environmental impact of public sector policies and programs and of investment projects (http://europa.eu.int/pol/env/index_en.htm).

On the Chinese side, the same attention is given to environmental protection, to be developed in accordance with Chinese demand for economic growth. Active public participation and capacity building are high on the Chinese agenda:

> The support and participation of public and social groups is essential to the achievement of sustainable development. The form and degree of their participation determine the rate at which the objectives of sustainable development are realized [...] New mechanisms are needed for public participation in sustainable development. It is necessary for the public to not only participate in policy-making related to environment and development, particularly in areas which may bear direct impact on their living and working communities, but also to supervise the implementation of the policies (http://www.acca21.org.cn/chnwp20a.html).

Urban regeneration programmes offer a unique opportunity to achieve a sustainable development strategy: they address the issues of environmental protection and economic growth while requiring the active participation of people. Land remediation and reuse and social and economic redevelopment of brownfield sites are urgent policies for local and national governments both in the West and in China. These policies may be quite expensive and they need the support of public and private sectors, together with the consensus of communities. Calculating and communicating the beneficial effects of cleaning up contaminated sites to the public may be decisive for the effectiveness and success of remediation policies and redevelopment plans, and of course it is necessary for improving public decision making. By referring to the Western experience and our own research we presented the reasons for and offered a detailed analysis of the ways to realize benefit-cost analysis for brownfields remediation and redevelopment projects, in the attempt to offer possible hints for similar projects in China.

Margherita Turvani and Stefania Tonin
Planning Department,
University IUAV, Venice, Italy

NOTES

[1] The Environmental Protection Law of the People's Republic of China is the cardinal law for environmental protection in China. The law has established the basic principle for coordinated development between economic construction, social progress and environmental protection, and defined the rights and duties of governments at all levels, all units and individuals as regard to environmental

protection. China has enacted and promulgated many special laws on environmental protection as well as laws on natural resources related to environmental protection. They include the Law on the Prevention and Control of Water Pollution, Law on the Prevention and Control of Air Pollution, Law on the Prevention and Control of Environmental Pollution by Solid Wastes, Marine Environment Protection Law, Forestry Law, Grassland Law, Fisheries Law, Mineral Resources Law, Land Administration Law, Water Resources Law, Law on the Protection of Wild Animals, Law on Water and Soil Conservation, and Agriculture Law (http://www.china.org.cn/e-white/environment/e-3.htm). To succeed in the application of the Ten Laws, China wants to commit to investigation of the major problems in the execution and enforcement of current laws and the collection and analyses of typical cases to serve as a basis for further action, investigation of the use of market mechanisms for enforcing environmental protection legislation. These measures could include environmental taxes, tradable permits, levies on pollution discharge ("polluter pays principle"), and economic incentives (http://www.acca21.org.cn/pp1-1.html).
[2] The Small Business Liability Relief And Brownfields Revitalization Act (2002).
[3] Every Chinese resident has a hukou designation as an urban or rural resident. Hukou is an important indicator of social status, and urban (chengshi) status is necessary for accessing urban welfare benefits, such as schools, health care or subsidized agricultural goods. Without urban hukou status it is very difficult to live in cities. By limiting access to the benefits of urbanization, the hukou system ostensibly served as the world's most influential urban growth management instrument. (Ding, Chengri and Gerrit Knaap http://www.lincolninst.edu/pubs/pub-detail.asp?id=793) For a discussion of the implication of the houko system for future rural productivity increase and urban life quality for citizen see http://www.cdrf.org.cn/2006cdf/pinglun5_en.pdf.

REFERENCES

Adler M.D., 2003. Fear assessment: cost benefit analysis and the pricing of fear and anxiety, *Joint Center for Regulatory Studies Working Paper*, No. 03-12.
Alberini, Longo A., Tonin S., Trombetta F. and Turvani M., 2005. The Role of Liability, Regulation And Economic Incentives In Brownfield Remediation and Redevelopment: Evidence from Surveys of Developers, *Regional Science and Urban Economics*, No. 35, pp. 327-51.
Brown L., 1995. *Who Will Feed China: Wake-Up Call for a Small Planet*, Washington DC, Worldwatch Institute.
Cao Y., Qian Y. and Weingast B. R., 1997. From Federalism, Chinese Style, to Privatization, Chinese Style, *Stanford Economics Working Paper* No. 97-049.
Chen Y., 2005. Regeneration and Sustainable Development in the Transformation of Shangai, *WIT Transaction on Ecology and Environment*, No. 81, WIT Press.
Chen Ding-jiang, You-run Li, Jing zhu Shen and ShanYing Hu, 2001. The Planning and design of Eco-Industrial Parks in China, *ICCP Conference*, Beijing.
Chengri Ding, 2003. Land Policy Reform in China: Assessment and Prospect, *Land Use Policy*, No. 20, pp. 109-120.
CCIED, http://eng.cciced.org/cn/company/create/page2102.htm?siteid=1&lmid=2102
Ding Chengri and Gerrit Knaap http://www.lincolninst.edu/pubs/pub-detail.asp?id=793
De Sousa C., 2003. Turning Brownfields into Green Space in the City of Toronto, *Landscape and Urban Planning*, Vol. 62(4), pp. 181-98.
EEA, 2000. Management of contaminated sites in Western Europe, *Topic Report* N. 13/1999, Copenhagen, Retrieved April, 26, from http://reports.eea.eu.int/Topic_report_No_131999/en
Eisen J.B., 1999. Brownfields Policies for Sustainable Cities, *Duke Environmental Law and Policy Forum*, No. 9(2), pp. 187-217.
Fox-Prezeworksi J., Goddard J. and de Jong M., 1991. *Urban Regeneration in a Changing Economy: An International Perspective*, Clarendon Press, Oxford.
Freeman A. M., 1993. *The Measurement of Environmental and Resource Values: Theory and Methods*, Resources for the Future, Washington DC.
Fujita M. and Thisse J.F., 2003. Does Geographical Agglomeration Foster economic Growth? An Who Gains and Loses from it?, *The Japanese Economic Review*, No. 54(2), pp. 121-145.
Fujita M., Krugman P. and Venables A., 2001. *The Spatial Economy: Cities, Regions and International Trade*, Cambridge, The MIT Press.

Henderson J., 2004. Issues Concerning Urbanization in China, *11th Five year Plan of China*, Washington DC World Bank

Hanson R., 2003. *Sustainable Industrialization in China and a Well-Off Society*, http://www.harbour.sfu.ca/dlam/issuepaper-03.htm#_ftn1

Hofman, B., 2006. *Comment on: Yang Xie's Development and Urbanization of China's Country Side in Reform and Opening Up*, Retrieved April, 26, 2006, from http://www.cdrf.org.cn/2006cdf/pinglun5_en.pdf

Improving the Legal and Administrative Systems Step by Step (n.d.). Retrieved April, 26, 2006, from http://www.china.org.cn/e-white/environment/e-3.htm

Jacobs J., 1985. *Cities and the Wealth of Nations*, New York, Random House.

Lin G.C., 2002. The Growth and Structural Change of Chinese Cities: a Contextual and Geographic Analysis, *Cities*, No. 19(5), pp. 299-316.

Mimi Li, 2003. *Urban Regeneration Through Public Space: a Case Study is Squares in Dalian, China*, PhD Thesis, Waterloo, Ontario, Canada.

OECD, 2000. *Urban Brownfields*, Retrieved May, 7, 2004, http://www1.oecd.org/tds/bis/brownfields.htm.

Oliver L., Ferber U., Grimski D., Millar K. and Nathanail P., 2005. *The Scale and Nature of European Brownfields*, Retrieved November, 20, 2005, from http://www.cabernet.org.uk/resourcefs/417.pdf

Pellow D.N., 1998. A Community based perspective on Brownfields: Seeking Renewal From the Bottom-Up, Evanston IL: North Western University, Department of Sociology, Institute for Policy Research.

Porter M. E., 1995. The Competitive Advantage of the Inner City, *Harvard Business Review*, No. 73(3), pp. 55-71.

Priority Programme for China's Agenda 21 – Priority 1 – Capacity Building for Sustainable Development. (n.d.). Retrieved April, 26, 2006, from http://www.acca21.org.cn/pp1-1.html

Probst K.N. and Konisky D. M., 2001. Superfund's Future: What Will it Cost? A Report to Congress, *Resources for the Future*, Washington DC.

Riddell R., 2004. *Sustainable Urban Planning*, Blackwell, Oxford.

RESCUE (Regeneration of European Sites in Cities and Urban Environments). 2004. *Guidance on Sustainable land use and urban design on brownfield sites*, Workpackage 4 – Deliverable D 4.1, Retrieved April 10, 2006, from http://www.rescue-europe.com

Sassen S., 2000. *Cities in a World Economy*, Thousand Oaks, Pine Forge Press.

Schelling T.C., 1968. The Life You Save May Be Your Own, in S. B. Chase Jnr (ed.), *Problems in Public Expenditure Analysis*, Washington DC, The Brookings Institution.

Small K., 2002. Chinese Urban Development: Introduction, *Urban Studies*, November, Special Issue.

Stone McGuigan J., 2000. The Potential Economic Impact of Environmental Liability: the American and European Contexts, Retrieved October, 10, 2004, from http://europa.eu.int/environment/liability/competitiveness_finalrep.pdf

Tonin S., 2006. What is the value of brownfields? A review of possible approaches, in Alberini, A., Rosato, P., Turvani, M., *Valuing Complex Natural Resource Systems – The case of the Lagoon of Venice*, Edward Elgar.

Turvani M., A. Chiabai, A. Alberini and S. Tonin, 2006. Public Support for Policies Addressing Contaminated Sites: Evidence from a Survey of the Italian Public, *3rd World Congress of Environmental and Resource Economists*, Kyoto (Japan), 3-7 July 2006.

Turvani M. and Trombetta F., 2006. Governing Environmental Restoration: Institutions and Industrial Sites Cleanups, in Alberini, A., Rosato, P., Turvani, M., *Valuing Complex Natural Resource Systems – The case of the Lagoon of Venice*, Edward Elgar.

Urban Institute, Northeast-Midwest Institute, University of Louisville, University of Northern Kentucky. 1997. The Effect of Environmental Hazards and Regulation on Urban Redevelopment, UI Project No. 06542-003-00.

US Environmental Protection Agency, 2005. Superfund Benefit Analysis, Partial Draft prepared for US Environmental Protection Agency Office of Superfund Remediation Technology Innovation, Retrieved April, 10, 2006 from http://www.epa.gov/superfund/news/benefits.pdf

Van-Camp L., Bujarrabal B., Gentile A. R., Jones R.J.A., Montanarella L., Olazabal C. and Selvaradjou S. K., 2004. *Reports of the Technical Working Groups Established under the Thematic Strategy for Soil Protection*. EUR 21319 EN/4, p. 872. Office for Official Publications of the European Communities, Luxembourg.

VIII. AGRICULTURE AND NATURAL RESOURCE MANAGEMENT

F. CAPORALI

ECOLOGICAL AGRICULTURE: HUMAN AND SOCIAL CONTEXT

Abstract. This contribution outlines how the science and philosophy of Ecology have brought about a new pattern of agriculture representation and management. Through the sequence of key-concepts like ecosystem, agro-ecosystem and sustainable agriculture, the emergence of a paradigm shift is being illustrated which identifies agriculture as a human activity system with multiple positive roles in society. Organic farming, as a practical way of doing agriculture defined by law, is being presented as the first institutional example world-wide of a human activity system explicitily oriented towards sustainability.

1. INTRODUCTION: THE SHIFT TOWARDS SUSTAINABILITY

Man, as any other biological species, needs to adapt to its natural environment in order to live, even if, more than the other biological species, man has the capability to modify its own environment to an extraordinarily large extent. This capability is today so pervasive and aggressive as to dramatically affect the life support processes for both man and the whole biosphere. Therefore, it is urgent to provide a cultural shift towards more balanced human behaviour, which can ensure socio-economic development without undermining its biophysical foundations. This new model or style of life is generally mentioned as sustainable development and it is now a crucial challenge for humanity (UN, 1992).

To pursue the goal of sustainable development big efforts are to be made by institutions and individuals, because the ordinary patterns of development currently running do not comply with this new strategy which is only an emergent one. In order to become a dominant one, it needs to gain consensus and support of every kind – cultural, political and economic.

Agriculture is a human activity system that more than others has to do with the natural environment and its resources, both abiotic and biotic ones. Because of its peculiar character of being a trophic link between man and nature, agriculture more than any other human activity needs to be sustainable, i.e. durable in the long term. If appropriately organized, agriculture can be a prototype of best practices for environmental management and can be regarded as ecological agriculture (Caporali, 2004).

2. THE NECESSITY FOR AN ECOLOGICAL UNDERSTANDING OF REALITY

Sustainable development is a human strategy which has a science behind it usually referred to as *Ecology*. This kind of science is different from most of the other ones in that it is founded upon a transdisciplinary paradigm (Jantsch, 1970), which aims at both integrating different fields of knowledge and developing attitudes for problem solving and multi-scale analysis (Giampietro, 2004). The main goal of Ecology is the search for connections, because reality is understood as consisting of a network of interrelated events being developed over a continuum of spatial and temporal scales. Its basic method is an input/output analysis approach, which has both philosophical and scientific roots. Since understanding is a human enterprise or competence, which uses knowledge (Boix Mansilla and Garner, 1997), focus will firstly be on the process of knowledge construction.

2.1. Two complementary (or competing?) paradigms for knowledge construction

Because reality is a complex issue, two main complementary (or competing?) approaches have been developed in order to know, explain and manage it. One approach views reality as composed of different parts – or components – and considers it as the sum of the parts. This is the mechanistic view of reality derived specially from the Cartesian philosophy and the scientific revolution of the 17th century onward. This approach is particularly suitable for anthropogenic designed objects and explains the seemingly endless success of technological applications, including the more recent biotechnological applications to organisms, being referred to as genetically modified organisms (GMOs) or transgenic organisms. This paradigm has privileged the study of reality through increasingly specialised disciplines, each deepening knowledge about single parts or aspects of reality. Daly and Cobb (1994) have labelled this dominance of a discipline-oriented approach "disciplinolatry", identifying it as "the overwhelmingly dominant religion of the university". However, their opinion is that

> the more successful and exclusive are disciplinary goals, the less the contribution of the discipline to true understanding. The result is an *information age* but little comprehension of our real condition.

The other approach views reality as a hierarchy of open, interconnected systems, being developed over spatial and temporal scales, where new properties arise when scaling up hierarchical levels – i.e. from a cell, to an individual and an ecosystem – and where the upper and earlier levels constrain the lower and later ones, which need to adapt themselves to the system they belong to (Allen and Star, 1982). This approach is called systems paradigm, where the concepts of system, process, hierarchy, openness, integration, emergence, coevolution, creativity, irreversibility and meaning are dominant (Jantsch, 1980). Ecology is the science that more than any other has drawn upon this systems paradigm.

Usually, the mechanistic paradigm is regarded as culturally dominant in today's society, i.e. strongly established in human institutions, while the systems paradigm is regarded as only an emergent one, i.e. present but hardly affecting human behaviour, at both individual and institutional level. Indeed, the current structure and functioning of schools and universities in Western societies largely comply with the mechanistic paradigm, as the disciplinary structure of both didactic curriculum and research departments testify (Jantsch, 1970; Daly and Cobb, 1994). Only more recently, the structure and functioning of institutions of higher education has shown some sign of change towards a systems approach, establishing curricula and research departments devoted to environmental issues, with emphasis on ethical aspects as well (Wals *et al.*, 2004). Looking at recent developments, it seems that the systems paradigm is starting to be nested upon the mechanistic one, instead of replacing it, which justifies the assumption that there are more elements of complementarity than of competitivity between the two paradigms.

2.1.1. Philosophical foundations of the systems paradigm

Science and philosophy were originally not separate fields of inquiry – which happened definitively only in the 18th century – but a unique body of speculative thought. Aristotle's *Physics* is the first systematic attempt of thought organization concerning the principles of knowledge. In the first book of *Physics* (*The method of the physical science*), Aristotle claims that true knowledge about things is acquired when the principles or causes of things are known. In the second book of *Physics*, he describes four causes as necessary in order to know and explain a thing: *efficient causes*, which go back to its origin; *formal causes*, dealing with its innate tendencies towards a formal expression; *material causes*, concerning the stuff a thing is made of; *final causes*, which have to do with future goals. Aristotle's illustration of the knowledge method is authentically systemic in that it unveils that each thing is a process, i.e. an event which happens in a spatio-temporal scale, whereby causality (or connections) is the dominant dimension and the becoming is the essence of reality. Moreover, Aristotle points out that the knowledge process must proceed from the whole towards its parts, because what is global is a "true" whole that is more knowledgeable.

In Fig. 1 is shown the model of representation of the knowledge process according to Aristotle's four causes; this model has many analogies with the current model of inquiry we adopt in systems analysis, which is called input/output analysis model. Aristotle's detailed description of the four causes in the process of knowledge is now recognized as a powerful epistemological tool for understanding reality and there is no surprise for statement as the following by Capra (2002):

> I find it fascinating that after more than 2,000 years of philosophy, we still analyse reality within the four perspectives identified by Aristotle.

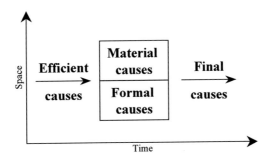

Figure 1. Representation of the knowledge process about one thing according to Aristotle's Physics.

At the very onset of conventional science in the 17th century, the scientific revolution started according to the Cartesian reductionist mechanical philosophy and the principle of analytical reduction which still characterises the Western intellectual tradition giving men a new myth (science) and a new powerful tool (technology) to modify reality. The exploitation of science in Western technology has largely created the modern world in a physical sense (Checkland, 1993), however, it has not helped in the search for meaning and therefore, in the search for appropriate management of reality by humans. Philosophers as Husserl, already in the 30's of the 20th century, pointed out at the failure of science (Husserl, 1954) and at the necessity of reconsidering the validity of a universal philosophy which entails every kind of knowledge, including science. The demand for better models of knowledge having greater descriptive and predictive power is still alive and the systems paradigm is a contribution to build up such new models. The first decades of the 20th century were crucial in the provision of philosophical contributions to the systems paradigm, especially by N.A. Whitehead (1927) and J.C. Smuts (1927).

The systems paradigm in philosophy is also referred to as process philosophy (Barbour, 1997), because A.N. Whitehead has been the most influential exponent of process categories, which give body to a systematic metaphysics that is consistent with the evolutionary, many-levelled view of nature. The basic components of reality are portrayed as processes of becoming, which Barbour (1997) gives a detailed account of, labelling Whitehead's process philosophy as "an ecological metaphysics". He emphasizes four points:

1) *the primacy of time,* because the starting point of process philosophy is becoming rather than being, and reality, along a temporal hierarchy, exhibits chance, creativity and emergence (production of novelty);
2) *the interconnection of events,* because the world is a network of interactions, where events are interdependent;
3) *reality as organic process,* because the world process implies temporal change and interconnected activity. The parts contribute to and are also modified by the unified activity of the whole. Each level of organization is affected by and influences the pattern of activity at the other levels;
4) *the self-creation of every entity,* because every entity is a new unity formed from an initial diversity, which contributes distinctively to the world.

As Barbour (1997) points out:

> Process thought is opposed to all forms of dualism: living and nonliving, human and nonhuman, mind and matter. Human experience is part of the order of nature.

Another contribution to the systems paradigm comes from J.C Smuts (1927), who elaborated a theory known under the name "holism", which has been defined as the synthetic tendency in the universe, i.e. the principle which makes for the origin and progress of wholes in the universe. Smuts develops "the concept of the whole as a means of tracing the evolution of reality". Smuts's philosophy is

> an attempt to reach the fundamental unity and continuity which underlie and connect matter, life and mind

through the ascending order of wholes along a temporal hierarchy. Smuts lists the most significant characteristics of wholes, which define "the great Ecology of the universe" as follows:

1) wholes are not closed, isolated systems; they have their fields in which they intermingle and influence each other;
2) the holistic universe is a profoundly reticulated system of interactions and inter-connections;
3) genetic relationships connect the entire holistic universe;
4) evolutionary process holds all the wholes together in one vast network of adaptations and harmonious co-ordinations;
5) the rise and self-perfection of wholes in the Whole is the slow but unerring process and goal of the holistic universe.

According to holism, reality is a progressive construction, which culminates with the human personality as the last emergence, whose essence is creative freedom that means self-determination in a universal holistic order.

2.1.2. The ecosystem concept as a scientific development of philosophical demands

If we look at the context in which the ecosystem theory has been generated, it is really astonishing to find out how strict connections it exhibits with the holistic philosophy. Indeed, Arthur George Tansley coined the term *ecosystem* and provided its basic description in a seminal paper published in the scientific magazine *Ecology* (1935) as the last outcome of a dialectical confrontation with the South African ecologist John Phillips, who based its idealist foundation for ecological research on the holistic philosophy of Smuts and defined plant communities as organisms (Golley, 1993; Jax, 1998; Anker, 2002). Already in a previous work of 1934, Tansley (2002) confirmed the validity of holistic concepts like: the temporal genetic series, inorganic matter – living organism – mind; the autopoietic character of cells and organisms; the teleological orientation of their behaviour; the belief in continuity throughout the universe; the emergence of mind and ethical values. All that is summarized in statements like:

> human beings belong to all four different planes of existence: the material, the biological, the psychical and the ethical;

> there seems to be no such thing in the universe as complete independence;

and

> I understand as evolution... the progressive appearance of integrated systems.

However, Tansley strongly rejected the idea of calling a plant community an organism. Given this context, it comes as no surprise the emergence of the ecosystem concept as a powerful heuristic tool, which was formulated in the following year (Tansley, 1935) with the intent of facilitating a better understanding of nature organization.

Tansley presented the ecosystem concept as follows:

420   F. CAPORALI

> But the more fundamental conception is, as it seems to me, the whole system (in the sense of physics), including not only the organism-complex, but also the whole complex of physical factors forming what we call the environment of the biome – the habitat factors in the widest sense.
> It is the systems so formed which, from the point of view of the ecologist, are the basic units of nature on the face of the earth.
> These ecosystems, as we may call them, are of the most various kinds and sizes. They form one category of the multitudinous physical systems of the universe, which range from the universe as a whole down to the atom.
> They all show organization, which is the inevitable result of the interactions and consequent mutual adjustment of their components.
> The whole method of science... is to isolate systems mentally for the purpose of study.

As put by Golley (1993) and Caporali (2004), *ecosystem* refers to a holistic and integrative ecological concept that combines living organisms and the physical environment into a system that is:

1) an element in a hierarchy of physical systems from the universe to the atom;
2) the basic system and the unit of study of ecology;
3) both a scientific and an epistemological model for representing the reality we live in.

From its origin in 1935 to nowadays, Tansley's ecosystem has remained a key concept in ecological sciences and thanks to its heuristic value that has been operationally accrued by other eminent ecosystem ecologists (Lindemann, 1942; Odum, 1953), it is now used as the main intellectual tool for expressing judgements about reality organization and management.

In a current vision taking into account both previous scientific and philosophical foundations, the ecosystem model as a unity of study or management can be symbolically constructed as a four-step process that culminates in a physically delimited reality as shown in Fig. 2 – the circle represents the space-time limit – continuously interconnected with the surrounding environment in terms of energy-matter-information input and output exchanges.

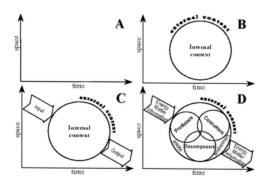

Figure 2. A four-step process for implementing a simple system model of a piece of reality (ecosystem). A) scale framing; B) boundary fixation: allocation of the system of interest into the scale frame; C) input/output exchanges; D) ecosystem structure and functioning.

3. AGROECOLOGY AS THE SCIENCE OF SUSTAINABLE AGRICULTURE

The systems paradigm is a very useful tool for studying and managing reality, since it is applicable to any kind of human activity. It is a method that reflects contents, i.e. that exhibits coherence between reality and its human representation, whereby ontology and epistemology coincide. In using it, it is important to remember its fundamental elements which are *hierarchy, emergence, communication and control* (Checkland, 1993). The systems paradigm applied to agriculture defines a new transdisciplinary science which is called *Agroecology*. When applied to agriculture, the systems paradigm is like a lens for focusing on rural reality at different levels of resolution along the sequence of the agroecosystem's hierarchical level which are considered in the enquiry (Tab. 1). In analogy with the ecosystem concept, the agroecosystem is both an ecosystem modified and used for agricultural purposes and the model that represents it.

Table 1. Application of the systems paradigm in agriculture: expression of the hierarchy concept

Agroecosystem's hierarchical level	Fields of study and management
Field system	Soil-plant-atmosphere relationships
Cropping system	Relationships between crops in a spatio-temporal scale
Farming system	Relationships between crops, livestock and management
Landscape system	Biodiversity. Land conservation practices and aesthetics
Regional system	Rural activity integration; socio-economic development
National and International system	Globalisation aspects (Market, Economy, Policy and Environment)

All this hierarchy is to be meant as a spatio-temporal continuum, an open, interconnected sequence or stratification of multi-layered agroecosytems, which are isolated only for a necessity of study and management. A disciplinary approach usually focuses only on components and processes of one level of the hierarchy and neglects the other ones, with great risks of incomplete or misleading understanding of the whole reality, which is definitely what counts in the long run. This strict disciplinary attitude is what Whitehead called "the fallacy of misplaced concreteness" being the cause of failure in understanding. In a transdisciplinary approach, as that used by *agroecology*, that risk is avoided because what is recommended is to try to establish coherence of knowledge, judgement and management within and among the different levels of the hierarchy that forms the organization of reality.

Starting from the upper hierarchical levels of Tab. 1, i.e. from international to regional systems, socio-economic stakeholders and processes constitute the driving force of the organization pattern and the agroecosystem can be represented by a model that identifies agriculture as a human activity system (Caporali, 2004). In this model, the human factor is dominant because it is the flow of information (economic and political, scientific and technical, cultural and ethical) which determine the kind of organization of the agroecosystem, i.e. the structure and functioning of the whole system. At these hierarchical levels of broader interaction, communication, trade and control, biophysical and socio-economic components of agroecosystems at local level are usually neglected and not included in the analysis, and therefore decisions at macro-economic scale are often taken that contrast in practice with the biophysical and socio-economic sustainability at local level.

If, according to an agroecological approach, the analysis along the sequence of hierarchical levels was completed proceeding from the regional system level to the field level, the risk of taking decisions against sustainability at local level could be avoided. Indeed, the model that represents the agroecosystem structure and functioning at farming system level (see paragraph 3.1) demonstrates that primary productivity, which is the only significant antientropic process on the earth due to plant photosynthesis, is sustainable only at local level (Fig. 3), i.e. it cannot occur everywhere but only where there are: a) favourable climate, b) appropriate farming conditions for sustaining soil fertility – which are driven mostly by biological processes – and c) human resources willing to reside on the spot. Therefore, every intervention that undermines this basis of local crop productivity is deemed to cause unsustainability of agriculture and rural abandonment.

Because of the above-mentioned reasons, *Agroecology* is currently intended as the science of sustainable farming and food systems, where humans and nature are integral parts (Francis *et al.*, 2003). *Agroecology* is the science that provides information, interpretation and knowledge in order to promote action in favour of agriculture sustainability. Its field of enquiry embraces production, processing, marketing, consumption, and their social and environmental impacts.

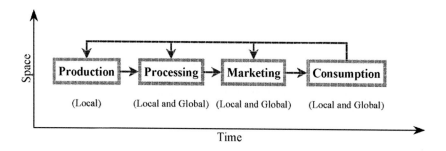

Figure 3. Agri-food chain structure.

Disseminating the systems paradigm in agriculture, which means to extend the perception of belonging to agriculture as a human activity system, is the first condition in order to achieve its re-orientation towards sustainability. Each stakeholder, both single individuals and institutions – especially consumers and their representatives – should be involved in a participatory communication network through any available means, from university to schools and from markets to mass media. Indeed, regulation and control of agroecosystem sustainability at international and national levels could only happen if a bottom up feedback process were promoted by responsible stakeholders. The philosopher Hans Jonas (1979) has laid the basis for the foundation of an ethics more appropriate for today's technological era focusing on a system paradigm and the principle of responsibility.

3.1. Design, implementation and management of sustainable agroecosystems

In the hierarchical sequence of agroecosystems, the farm level is a crucial one, because the farm is the management unit with a biological base, easily identifiable because of its boundaries, which represents the meeting point between human interests and the natural environment (Caporali et al., 1989). The identification of a farm as an agroecosystem (Caporali, 2004) permits a clear perception of the impact of human action on the structure and functioning of agricultural systems. Above all, it promotes awareness of the fact that:

1) the farmer is the direct creator of the agroecosystem;
2) society has great possibilities for orienting farmers' decisions and the structure and functioning of agroecosystems through the definitions of the economic system;
3) the structure and functions of agroecosystems can be modified in a short span.

Today's advanced societies see agriculture as a potential multifunctional activity, which responds to the double purpose of production and protection, yielding both social and environmental benefits.

A plan of action promoting a productive and protective agriculture must be well founded on a solid scientific base, with contributions from the international scientific community. Recently in Europe (2004), a transnational programme has been approved at academic level as a Socrates Thematic Network under the title "Redefining the curricula for the multifunctional rural environment – agriculture, forestry and the rural society" (MRENet). The MRENet aims at supporting the development of a European dimension for teaching and learning disciplines related to the multifunctional rural environment by redefining the curricula to reflect environmentally sustainable, ethically defensible and risk conscious agricultural and forestry industries that foster the wellbeing of both rural communities and the biophysical and socio-cultural environment in which they live.

The farming level is the preferred target of intervention of agricultural policy for a sustainable agriculture, with agreements, incentives and subsidies assigned by the civil society directly to farmers. In Europe, this kind of policy started in 1985 with the EEC Regulation 797/85 concerning the types of environmentally compatible agriculture and it still exists. The Agenda 21 (UN, 1992), which is the main outcome of the United Nation Conference in 1992, devotes chapter 14 to "Promoting sustainable agriculture and rural development" and chapter 32 to "Strengthening the role of farmers" because "a farmer-centred approach is the key to the attainment of sustainability in both developed and developing countries". Moreover, the decentralisation of decision-making towards local and community organization is seen as a key-strategy in changing people's behaviour and implementing sustainable farming plans.

Behind these initiatives are the science of agroecology and the concepts of ecosystem and agroecosystem. The concept of ecosystem, which has been originally developed in ecology and subsequently applied to farming to originate the concept of agroecosystem, has formed the basis for the conception of an ecological agriculture and has laid the foundations for its design and practical application.

The advantage of considering a farm as an agroecosystem was realized and supported in Italy by the agronomist Alfonso Draghetti who, as early as in 1948, published a volume titled "Farm Physiology Principles". He distinguished two fundamental pathways for the transfer of energy and nutrients in the functional processes of a farm, named "small" and "large" circulation respectively (Fig. 4).

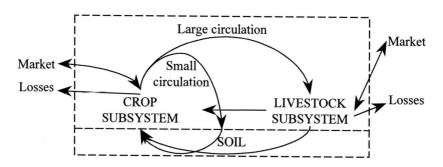

Figure 4. Small and large circulation patterns of energy-materials in a mixed farm (modified after Draghetti, 1948).

Through the small circulation, the exchange of energy and material between crops and soil (soil-crop sub-system) is performed; through the large circulation involving livestock transformation (livestock sub-system) the food chain of grazing and detritus is completed at farming level, allowing the system to sustain itself. The level of self-sustainability depends on the level of energy-matter re-cycling on the farm and it is strictly connected with the type of farming. This input/output functional scheme of a farming system constitutes a reference picture for designing, monitoring and managing sustainable agroecosystems.

A way of developing such agroecosystems is to imitate the structure and function of natural ecosystems, which perform maximum utilization of native resources and processes, both physical and biological, through a more complicated structure and functioning. Hence, sustainable farming systems need to be arranged in tune with the principles of ecological development in a way that a farm results in an agroecosystem in which the various components (cropping systems, livestock systems, soil and climate) are so fully integrated as to lead the farm to serve as:

1) a "solar plant", capable of converting solar energy to biomass throughout the whole year;
2) a rainfall water catchment area in which infiltration is enhanced over water run-off and soil erosion;
3) an area where atmospheric nitrogen fixation is enhanced by the frequent and intensive use of legume crop rotation, intercropping, cover cropping and green manuring;
4) a mixed farm, which combines crop growing and livestock husbandry so as to attain the maximum integration of grazing and detritus chains and increased soil fertility;
5) an integrated biological community in which weed and pest control is a prevention measure due to the diversity inside (crop rotation, intercropping, etc.) and outside (hedges, tree rows, etc.) the cultivated field.

The overall goal of these principles is to develop a multifunctional agroecosystem that is productive, protection-driven, and healthy, relies on the minimal use of external fossil fuel energy and yields quality and contaminant-free food.

3.2. Measuring agriculture sustainability

Agricultural researchers have long recognized the importance of the sustainability of agricultural systems and the need to develop appropriate ways to measure sustainability. Lynam and Herdt (1989) emphasized that agricultural researchers should:

1) recognize the importance of the sustainability of agricultural systems;
2) devise appropriate ways of measuring sustainability;
3) empirically examine the sustainability of some well-defined cropping or farming systems;
4) define the externalities present in such systems;
5) develop methods to measure those externalities.

Webster (1997) stressed that conventional farming systems are developing an undesirable rural development in the social, economic and biophysical sense, and that there is a need to move from conventional farming systems to sustainable farming systems. Sustainability should involve a reduction in external inputs and a move towards internal self-sufficiency. There is a general agreement that negative environmental impact is lower in agroecosystems which employ less auxiliary energy (Wagstaff, 1987). In general there is a shared awareness that obtaining acceptable indicators for sustainability assessment is an urgent challenge for both researchers and decision-makers. The search for agroecosystem performance indicators (APIs) is an urgent task to develop understanding and to facilitate decision-making processes. The aim to develop APIs has both an epistemological and a practical significance, representing, respectively: a) an efficient instrument of inquiry for studying agroecosystem functioning and performance, according to an input/output approach; and b) a relevant knowledge base for both the design of sustainable agroecosystems and decision-making processes (Tellarini and Caporali, 2000).

According to the above-mentioned Draghetti's scheme (1948), small and large circulation, activated by solar energy, constitutes the motor of the entire production process. This motor is also fuelled by inputs of auxiliary energy-matter from the outside and produces outputs that can be re-used on the farm (internal transfers) or leave the system definitively. The market is regarded as the external socio-economic environment in which the farm is immersed. With this kind of model, it is not only possible to measure the total input and output, but also the internal movements from one sub-system to another, and even within each sub-system. Since the flows within the farm and between the farm and the external environment can be expressed as energy or as monetary values or as nutrients, all direct APIs can be calculated for each of these flows. By relating each type of output to each type of input, it is possible to obtain a considerable amount of information on both the circulation of energy-matter (and monetary values) within the farm and the efficiency with which these resources are used, both at the level of individual sub-system and the farm as a whole. It is also possible to evaluate the level of energy-matter dependency of the farm (or of each single sub-system) both on the context external to the farm and on non-renewable inputs. In other words, the sustainability level of the farm's agricultural activity. Applications of this kind are now of current use and of great interest, especially in documenting elements of both structural and functional diversity with a value for agroecosystem sustainability at farming system level (Tellarini and Caporali, 2000; Rygby *et al.*, 2001; Caporali *et al.*, 2004; Thomassen and de Boer, 2005).

4. THE ASSETS OF ORGANIC AGRICULTURE

Organic agriculture is the most advanced outcome of the societal demand for a sustainable agriculture. It is defined by law through: a) a body of technical rules (standards) to be adopted at the farm level; b) administrative procedures permitting control of the production processes by an entrusted institution; c) certification and labelling of products before marketing. Since 1991, organic crop production in the European Union has been legally defined by EC Regulation 2092/91. Since this date, all produce sold with an organic label has had to comply with this legislation (Lampkin, 1990). Organic agriculture is practised in almost all countries and its share of agricultural land and farms is continuously increasing.

Two main sources at the international level can be mentioned to provide a definition of organic agriculture: one is that of the FAO/WHO Codex Alimentarius (2001):

> Organic agriculture is a holistic production management system, which promotes and enhances ecosystem health, including biological cycles and soil biological activity. Organic agriculture is based on minimizing the use of external inputs, avoiding the use of synthetic fertilisers and pesticides.... The primary goal of organic agriculture is to optimise the health and productivity of interdependent communities of soil life, plants, animals and people.

The other one is that of the International Federation of Organic Agriculture Movements (IFOAM) (2002):

> Organic agriculture is a whole system approach based upon a set of processes resulting in a sustainable ecosystem, safe food, good nutrition animal welfare and social justice. Organic agriculture therefore is more than a system of production that includes or excludes certain inputs.

Organic agriculture principles are consonant with principles of ecological agriculture, as expressed in paragraph 3.1, because the ecosystem approach is the base of designing and management adopted by organic farmers who "need to restore the natural ecological balance because ecosystem functions are their main productive input" (FAO, 2002). Maintenance of soil fertility for crop production and maintenance of biological balances for crop protection are the two pillars of organic agriculture, which mostly depend upon a technical management being based on biodiversity and re-cycling.

According to Caporali (2004), characters and assets of organic agriculture can be summarized as follows:

a) it is scientifically founded and the founding discipline is agroecology;
b) it promotes "the culture of life". Life is interpreted as a process of necessary interdependence among all the components of the agroecosystem;
c) it promotes biodiversity as a key to autonomy and sustainability;
d) it promotes the re-establishment of environmental and social balances – food, socio-economic and health-related balances such as: more favourable balances of population density; a better distribution of livestock on the territory; the improvement of the agricultural landscape; the improvement of the functionality of the agroecosystem; the maintenance of the quality of natural resources as well as agricultural products; the enhancement of local economy;
e) it has an educational value. An organic farm is a practical case of management of life scenario in accordance with ecological principles and therefore it has potential for disseminating an ecological culture by experiential learning;
f) it is a guarantee of the origin of food. Conscious consumers share the product, the process and the environment of production and all their effects (health-related, social and environmental). Organic agriculture meets the concept of food traceability, i.e. the story of its path from the field to the table, which is becoming an important subject on the policy of the quality and security of food.

For the above mentioned characters and assets, organic agriculture is the first human activity system consciously and legally devoted by the civil society to implementing sustainability in practice and therefore, to operating for the common good.

5. CONCLUSIONS

The current ecological perspective of agriculture calls for a commitment to responsibility by all stakeholders of this human enterprise in order to promote agricultural systems able to provide multiple socio-economic and environmental benefits.

Agroecosystem health is now emerging as a potential goal, which requires an understanding of the scientific and philosophical roots on which the concept of sustainability is based. New agroecology-oriented curricula at university level should drive the expected cultural shift toward more systemic teaching and learning attitudes, because no real societal change can intervene and last in practical activity like agriculture without a previous correspondent change in the training process of the human resources involved. Because the human component is today the most important driving force in ecosystem dynamics, it is of paramount importance to work firstly for mind health, if we really want to achieve both healthy individuals and institutions in a healthy environment.

Fabio Caporali
Crop Production Department,
University of Tuscia, Viterbo, Italy

REFERENCES

Allen T.F.H. and Starr T.B., 1982. *Hierarchy. Perspectives for ecological complexity*. The University of Chicago Press, Chicago.
Anker P., 2002. The Context of Ecosystem Theory. *Ecosystems*, 5, pp. 611-613.
Barbour J.C., 1997. *Religion and Science*. Harper, San Francisco.
Boix Mansilla V. and Gardner H., 1997. Of Kinds of Disciplines and Kinds of Understanding. *Phi Delta Kappan*, January 1997, pp. 381-386.
Caporali F., 1999. Ecosystems controlled by Man. *Frontiers of Life*, Vol. IV, 519-533. Academic Press, New York.
Caporali F., 2004. *Agriculture and Health. The Challenge of Organic Agriculture*. EDITEAM, Cento (FE), Italy.
Caporali F., Nannipieri P., Paoletti M.G., Onnis A., Tomei P.E. and Tellarini V., 1989. Concept to sustain a change in farm performance evaluation. *Agriculture, Ecosystems and Environment*, 27, pp. 579-595.
Caporali F., Mancinelli R. and Campiglia E., 2004. Indicators of Cropping System Diversity in Organic and Conventional Farms in Central Italy. *International Journal of Agricultural Sustainability*, 1, pp. 67-72.
Capra F., 2002. *The Hidden Connections*. Doubleday, New York.
Checkland P.B., 1993. *Systems Thinking, Systems Practice*. John Wiley and Sons, London.
Daly H. E. and Cobb B.Jr., 1994. *For the Common Good*. Beacon Press, Boston.
Draghetti A., 1948. *Principi di fisiologia dell'azienda agraria*. Istituto Editoriale Agricolo, Bologna.
FAO/WHO Codex Alimentarius Commission, 2001. *Guidelines for the production, processing, labelling and marketing of organically produced foods*. CAC/GL 32-1999-Rev.1-2001, Rome.
FAO, 2002. *Organic agriculture, environment and food security*, Rome.
Francis C., Lieblein G., Gliessman S., Breland T.A., Creamer N., Harwood R., Salomonsson L., Helenius J., Rickerl D., Salvador R., Wiedenhoeft M., Simmons S., Allen P., Altieri M., Flora C. and Poincelot R., 2003. Agroecology and Agroecosystems: The Ecology of Food Systems. *The Journal of Sustainable Agriculture*. 22, pp. 99-118.
Giampietro M., 2004. *Multi-Scale Integrated Analysis of Agroecosystems*. CRC Press, New York.

Golley F. B., 1993. *A History of the Ecosystem Concept in Ecology*. Yale University Press, New Haven.
Husserl E., 1954. *Die Krisis der europaischen Wissenschaften und die transzendentale Phanomenologie*. Band VI, Haag, Martinus Nijhoff.
IFOAM, 2002. *Basic Standards for Organic Production and Processing* (available at http://www.ifoam.org/about_ifoam/standards/index.html).
Jantsch E., 1970. Inter- and Transdisciplinary University: a Systems Approach to Education and Innovation. *Policy Sciences*, 1, pp. 403-428.
Jantsch E., 1980. *The self-organizing universe. Scientific and human implications of the emergent paradigm of Evolution*. Pergamon Press, Oxford.
Jax K., 1998. Holocen and Ecosystem – On the Origin and Historical Consequences of two Concepts. *Journal of the History of Biology*, 31, pp. 113-142.
Jonas H., 1979. *Das Prinzip Verantwortung. Insel Verlag*, Franfurt am Main.
Lampkin N., 1990. *Organic Farming. Farming Press*, Ipswich, UK.
Lindemann R.L., 1942. The Trophic-Dynamic Aspect of Ecology. *Ecology*, 23 (4), pp. 399-418.
Lynam K.J. and Herdt R.W., 1989. Sense and sustainability: sustainability as an objective in international agricultural research. *Agricultural Economics*, 3, pp. 381-398.
Odum E.P., 1953. *Fundamentals of Ecology*. W.B. Saunders, Philadelphia.
Rygby D., Woodhouse P., Young T. and Burton M., 2001. Constructing a farm level indicator of sustainable agricultural practice. *Ecological Economics*, 39, pp. 463-478.
Smuts J.C., 1927. *Holism and Evolution*. MacMillan, London.
Tansley A. G., 1935. The use and abuse of vegetational concept and terms. *Ecology*, 16, pp. 284-307.
Tansley A.G. (prepared by P. Anker), 2002. The Temporal Genetic Series as a Means of Approach to Philosophy. *Ecosystems*, 5, pp. 614-624.
Tellarini V. and Caporali F., 2000. An input/output methodology to evaluate farms as sustainable agroecosystems: an application of indicators to farm in Central Italy. *Agriculture, Ecosystems and Environment*, 77, pp. 111-123.
Thomassen M.A. and de Boer I.J.M., 2005. Evaluation of indicators to assess the environmental impact of dairy production systems. *Agriculture, Ecosystems and Environment*, 111, pp. 185-199.
UN (United Nations), 1992. *Agenda 21. United Nations Conference on Environment and Development*. Rio de Janeiro, Brazil.
Wagstaff H., 1987. Husbandry methods and far systems in industrialized countries which use lower levels of external inputs: a review. *Agriculture, Ecosystems and Environment*, 19, pp. 1-27.
Wals A.E.J., Caporali F., Pace P., Slee B., Sriskandarajah N. and Warren M., 2004. *Education and Training for integrated Rural Development. Stepping Stones for Curriculum Development*. Elsevier, Overhei. Reed Business Information, Netherlands.
Webster J.P.G., 1997. Assessing the economic consequences of sustainability in agriculture. *Agriculture, Ecosystems and Environment*, 64, pp. 95-102.
Whitehead A.N., 1927. *Process and Reality. An Essay in Cosmology*. The Free Press, New York, 1927.

M. L. GULLINO, A. CAMPONOGARA AND N. CAPODAGLI

SUSTAINABLE AGRICULTURE IN THE FRAME OF SUSTAINABLE DEVELOPMENT: COOPERATION BETWEEN CHINA AND ITALY

Abstract. After a short introduction on the importance of the agro-food sector in the economies of most societies, this chapter focuses on the role of agriculture in protecting the environment and enhancing biodiversity. The importance of a shift toward agricultural systems, which are more complex in terms of biodiversity in emerging countries such as China, where most of the population (about 60%) lives in rural areas, in poor conditions and still relies on agriculture as the main source of income, is discussed. In China the problems caused by shifting to more intensive and polluting agricultural production patterns to address the demand for food of the growing population and by the need of higher incomes of rural communities are serious. The experience gained in the framework of the Sino-Italian Cooperation Program for Environment Protection jointly launched by the Italian Ministry for the Environment Land and Sea (IMELS) and the State Environment Protection Administration of China (SEPA) in the year 2000, through the implementation of several cooperation projects in different rural areas of China, is critically discussed.

1. INTRODUCTION

Sustainable agricultural development is an important goal in economic planning and human development worldwide. Agriculture plays a significant role in protecting the environment and enhancing biodiversity (Altieri and Merrick, 1987; Paoletti, 2001; Marshall and Moonen, 2002) when it is carried out in a sustainable manner and takes into account its genetic resources. The three dimensions (economic, environmental and social) of sustainable development are highly visible in the agro-food sector: economic, environmental and social issues are probably more acute in this sector than in most of the other economic sectors. As a consequence, the agro-food sector is a major component (of course not the only one) of the overall issue of sustainable development and it is a highly relevant example of the main issues, which this topic at large can raise (Viatte, 2001).

At present, agriculture faces the need to meet new challenges both in highly industrialized countries as well as in developing ones. Such challenges are represented by sustainable growth, social integration of rural communities and proper use of the advantages deriving from emerging global markets. These markets become more demand-driven and consumer-oriented and they have a strong tendency towards openness and loss of trade barriers. These challenges go along with radical changes in the relation between society and agriculture.

In industrialized societies, the need for food security, sufficient foreign currency, and maintenance of economic basis in rural regions is largely met. Consequently, there is no longer justification for the traditional trade barriers or the vigorous promotion of agricultural technology by governments. Specialization and increased production no longer seem to be the main goals of agricultural policy. Other competing societal values, needs, and interests emerge in the political arenas of many states. Since agriculture has been often perceived as a major source of risk in our risk society, agriculture and the food industry are engaged in a quest to redefine and normalize their relation with society. Today people recognize that agriculture not only produces food, but also protects the diversity of life, the landscapes, the territory, and the waters, prevents disasters, promotes the vitality of rural areas (favoring employment, strengthening farm-related economies, and maintaining local culture), alleviates their poverty, and offers recreation possibilities to urban people (Negri, 2005).

Agriculture is an elemental engine of economic growth in the developing world. Today the world is not (food) secure in terms of access to food. Most populations in the developing world are increasing rapidly: by the year 2020 there will be an additional 1.5 billion people to feed. Improvements in yields on a sustainable basis will be needed to meet the food demands of this growing population. Local gains in productivity will not only increase food security for the poor, they will also increase farmer incomes and allow them greater opportunity to break the cycle of poverty. A new type of agriculture is needed which is ecologically sound and meets the food needs of the poor.

2. AGRICULTURE AND ENVIRONMENT

Environmental trends in agriculture have been analyzed in depth, in each individual country, as well as at the international level. There are both positive and negative features and aspects. Presently, agriculture mostly has a productive role and feeds the world through intensive, not sustainable agricultural systems (Matson *et al.*, 1997; Rasmussen *et al.*, 1998). On the one hand, in many industrialized countries there are still regions with high pollution levels resulting from agriculture, such as nitrogen and pesticide loadings in water, and threats to natural resources such as land, water and biota. Two issues merit highlighting because they have a very strong societal character, and hence a marked ethical dimension: one relates to the use of water, the other to biodiversity (Falkenmark and Rockstrom, 2004; Thrupp, 2004). On the other hand, there have been improvements in the efficiency of the use of inputs as a result of more targeted farm practices. As the world enters into an era in which global food production is likely to double in order to continue to feed a growing population, human needs related to food production must be met without compromising the health of agro-ecosystems. As a consequence, the protective role of agriculture must prevail over the productive one. In order to minimize the environmental impact, a shift toward agricultural systems that are more complex in terms of biodiversity and productive strategies is critical (Tilman, 1999).

In past decades, the management of natural resources by agriculture was mainly oriented at improving on-farm productivity. While such approaches may have fulfilled their primary objective – e.g. soil conservation – other objectives, such as preventing polluted run-off, if considered at all, were paid less attention.

An ecosystem approach refers to an integrated strategy for the management of land, water and living resources that promotes conservation and sustainable use, by both current and potential future users. It stresses the application of scientific methods focused on levels of biological organizations that encompass the essential processes, functions and interactions among organisms and their environment. Humans are an integral part of any ecosystem approach. Adopting an ecosystem approach implies a change not only in the institutions but also in their objectives. In the case of agriculture, such an approach requires considerations of alternative uses of resources, as well as substitutes for the goods or services that they render, such as energy crops, environmental amenities and services. The question is not, therefore, how should we maintain resources in existing agricultural activities, but how can society ensure that resources are allocated to their highest value use in contributing to sustainable development.

3. SUSTAINABLE AGRICULTURE FOR ENVIRONMENTAL PROTECTION

Sustainable agriculture is defined as

> an integrated system of plant and animal production practices having a site-specific application that will, over the long term: satisfy human food and fiber needs; enhance environmental quality and the natural resource base upon which the agricultural economy depends; make the most efficient use of nonrenewable resources and on-farm resources and integrate, where appropriate, natural biological cycles and controls; sustain the economic viability of farm operations; enhance the quality of life for farmers and society as a whole" (US Congress, 1990). An early definition by Gips defines sustainable agriculture as "ecologically sound, economically viable, socially just, and humane" (Appleby, 2005).

Humaneness of sustainable agriculture is a very important feature.

Sustainable agriculture uses ecological principles in combination with traditional and modern technologies so as to combine higher productivity with environmental friendliness and social and cultural sensitivity (Caporali, 2007). Implementing sustainable agriculture means working with natural systems to prevent pest outbreaks and other problems, rather than waiting to treat them, once they have occurred (Bird, 2003; Cook, 2000). Sustainable agriculture relies on diverse cropping patterns, integrated crop-livestock systems, new and appropriate crop varieties, organic and inorganic inputs, use of biological control and, where appropriate, comparatively safe, selective pesticides. Agricultural biodiversity (agrobiodiversity) is a fundamental feature of sustainable agriculture: it includes not only a wide variety of species and genetic resources, but also refers to the ways in which farmers can use biological diversity to produce and manage crops, land and water (Altieri, 1991). Sustainable agriculture does not necessarily mean low-tech approach. Although low input, natural methods fit best with sustainable agriculture, cutting edge technology might be adopted. For instance, biotechnology may provide useful tools (i.e. pest-resistant varieties, diagnostic tools...). Similarly, adopting no-tillage agriculture, although is going back to a more natural method of growing crops without plowing, requires specialized seeding equipment and most commonly involves the application of herbicides. Although agroecology has an inherent bias towards more natural or cultural means of agronomy, an agroecological approach does not preclude the technological innovations we have at our disposal in the 21st century. For instance, electronic tools such as geographic information and global positioning systems have been recommended to develop an understanding of interactions among complex and interrelated components of agroecosystems (Ellsbury et al., 2000).

All farming systems, from intensive conventional farming to organic farming, have the potential to contribute to sustainability. However, whether they do in practice depends on farmers adopting the appropriate technology and management practices in the specific agro-ecological environment within the right policy framework. There is no unique system that can be identified as sustainable and no single path to sustainability. There can be co-existence of more intensive farming systems with more extensive ones that overall provide environmental benefits, while meeting food demands. However, most sustainable farming systems – even extensive systems – require a high level of farmers' skills and management to operate.

Modern biotechnology holds the potential to provide new tools for farmers in developing countries to increase yields, produce crop resistant to droughts, salinity, pests and diseases, and produce new crop products of greater nutritional value. It has also the potential to reduce unfavorable impacts on the environment by reducing the use of pesticides, to reduce the use and costs of inputs and, hence, to increase farmer income. Moreover, it provides means very useful for identifying the types of organisms that occur in microbial communities (Wang and Dick, 2004). However, while biotechnology has the potential to help, it can only be part of the solution (Veltkamp, 2001).

4. AGRICULTURE IN CHINA

China is a large agricultural country. The rural areas greatly contributed during the reform period to boost China's economic growth and social stabilization. In the early 1980s, the conversion from the collective system to the "Household Responsibility System" gave farmers autonomy in decision making and the right to bear profit or losses from their decisions. In the late 1980s, the promotion of the establishment of Township and Village Enterprises provided a second contribution to China's growth and development, while avoiding massive migration from rural to urban areas. Nowadays the agricultural sector is declining, but still represents an important element of China's economy. Agriculture accounts for almost 15% of the GDP and above 40% of employment. Even though urbanization process goes at high speed, the 60% of the total population (1.3 billion people) keeps living in rural areas (OECD, 2005). Notwithstanding the still significant economic performance of the sector, China's agriculture faces serious challenges from both global economic integration and global changes in the natural environment. As a developing country with 9.6×10^6 Km^2 of land, China must acquire a good understanding of sustainable development, develop scientific methods for evaluating the sustainable capability of its agricultural lands, and deal with the issues of regional differences and imbalance in levels of agricultural development in order to devise strategies to achieve the goal of sustainable development across the country (OECD, 2005; Xu et al., 2006). The Chinese agriculture is living a challenging momentum. A set of interacting trends is making the conversion of conventional agricultural production patterns into sustainable ones really difficult.

China is relatively scarce in agricultural land and water, since it has only the 10% of the world's arable land and its water resources per capita are around one quarter the world average. Overgrazing and misuse of water for irrigation are converting grassland and cropland into desert, mainly in the northern regions of China (i.e. Xinjiang and Inner Mongolia) and the North China Plain where grain production and grazing are more concentrated. The number of cattle, sheep and goats tripled between 1950 and 2003, while the protection effect of small vegetation against soil erosion by winds has been progressively reducing. In around fifty years, the annual loss rate of arable land due to desertification has more than doubled passing from 156,000 hectares in 1950 to the current 360,000 hectares (Wang, 2004). The rapid industrialization that drives up demand contributes to shrink the cropland area. Cropland is being converted to non-farm uses at a record date, including industrial and residential construction and the paving of land and roads, highways and parking lots. Similar impressive trends are observed with regard to water tables. Rivers, wells and aquifers, mainly in northern China, are getting dry – some have already disappeared – with heavy recoil on agricultural production. In the North China Plain, the over pumping of water from aquifers also due to backlog irrigation techniques (e.g. flooding and furrow) set the depletion rate of groundwater levels at 1 to 3 meters a year.

Population is increasing at fast rate, nearly 15 million people annually. The efforts in tackling the higher demand of food are undermined by the decreasing trends in grain production. The phenomenal rise in China's production from 90 million tons in 1950 to 392 millions tons in 1998 came to a halt. In 1998 it turned downward, falling to 322 million tons in 2003 (USDA, 2004). During the period 1992-2003, agricultural trade has shown a large increase in cereal imports in 1995 and 1996 and an increasingly high level of oilseeds imports since 1997. These decreasing trends are also due to radical dietary changes affecting both urban and rural population. With the improvement of welfare throughout the country and the general increase of per capita income, population started consuming fewer carbohydrates and more proteins (meat, fruits, vegetables, eggs, fish). Protein intake from animal sources increased 4.4 times from 1997 to 2001. Protein consumption is now higher than the world average and just below the levels in Japan and South Korea. Measured in energy terms, average food consumption in China is high, at 2,951 calories per capita per day in 2002, compared to the world average of 2,804 (2,761 in Japan and 3,058 in South Korea) (OECD, 2005). Reflecting changes in consumer demand, the composition of primary production continues to shift from crops to livestock and fish production. In 2003, crops accounted for 50% of the total agricultural production, livestock for 32%, fishery for 14% and forestry for 4%. While cereals remain the key crop, their share in total crop production and in the area sown declined quite substantially between 1990 and 2003, with other crops, such as vegetables and fruits, becoming more important and profitable (OECD, 2005).

The fast urbanization is cause of increasing social conflicts in rural areas. Agricultural incomes stopped growing in real terms after 1996 and grew slowly from 1985 onwards, while incomes in urban areas continued to rise rapidly. The widening gap between rural and urban in terms of incomes and access to most social services is provoking a massive migration of rural workers – mainly young people – to urban areas. The actual number of people employed in agriculture declined between 1990 and 2000 by 28.5 million. The number of "floating" workers, workers with jobs in cities but registration in the countryside, is approaching 100 million. Over the next decade, another 100 million are likely to leave rural areas and the process probably will not stop until the rural workforce has been reduced to perhaps 10% of the total national workforce (Perkins, 2004). The migration of young farm workers to cities may halt the ongoing conversion to high value crops (i.e. vegetables, fruits, and grape) usually more labor-demanding. During the last decade the area of fruits and vegetable productions expanded by an average of 1.3 million hectares per year. In a country where 60% of the population lives in rural areas, in poor conditions, and still relies on agriculture as a main source of income, the shifting to high-value crops is the only option for achieving a higher quality of life (Brown, 2005).

China's challenge to individuate sustainable solutions to social and economic pressure put on agriculture is even greater due to the threat posed by agriculture on environment and human health. China's agriculture represents one of the most polluting production sectors. In 2003 China used 44.12 million tons of chemical fertilizers and 1.33 million tons of pesticides, reaching the top of the world consumers list. Application of chemical fertilizers has tremendously increased in the last decades. The total fertilizer consumption increased by more than 500% between 1980 and 1988, while domestic consumption of pesticides nearly doubled from 730,000 tons in 1990 to 1.32 million tons in 1999. In 2003 the total consumption of fertilizers and pesticides was registered at 44.20 and 1.33 million tons respectively. Estimations calculate around 135 million hectares of cropland highly polluted because of accumulation of heavy metals, nitrogen, phosphorus and other chemical compounds. Each year nearly 1.7 million tons of nitrogen are released into the soil from fertilization of grains and horticultural crops at average doses often beyond 150-200 kg per hectare. In the areas of intensive agriculture where up to 5 cropping cycles can occur in the same year on the same soil, the total load of nitrogen released into the soil can reach 1,500 kg per hectare. Around 50% of the nitrogen released reaches the basins of the Yellow River, the Yangtze River and the Pearl River – the three longest rivers of China – that flow throughout the main Chinese agricultural areas. Serious problems of water eutrophication and alteration of natural ecosystem have been observed along the 13,000 km aggregated length of the three rivers (CCICED, 2004).

Sanitary problems of agricultural produce worsen the general situation. About 10% of cereals, more than 20% of animal products and nearly half of the fruit and vegetable production are low quality. The rapid increase of animal production without a proper improvement of sanitary and veterinary conditions raised the degree of contamination of food by viruses, bacteria and other pathogens. The exceeding reliance on inorganic fertilizers drastically reduced the protein content of Chinese agricultural products, staple food of the 900 millions rural people. After China's accession to the WTO, safety concerns on Chinese products have been raised as potential barrier to the prospected increasing export of agro-food to Western countries, Europe in particular, because of its high quality and sanitary standards.

In the effort to reconcile economic and social needs and environmental protection China is undertaking countermeasures towards the promotion of sustainable agricultural practices. The attention paid to activities in the agro-environmental sector has been increasing over time due to the high social and economic priority attached by the Chinese authorities to the modernization of agriculture that must be pursued in a sustainable manner, addressing at once food security, environment protection, economic development, and good management of natural resources. On the one side, China is committed to comply with the multilateral environmental conventions and protocols with direct impact on the agricultural sectors (e.g. the Montreal Protocol on Substance that Deplete the Ozone Layer, the Stockholm Conventions on Persistent Organic Pollutants, the Convention on Biological Diversity, the Convention to Combat Desertification, the Framework Convention on Climate Change); on the other side it is pushing the adoption of agricultural practices and technologies with low impact on the environment. Fifty-one special agricultural zones have been established by the Chinese government in order to concentrate the economic and technological investments. The Tenth Five-Years Plan (2001-2005) allocated more than 950 million dollars to the improvement of agricultural infrastructures, the promotion of basic and applied research, the conversion to innovative technologies, the training of extension agents and growers. While priority is given to boosting grain production and to stop the increasing reliance on grain import, particular attention is paid to keep the positive trends registered during the last 10 years in fruit and vegetable production. Indeed, the area invested in fruits and vegetables passed from 10 million hectares in 1991 to 26 million hectares in 2003, mainly in response to a rapid growth in domestic demand and in the export market. In a country like China where the average farm surface is around 0.65 hectares, the shift to higher value crops remains the only solution to increase salaries. Although modernization is occurring mainly in the traditional agricultural sectors, other emerging sectors such as organic farming are drawing attention of policy makers as promising both in terms of export and sustainability. Nowadays organic farming in China represents only 0.4% of the total agricultural area, far below the European average of 3.5%. However organic food production is estimated to increase by 10 times over the next decade. The value of trade in exports of organic products that jumped from 300,000 dollars in 1995 to 120 million dollars in 2003 is expected to further increase in the future at an annual rate of 30% (Xie *et al.*, 2005).

5. THE AGRICULTURE COOPERATION PROJECTS IN CHINA: RESULTS, PROBLEMS AND PERSPECTIVES

The Chinese government has been paying a great deal of attention to the concept of sustainable development by listing it as a significant goal in the 21st Century China Agenda (National Planning Committee, 1994). In the past decades, considerable research has been carried out by Chinese researchers and governmental agencies as well as by some international agencies and foreign researchers, to address the issues that China is facing (Chinese Academy of Science, 1999, 2000, 2001; Fisher and Sun, 2001; Gong and Lin, 2000; Guo and Li, 1999; Liao and Liu, 2000; Yang and Cai, 2000; Xu et al., 2001). However, research and development investments in the agricultural area have fallen far below an optimum level. Agricultural research as a percentage of agricultural GDP actually declined from 0.49% to 0.38% from the late 1970s to the mid-1990s. This, coupled with the rapid increase in the number of agricultural research scientists during the reform period, reduced the real research funds per scientist by 25-30% during the 1990s (Nyberg and Rozelle, 1999). This might slow the march of Chinese agriculture towards sustainability, when not supported in the long term by continuous technology innovation and development of qualified human resources. In this context, foreign assistance is regarded by China as a fundamental means to channel additional funds and to accelerate agricultural modernization. According to the data of the Chinese Ministry of Commerce, foreign direct investments in agriculture almost doubled from 2002 to 2003, passing from more than 0.7 billion US dollars to around 1.4 billions, and stabilized at 1.2 billions in the year 2004. International multilateral institutions (e.g. the World Bank, IFAD, FAO, UNDP, the Asia Development Bank, etc.) represented the first channel of foreign capitals, followed by bilateral technical and economical cooperation between governments, and collaboration at enterprise level. A growing trend of projects linking agriculture to environmental protection and sustainable development at large has been registered during the recent years. Agriculture is not addressed anymore as a stand alone sector. On the contrary its deep interconnections to the societal, economical and environmental aspects of sustainable development projects objectives are recognized, shifted from agriculture *per se* to agro-environment, intended as a complex dimension where food production "internalizes" the principles of environmental protection and sustainable development. Particularly in emerging countries like China that achieved acceptable levels of food production, cooperation in agriculture is not regarded anymore as a matter of food security, but as a fundamental occasion to develop sustainable models of production, ensuring protection and conservation of natural resources and improvement of quality of life in rural areas.

The agro-environmental projects implemented within the Sino-Italian Cooperation Program for Environmental Protection, a framework program jointly launched in 2000 by the Italian Ministry for the Environment Land and Sea and China State Environmental Protection Administration for the realization of demonstration and technology transfer projects, reflect this changing trend.

Under such umbrella, all agro-environmental projects (Tab. 1) respond to the primary goal of reducing China's reliance on a massive use of fertilizers and pesticides that is posing serious threats to global environment and causing exceptional phenomena of soil erosion and water pollution within the Chinese borders. While the general effort is to enable China's compliance with the obligations set by the Multilateral Environmental Agreements and with the Millennium Development Goals, project objectives and instruments of implementation have to be site specific and to properly address the particular social and economic needs of the area of intervention. In large developing countries like China, with an extremely diversified agricultural sector in terms of climate, levels of infrastructure and mechanization, economic and social conditions, it is furthermore important avoiding generalized approaches.

Since the launch of the Sino-Italian Cooperation Program, significant investments have been made for the phasing out of methyl bromide, a highly toxic fumigant used in the horticultural sector for pre-plant soil disinfestation and banned by the Montreal Protocol because of its implication in the ozone layer depletion (Gullino et al., 2003). Pilot activities aimed at demonstrating the technical and economic feasibility of innovative and low environmental impact techniques for soil disinfestation started in 2001 in Shandong and Hebei provinces with the final objective to individuate solutions replicable in other areas of China. The positive project outcomes eventually contributed to the definition of China's MB National Phase-out Plan under the framework of the Multilateral Fund of Montreal Protocol. While the choice of target areas went quite automatically to Hebei and Shandong provinces because of their high methyl bromide consumption and characterized by an expanding horticultural sector, the selection of target technologies took fully into account the local level of infrastructure, mechanization, availability of agricultural inputs and know-how. Solutions like soil steam pasteurization and soilless cultivation systems were ruled out in favor of cheaper alternatives, easier to apply and less energy consuming. Soil solarization, the use of grafting on resistant rootstocks and the application of less harmful chemicals at reduced dosages *via* drip irrigation (e.g. metham sodium, chloropicrin, 1,3-dichloropropene), tested on tomatoes and strawberries, resulted in higher acceptance by local growers because while providing level of treatment effectiveness comparable to MB, they require lower investment costs and small changes to fit within the traditional cultural practices (Cao et al., 2002a; Cao et al., 2002b). For these reasons, they were registered by the Chinese Ministry of Science and Technology as successful cases suitable to Chinese agriculture for the control of soilborne pathogens and as effective alternatives to the use of methyl bromide.

The same approach, avoiding pursuing short-term objectives due to the pressure for profit was adopted in different regions, in order to ensure the reproducibility and long-term sustainability of transferred technologies. In Xinjiang and Inner Mongolia, for instance, which are Chinese western regions characterized by poor social conditions, a very low level of infrastructure and scarce capacity of farmers in managing modern cropping systems, the choice went to very basic and low cost technologies. Drip irrigation systems resulted as a win-win solution to the serious problems of desertification, soil erosion and pollution affecting the two regions. Used in substitution of the locally adopted flood irrigation and also for the distribution of fertilizers at reduced dosages, drip irrigation systems permitted to achieve significant reduction in the use of water and fertilizers (5-6 times less compared to common practices) on tomatoes, pumpkins, cabbages, grapes and corn. These were promising outcomes for regions like Xinjiang, formerly one of the poorest regions of China and now preparing to be one of the main agricultural production areas of the country. Different considerations should be made on the use of starch-based biodegradable plastic films used in replacement of traditional polyethylene mulching films, which had been causing "white pollution" problems in the area. The agronomic and environmental performance was satisfactory: a good control of weeds and a complete degradation of the film few months after the end of the cropping cycle were observed. However, since the cost of biodegradable plastic is still too high in comparison with traditional plastic films their actual transfer into practice must be considered in the very long term (Gullino et al., 2002). Differently from Xinjiang and Inner Mongolia, in Chongnming (Shanghai) the implemented project considered more specialized technologies and complex cropping systems. The Shanghai area is the most advanced in China as for technology, know-how, foreign trade and capital turnover. The expectation in terms of technology transfer is quite high.

In Chongming Island, the third biggest Island of China after Taiwan and Hainan a few kilometers from Shanghai, IMELS and the Shanghai Municipal Government are implementing a project meant to convert the traditional local agricultural systems into organic farming production. Chongming Island is the world's largest alluvial island. Its coastal wetland and tidal flats provide many important ecological services including buffers against tidal surges and staging areas for migratory birds. Due to its extraordinary resources, scenic qualities, and its proximity to the city of Shanghai 45 km away, the island is also an attractive tourist destination, and it supports important agricultural and fisheries economies. The aim is to develop environmentally friendly green food production not only to increase potential for higher income for local growers looking with interest to foreign markets, but also to enable the production of healthy food and the promotion of a safe environment for national eco-tourists visiting Chongming Island in the future. In particular the project aims at stopping soil salinization processes, to reduce the use of chemical fertilizers (currently far over the national safety limit of 225 kg/ha) and pesticides. In two-years field experimental trials on tomatoes, watermelons, pumpkins, horse beans and other horticultural crops, technical and economical feasibility of the use of tolerant and resistant varieties, grafting on resistant rootstocks, biodegradable mulching films in combination with the use of fertigation and environmental monitoring systems, and integrated pest management also based on the use of biological products will be evaluated. The project goes beyond the merely environmental concerns and strengthens the role of rural areas as multifunctional dynamic systems. This is an important aspect in China, since the present economic growth, urbanization and increased leisure time, also increase the demand for tourism and recreation activities in rural areas. Considering the future development plan of Chongming Island as the first Ecological Recreational Island of China, the organic production of high value crops (vegetable and fruits) is regarded as a means to link higher income and market opportunity to environment protection (Gullino et al., 2006).

The selection of vegetables and fruits as target crops represents a constant in all the agro-environmental projects of the Sino-Italian Cooperation, due to environmental and economic reasons. While grains remain the key crop in China, their share on total crop production and on the area sown declined quite substantially between 1990 and 2003 as other crops, like fruit and vegetables, became more profitable and the government relaxed most of the policy measures, which had previously forced farmers to produce cereals (OECD, 2005). However, if on the one hand changes in the domestic demand and emerging export opportunities guided the impressive increase in vegetable and fruit production and provided farmers with more margin for higher income, on the other hand the shifting of Chinese agricultural production from grains to horticultural crops, notoriously demanding higher amount of chemical inputs, is likely to worsen the increasing consumption trend of fertilizers and pesticides. It is therefore urgent to take effective and immediate actions for the development of a horticultural sector able to take environmental externalities into account.

Next to the agricultural areas interested by the project in Inner Mongolia, reforestation activities are implemented as a means to protect soil from erosion and desertification, and to promote a sustainable use of non-agricultural lands. The "Windust" project is developing an overall strategy for controlling dust and sand storms originating in the Alashan desert and plaguing Beijing in springtime. As land use change of natural areas to agriculture with high-energy demanding species (e.g. corn, cotton) occurred in the area accelerating soil degradation and increasing the load of topsoil eroded by wind, a new model of farm is promoted by the project. The farm will combine together the utilization of wind energy, innovative technologies improving soil capacity to retain water, the breeding of less grassland-deteriorating animal species and the conversion of agricultural areas into forest reserves. The project is also intended to create new market opportunities by building up nurseries of saxoul (*Haloxylon ammodendron*), a local C4 shrub that will be eventually utilized for forestation activities. In order to increase the economic feasibility of the nurseries, desert ginseng, which has a great importance in the Chinese traditional medicine and a high economic value, will be grafted onto the saxoul's root system in a sort of double cropping.

The lack of technical and scientific know-how, and capacity in managing innovative cropping systems emerged as one of the major barriers towards the actual adoption of transferred technologies. The results from the Sino-Italian agro-environmental projects do not surprise as they reflect the general Chinese situation. Although there is an urgent need to create new profiles of researchers and extension agents supporting in the long term the conversion to sustainable agricultural practices, investments in agricultural research and education are at minimum levels while the Chinese extension system is hardly seeking a new balance after conversion from the central planning systems to the "Household Responsibility Systems", which claimed individual farmers as the basic production unit, instead of the production brigade of the previous collective systems. The challenge is to re-orient scientific and technical capabilities towards the market and industry requirements. Again, international cooperation programs play a fundamental role in filling the educational gap.

All demonstration projects implemented within the Sino-Italian cooperation projects provide back-to-back with technology transfer a full package of training activities particularly tailored on the specific needs of local farmers and technicians, but involving also academic institutions and private companies in the attempt to establish cross-sectoral partnerships. Particular emphasis is given to the scientific collaboration with academic institutes and research centers. At the micro level, there is a need for China's higher education institutions to learn how to identify, develop, and implement research and extension programs well adjusted to the global and domestic scenario. At the macro level, there is a need for China's policy makers to consult experts in order to formulate appropriate sustainable development strategies and policy. In this perspective, strengthening capacity and efficiency of the role of universities towards government, industry and market operators is strategic for the future sustainable development of Chinese agriculture and the promotion of innovative "green" technologies. Several Sino-Italian cooperation projects are pursuing this objective.

Joint research programs have been established between Italian and Chinese universities with the aim of making Chinese scientists and technicians more acquainted with some of the modern techniques targeted by the demonstration projects. The use of biocontrol agents and grafting on resistant rootstocks within integrated pest management systems is addressed by the project "Sustainable plant protection in respect of the environment: modern techniques for the control of plant pest and diseases of horticultural crops in China". Within the Scientific and Technological Cooperation Program between the Italian Ministry of Foreign Affairs and the Chinese Ministry of Science and Technology, a replicable model of research and semi-commercial scale application of innovative techniques and technologies for organic agricultural waste composting, as well as of the use of biodegradable plastics as a means to reduce production of non-compostable agricultural waste, is being developed under the project "Innovative techniques for reduction and recycling of agricultural wastes". A project co-financed by the European Commission in the framework of the Asia-Link Programme will promote sharing of technical, scientific, economical and ethical knowledge on organic farming, while a project co-financed by the Italian Ministry of Education, University and Research will permit research on molecular diagnostic techniques for plant pathogens, prevention of food contamination by micotoxins, recycling of industrial organic wastes for compost production and influence of climate change on the diffusion of plant diseases. The involvement of the private sector is particularly sought. Each project is implemented on a participatory basis, stimulating the creation of broad partnerships involving all relevant stakeholders, from governmental agencies to NGOs, from academic institutes to private companies. The model followed is that of Type II Partnership emerged from the 2002 Johannesburg World Summit on Sustainable Development as a means to promote a full integration of public and private sectors at large in the multilateral and bilateral cooperation programs, both in technical and financial terms. It is worth underlining, as an example, that the academic partnership between Italian and Chinese academic institutes, consolidated over 5 years of activities within the framework of the Sino-Italian Cooperation Program, eventually showed high capacity to convey human and financial resources from private and public sectors into the development of broader mobility programs. Only in the year 2005, thanks to the exploitation of different sources of co-financing, the number of students and researchers visiting Italian firms and research institutes for study and research activities on sustainable agriculture within MSc and PhD programs jumped from few units to over 50 people, and the period of staying from few weeks to 3 years.

Nevertheless, it is still quite difficult to plan the involvement of the private sector in long-term programs and give cooperation projects also a market perspective, facilitating the introduction and commercialization of environmentally friendly innovative technologies in China. Due to the fragile Chinese regulation framework on intellectual property protection, the involvement of Italian private companies is often limited to stand-alone interventions within each single project (e.g. field visits, lectures during seminars and trainings, short-term internships and technology procurement). Even though there is the possibility to piggy-back on a governmental program like the Sino-Italian Cooperation Program for Environmental Protection and create effective synergies mutually benefiting both sides, the skepticism showed by private firms about a not yet conducive Chinese regulatory system represents a great barrier against an effective technology transfer. First signals of potential cooperation fully involving Chinese and Italian private companies in the development of innovative technologies for agriculture have been shown by the project "Sustainable plant protection in respect of the environment: modern techniques for the control of plant pest and diseases of horticultural crops in China". The project, co-financed by the Italian Ministry of Production Activities and the Italian Trade Commission with the aim to stimulate partnership between SMEs and universities, is opening the possibility to develop co-patents of biocontrol products for pest and disease control. Same opportunities are emerging in Chongming Island for the creation of agricultural waste composting facilities for the production of biogas and organic fertilizers.

Table 1. Projects on sustainable agriculture implemented within the Sino-Italian Cooperation Program for Environmental Protection and other cooperation initiatives

Project title	Period (Duration)	Funded by	Institutions/companies involved
Alternatives To The Use of Methyl Bromide in Soil Fumigation	2001-2003 (24 months)	IMELS	SEPA/FECO; CAU; CAAS; AGROINNOVA
Strengthening Technology and Capacity of Sustainable Agriculture in China	2002-2005 (36 months)	IMELS	SEPA/FECO; Chinese Research Academy of Environmental Sciences; AGROINNOVA
Sustainable plant protection in respect of the environment: modern techniques for the control of plant pests and diseases	2005-2007 (24 months)	MAP, ICE, CRUI	AGROINNOVA; Ce.R.S.A.A.; CAU, Intrachem Bio Italia S.p.A., Nuovo Centro S.E.I.A. S.p.A.

of horticultural crops in China			
Organic Farming Systems and Techniques for the Promotion of "Green" Agriculture in Dongtan Chongming Island	2005-2008 (36 months)	IMELS	AGROINNOVA; Shanghai Environmental Protection Bureau; Shanghai Academy of Environmental Sciences; SIIC Dongtan Investment & Development (Holdings) Co., ltd.; Ce.R.S.A.A.
Sino-Italian Cooperation Project (WINDUST) to Combat Dust Storms in Northern China	2004-2007 (36 months)	IMELS	Tuscia University; AGROINNOVA; CETMA Consortium; Beijing Environmental Protection Bureau; Beijing Capital Group ltd.
Organic Farming: Social, Ethical, Economical, Scientific and Technical Aspects in a Global Perspective	2005-2008 (36 months)	European Commission – Asia Link Programme	AGROINNOVA; Tuscia University; University of Bonn; University of Wageningen; CAU; ZJU; Qinghai College of Animal Husbandry and Veterinary Medicine; NEAU
Innovative techniques for the reduction and recycling of agricultural wastes	2006-2009 (36 months)	MAE, MOST	AGROINNOVA; CAU
Technological innovations in crop protection to enhance food quality in China	2006-2008 (36 months)	MIUR, IMELS, University of Torino	AGROINNOVA; CAAS; CAU

IMELS Italian Ministry for the Environment Land and Sea, *SEPA/FECO* Foreign Economic Cooperation Office of the State Environmental Protection Administration, *CAU* China Agricultural University, *CAAS* Chinese Academy of Agricultural Sciences, *AGROINNOVA* Centre of Competence for the Innovation in the Agro-Environmental Sector of the University of Torino, *MAP* Italian Ministry for Production Activities, *ICE* Italian Trade Commission, *CRUI* Conference of Rectors of Italian Universities, *Ce.R.S.A.A.* Regional Center for Agricultural Experimentation and Extension of the Chamber of Commerce of Savona, *ZJU* Zhejiang University, *NEAU* Northeast Agricultural Universities, *MAE* Italian Ministry of Foreign Affairs, *MOST* Chinese Ministry of Science and Technology, *MIUR* Italian Ministry of Education, University and Research.

6. CONCLUDING REMARKS

A study carried out by Xu et al. (2006) by zoning China into nine regions and 22 sub-regions, depicted Chinese agriculture in the year 2000, and showed that 16 provincial units have reached a level of sustainable development. Of the 16 provincial units, Shanghai, Guangdong, Beijing and Zheijang show strong capacity for sustainable development; Jiangsu, Liaoning, Fujian, Tianjing and Shandong show a moderate capacity for sustainable development. The rest (Heilongiiang, Jilin, Hebei, Henam, Hunam, Guangxi) exhibit a weak capacity for sustainable development. Fifteen provincial units have not yet reached a level of sustainable agricultural development (Xu *et al.*, 2006). The regions that have reached a level of sustainable development are all located in China's eastern zone or in the middle transition zone. The provincial units with better conditions in the agricultural resources supporting system are mainly located in Southern China and Xinjiang, whereas the provincial units with less favorable conditions are Chongqing, Gansu, Tianjin, Qinghai, Shanxi, Guizhou and Shanghai: in this case relative advantage declines from eastern China via mid China to western China (Xu *et al.*, 2006).

According to Xu *et al.* (2006), the situation depicted for the year 2000 suggests that Eastern China needs to elevate the level of sustainable agricultural development by increasing the input of science and technology, and adjusting its internal structure to develop intensive, commercial and green agriculture. In mid and western China, efforts must be done to transform their unsustainable status into sustainable development, and to improve the capacity of sustainable development: such regions occupy two thirds of the total area of China, therefore achieving sustainable agriculture in those regions is very important to ensure national food security.

Most of the agricultural projects carried out within the Sino-Italian cooperation program interested areas needing support. Being such projects fully integrated within a broader sustainable development program, they were able to involve all stakeholders in both project preparation and implementation. Particularly, governmental Institutions, Academic institutions, public research centers and private companies have always been partnered with the aim to create a long lasting network of local and international researchers and experts supporting the development and the adaptation of sustainable farming systems as well as the design of a new regulatory framework supporting the adoption of innovative technologies.

In all cooperation projects training and information activities have been important to enable the actual transfer into practice of target sustainable agricultural technologies and practices. District workshops and seminars have been organized in order to keep people involved in project activities continuously aware of all progress as well as to inform stakeholders on scientific, technical and economic feasibility of upcoming new techniques and systems.

Education and training are part of any strategy to promote the adoption of appropriate technologies and farm management practices (Francis, 2004): all the projects implemented included exchange of MS and Ph.D. students as well as of researchers among the Research Institutions involved.

The experience gained through the implementation of cooperation projects in rural areas of China shows the strategic role that sustainable agriculture plays towards the promotion of sustainable development (Gullino *et al.*, 2006). Moreover, the projects implemented respond to the need of creating job opportunities alternative to the agricultural sector. In this respect, agro-tourism is a valued option protecting the rural environment, sustaining small-sized enterprises and providing income and job opportunities. In China, as well as in other parts of the world, individual farmers are discovering profitable niche markets selling to consumers, concerned about the environment and animals' good treatment, goods such as organic food, free range products, and food from certified production systems (Appleby, 2004). Local, small-scale, indigenous farming systems can be helped to become more productive while improving sustainability (Pretty and Hine, 2001). China offers great opportunities in such respect.

In conclusion, the synergic role provided by the different projects in promoting environment protection and sustainable agriculture at once, will represent a strong effective effort toward the rural development of different regions. Such broader approach recognizes the multi functionality role of agriculture accordingly with the objectives of Agenda 21 and should represent an example for similar cooperation initiatives likely to be implemented.

In many respects, the pressures of the agro-environment in China follow the same trends in the structural adjustment of agriculture observed in other emerging economies, in particular where policies of increasing food self-sufficiency have been pursued. However, given the rapid rate of past and projected economic growth, the sheer size of the country, and the vulnerability of much of the land and water resources, including climate variability, the pressures on the environment pose a potential constraint on future agricultural growth, with international trade implications. Cooperation between the EU and China is therefore essential in order to guarantee an improvement of agricultural productivity and farmer incomes while protecting the environment in China, also adopting production and quality standards in line with the EU ones.

ACKNOWLEDGMENT

Work carried out in the frame of the Framework Agreement between AGROINNOVA and Italian Ministry for the Environment Land and Sea on "Sustainable agriculture".

The authors thank Prof. Angelo Garibaldi for critically reading the manuscript. Thanks are due to all the Italian and Chinese colleagues and co-workers who contribute to the implementation of the projects in China.

Maria Lodovica Gullino and Andrea Camponogara
Centre of Competence for the Innovation in the Agro-Environmental Sector (AGROINNOVA), University of Torino, Italy

Nevio Capodagli
Centre of Competence for the Innovation in the Agro-Environmental Sector (AGROINNOVA), University of Torino, Italy
Project Management Office of the Sino-Italian Cooperation Program for Environmental Protection, Beijing, China

REFERENCES

Altieri M., 1991. Traditional farming in Latin America. *The Ecologist, 21*, pp. 93-96.
Altieri M.A. and Merrick L.C., 1987. *In situ* conservation of crop genetic resources through maintenance of traditional farming systems. *Economic Botany, 41*, pp. 86-96.
Appleby M.C., 2004. Alternatives to conventional livestock production methods. In G.J. Benson and B.E. Rollin (eds.). *The well-being of farm animals: challenges and solutions*, pp. 339-350, Blackwell, Ames, IA.

Appleby M.C., 2005. Sustainable agriculture is humane, humane agriculture is sustainable. *Journal of Agricultural and Environmental Ethics, 18*, pp. 293-303.
Barret G.W. and Odum E.P., 2000. The twenty-first century: the world at carrying capacity. *BioScience, 50*, pp. 363-368.
Barrett G.W. and Peles J.D., 1994. Optimizing habitat fragmentation: an agrolandscape perspective. *Landscape and urban planning, 28*, pp. 99-105.
Bird G.W., 2003. Role of integrated pest management and sustainable development. In K.M. Dakono and D. Mota-Sanchez (eds.). *Integrated pest management in the global arena*, pp. 73-95, Wallingford, UK: Maredia, CAB International.
Brown L. R., 2005. Reversing China's harvest decline. Earth Policy Institute (eds.), *Outgrowing the Earth*, pp. 133-155, New York: W. W. Norton and Company Inc.
Cao A., Guo M., Cao Z., Zheng C., Gullino M.L., Camponogara A. and Minuto A., 2002. Sustainable practices for soil disinfestation: a project between Italy and China. *Proceedings 2nd International Conference on Sustainable agriculture for food, energy and industry*, Beijing, 2002, Vol. 2, pp. 1492-1500.
Cao Z., Yu Y., Chen G., Minuto A., Camponogara A. and Gullino M.L., 2002. Ecological studies on nematodes in alternative technologies to the use of methyl bromide in soil fumigation. *Proceedings 2nd International Conference on Sustainable agriculture for food, energy and industry*, Beijing, 2002, Vol. 1, pp. 85-92.
Caporali F., 2007. Ecological agriculture: human and social context. In C. Clini, M.L. Gullino and I. Musu, (eds.), *Sustainable development for Environmental Management,* Springer, Dordrecht, The Netherlands, pp. 415-429.
CCICED – Agricultural and Rural Development Task Force, 2004. China's Agricultural and Rural Development in the New Era: Challenges, Opportunities and Policy Options. Policy briefs.
Chinese Academy of Science. Sustainable Development Research Group of China, 1999. *Report on Chinese sustainable development strategy*, 1999. Science Press, Beijing.
Chinese Academy of Science. Sustainable Development Research Group of China, 2000. *Report on Chinese sustainable development strategy*, 2000. Science Press, Beijing.
Chinese Academy of Science. Sustainable Development Research Group of China, 2001. *Report on Chinese sustainable development strategy*, 2001. Science Press, Beijing.
Clements D.R. and Shrestha A., 2004. New dimensions in agroecology for developing a biological approach to crop production. In D.R. Clements and A. Shrestha (eds.) *New dimensions in agroecology*, pp. 1-20, Food Products Press, Haworth Press Inc., N.Y.
Cook R.J., 2000. Advances in plant health management in the twentieth century. *Annu. Rev. Phytopathol., 38*, pp. 95-116.
Ellsbury M.M., Clay S.A., Fleischer S.J., Chandler L.D. and Schneider S.M., 2000. Use of GIS/GPS systems in IPM: progress and reality. In G.G. Kennedy and T.B. Sutton (eds.). *Emerging technologies for integrated pest management: concepts, research, and implementation*, pp. 419-438, St Paul, MN, USA: APS Press.
Falkenmark M. and Rockstrom J., 2004. *Balancing water for humans and nature*. Earthscan, UK.
Fisher G. and Sun L., 2001. Model based analysis of future land use development in China. *Agriculture, Ecosystems and Environment, 85*, pp. 163-176.
Flora C.B., 2001. Shifting agroecosystems and communities. In C. Flora (ed.). *Interactions between agroecosystems and rural communities*, pp. 5-13, Boca Raton, USA.
Francis C.A., 2004. Education in agroecology and integrated systems. *Journal of Crop improvement, 11*, pp. 21-43.
Gong J. and Lin H., 2000. Sustainable development for agricultural regions in China. Case studies. *Forest Ecology and Management, 128*, pp. 27-38.
GSICI (Group of Shanghai Island Comprehensive Investigation), 1996. *Report of Shanghai Island Comprehensive Investigation*. Shanghai Scientific and Technological Press, Shanghai.
Gullino M.L., Camponogara A., Clini C., Yi L., Guanghui X. and Xiaoling Y., 2002. Sustainable agriculture for environment protection: a Sino-Italian Cooperation Program. *Proceedings 2nd International Conference on Sustainable agriculture for food, energy and industry*, Beijing, 2002, Vol. 1, pp. 948-952.
Gullino M.L, Camponogara A., Capodagli N., Xiaoling Y. and Clini C., 2006. Sustainable agriculture for environment protection: cooperation between China and Italy. *Journal of Food, Agriculture & Environment, 4*, pp. 84-92.

Gullino M.L., Camponogara A., Gasparrini G., Rizzo V., Clini C. and Garibaldi A., 2003. Replacing methyl bromide for soil disinfestation. The Italian experience and implication for other countries. *Plant Disease, 87*, pp. 1012-1021.

Guo H. and Li J., 1999. Regionalization of rural economics in China. *Regional development research in rural economics in China*. Science Press, Beijing.

Liao Z. and Liu Y., 2000. Comprehensive indexes and spatial distribution characteristics of the regional sustainable development of China. *Acta Geographica Sinica, 55*, pp. 139-150.

Marshall E.J.O. and Moonen A.C., 2002. Field margins in Northern Europe: their functions and interactions with agriculture. *Agriculture, Ecosystems and Environment, 89*, pp. 5-21.

Matson P.A., Parton W.J., Power A.G. and Swift M.J., 1997. Agricultural intensification and ecosystem properties. *Science, 277*, pp. 504-509.

National Planning Committee. National Science and Technology Committee. 1994. *China 21st Century Agenda White Book of China 21st Century Population, Environment and Development*. China Environmental Science Press, Beijing.

Negri V., 2005. Agro-biodiversity conservation in Europe: ethical issues. *Journal of Agricultural and environmental Ethics, 18*, pp. 3-25.

Nyberg A. and Rozelle S., 1999. *Accelerating China's rural transformation*. Retrieved April, 2006. From The World Bank: http://www-wds.worldbank.org/external/deFault/WDSConTenTServer/WDSP/IB/1999/11/19/000094946_99111006010783/Rendered/PDF/multi_page.pdf.

OECD, 2005. *OECD Review for Agricultural Policies – China*. OECD Publishing, Paris.

Paoletti M., 2001. Biodiversity in agroecosystems and bioindicators of environmental health. In M. Shiyomi and H. Koizumi (eds.). *Structure and function in agroecosystems design and management. Advances in agroecology*, pp. 11-41, CRC Press, Boca Raton, USA.

Perkins D., 2004. *Declining Growth in Farm Output and Employment. Implications for China's Economy and Society.* Retrieved April, 2006, from Harvard University Website: http://post.economics.harvard.edu/faculty/perkins/papers/farmoutput.pdf

Pretty J. and Hine R., 2001. *Reducing food poverty with sustainable agriculture*. UK Department of International Cooperation.

Rasmussen P.E., Goulding K.W.T., Brown J.R., Grace P.R., Janzen H.H. and Korschens M., 1998. Long term agroecosystems experiments: assessing agricultural sustainability and global change. *Science, 282*, pp. 893-896.

Thrupp L.A., 2004. The importance of biodiversity in agroecosystems. In D.R. Clements and A. Shrestha (eds.). *New dimensions in agroecology*, pp. 315-337, New York: Food Products Press, Haworth Press Inc.

Tilman D., 1999. Global environmental impacts of agricultural expansions. The need for sustainable and efficient practices. *Science, 96*, 5996-6000.

U.S. Congress, 1990. Food, agriculture, conservation and trade act. US Congress Papers, Title XVI, Research, Subtitle A, Section 1602.

U.S. Department of Agriculture, 2004. *Production, Supply and Distribution (electronic database)*. Retrieved March, 2005, from USDA web site: http://www.fas.usda.gov/psd.

Veltkamp E., 2001. Use of modern biotechnology in food and agriculture: views from the EU-US Biotechnology consultative Forum. *Proc. Third Congress of the European Society for Agricultural and Food Ethics*, pp. 93-96.

Viatte G., 2001. Agriculture and sustainable development: a societal and policy challenge. *Proc. Third Congress of the European Society for Agricultural and Food Ethics*, pp. 97-102.

Wang P. and Dick W.A., 2004. Microbial and genetic diversity in soil environments. In D.R. Clements and A. Shrestha (eds.). *New dimensions in agroecology*, pp. 249-287, New York: Food Products Press, Haworth Press Inc.

Wang T., 2004. Progress in sandy desertification research of China. *Journal of geographical sciences, 14*, pp. 387-400.

Xie B., Li T., Zhao K. and Xi Y., 2005. Impact of EU organic-certification regulation on organic exports from China. *Outlook on Agriculture, 34*, pp. 141-147.

Xu J., Chen B. and Zhang X., 2001. Ecosystems productivity regionalization of China. *Acta Geographica Sinica, 56*, pp. 401-408.

Xu X., Hou L., Lin H. and Liu W., 2006. Zoning of sustainable agricultural development in China. *Agricultural Systems, 87*, pp. 38-62.

Yang Y. and Cai Y., 2000. Sustainable evaluation on rural resources, environment and development of China. The SEEA method and its application. *Acta Geographica Sinica, 55,* pp. 596-606.

Zimdhal R.L., 2002. Moral confidence in agriculture. *American Journal of alternative agriculture, 17*, pp. 44-53.

S. DALMAZZONE

ECONOMICS AND POLICY OF BIODIVERSITY LOSS

Abstract. The first part of the chapter provides concise definitions of biodiversity. It explores the different meanings with which the term is used – genetic diversity, species diversity, intra-specific diversity, ecosystem diversity, functional diversity. It then considers the ecological services connected to biodiversity and the indices that can be used to quantify it, which may focus alternatively on species richness or on species diversity. The central part of the chapter focuses on the economics of biodiversity loss, including the economic value of biodiversity, biodiversity as a public good, use and non-use values, biodiversity and risk in agro-ecosystems. After analysing the social and economic factors driving the loss of biological diversity, the final sections discuss the available conservation policies: incentives, property rights, price reforms, protected areas, safe minimum standards, and international agreements such as the Convention on Biological Diversity and the Biosafety Protocol.

1. INTRODUCTION

According to the definition adopted in the Convention on Biological Diversity opened for signature, at the same time as the Kyoto Protocol, at the 1992 United Nations Conference on Environment and Development in Rio de Janeiro, biodiversity is

> the variability among living organisms from all sources, including, *inter alia*, terrestrial, marine, and other aquatic ecosystems, and the ecological complexes of which they are part: this includes diversity within species, between species and of ecosystems. [art. 2][1]

One of the virtues of this definition is that it makes it explicit that the expression can be used to refer to a few distinct concepts. It may point to the number of different species that are present, for instance, within a region or a country, to the variety of ecosystem types, or to the genetic variability within a single species. Measures of biodiversity exist for each specific meaning. Species diversity is a measure of the number of and differences between taxonomically distinct species. Ecosystem diversity is a measure of the differences between distinct ecosystems. Intraspecific diversity is a measure of the variation among individuals or populations of the same species, reflected in the composition of their gene pool; examples are the diversity between different dogs or varieties of rice.

The distribution of biodiversity on Earth is not even. The areas richer in endemic species are typically in the tropics, the poorer in the polar regions. Biodiversity also varied dramatically over time in Earth history, rising and falling in cycles (Rohde and Muller, 2005). The current documented extinction rates of a wide range of species, however, is approaching 1,000 times the background rate and may climb to 10,000 times the background rate during the next century, if present trends of habitat destruction continue. At this rate, one-third to two-thirds of all species of plants, animals, and other organisms would be lost during the second half of the next century, a loss comparable with those of past mass extinctions, the fifth and last of which, 65 million years ago, coincided with the disappearance of dinosaurs and lead to a permanent change in the character of life on our planet.

The drivers of change affecting biodiversity in our time are habitat destruction, the introduction of invasive species, overexploitation of natural resources, climate change, and pollution. More land was converted to agriculture since 1945 than in the 18th and 19th centuries combined. More than half of all the synthetic nitrogen fertilisers (first made in 1913) ever used were applied after 1985 (Land Institute, 2005). In the last three centuries, the global forest area has been reduced by approximately 40%, with three quarters of this loss occurring during the last two centuries (Millennium Ecosystem Assessment, 2005). Tropical forests are disappearing due mainly to logging, mining, hydropower and the need of land for agriculture. Temperate and northern old-growth forests are being destroyed by the timber and paper industries. It is economic activity, in a variety of forms, that is driving biodiversity loss. Thus, the causes, the consequences and the available policy options cannot be properly investigated without an understanding of the economic forces at work behind the current process of environmental change.

2. THE MEASUREMENT OF SPECIES DIVERSITY

Given the multifaceted character of the biodiversity concept, its quantitative measurement is a complex issue. A large number of different indices have been developed, at the species, ecosystem and intraspecific level, each capturing different aspects of the problem. There are two types of measures of diversity at the species level: measures based on the absolute numbers of species present (or species richness), and measures based on relative abundance.

The simplest index (α-*diversity*) is given by the number of species occurring within an area of a given size, giving equal weight to each species. It is a within-area measure of species richness.

A second measure of within-area diversity, but aimed at a larger scale such as a whole region, country or even continent, is called γ-*diversity*. It is often used to deal with biodiversity at the landscape level.

A third measure, β-*diversity*, indicates the rate at which the mix of species changes along a given habitat or geographic gradient. It is a measure of between-area diversity. It is often used also to compare the diversity of discrete sites or habitats, in which case it is not a rate measure but rather a measure of community dissimilarity[2]. β-diversity can be defined also as the ratio of the γ-diversity of a region to the average α-diversity of local areas within the region ($\beta=\gamma/\alpha$) (Bisby, in Heywood 1995).

For conservation as well as for economic purposes it can be vital to account for the uniqueness of the species that enter a diversity index. Species without close relatives may have a higher conservation value than those with close relatives. Closeness may be defined by several different criteria, for instance genetic similarity or functional features. Measures of relative abundance allow the assessment of biodiversity to go beyond species richness, and to provide insights on ecological diversity based on how even are the populations of species within an ecological community. A community of N species is said to be diverse if all of the species have relative abundances close to $1/N$. An ecological community with a dominant species that accounts for a large fraction of the community's biomass is relatively less diverse than a community with the same total number of species but represented in equal shares. Ecological diversity increases therefore with the number of species N, but also with evenness of species abundance (Polasky, 2006). An example of index of relative abundance is the Shannon-Weaver index,

$$SW = - \sum_{i=1}^{N} p_i \ln p_i, \qquad (1)$$

where p_i is the proportion of individuals belonging to the i-th species and $\ln$ is the natural logarithm. It captures the relative rarity of species: giving more weight to rare species, it provides a measure of the information content in the system.

Another important example is the Simpson's index,

$$D = \sum_{i=1}^{N} n_i (n_i - 1)/N(N-1), \qquad (2)$$

where n_i is the number of individuals in a particular species and N the total number of organisms of all species. It captures the dominance of species, that is, it gives the probability (in values between 0 and 1) of any two individuals drawn at random from a large community belonging to the same species. In this way, it gives more weight to the abundance of the most common species. To overcome the counterintuitive nature of the index, whose value decreases with increasing diversity, it is sometimes expressed as $1-D$, in which case it gives the probability that two individuals drawn at random belong to different species, with 1 indicating maximum diversity. Alternatively, it can be expressed as $1/D$, with values ranging between 1 (a community containing only one species) and the number of species in the sample.

Finally, there are indices of taxic or phylogenetic diversity, emphasizing evolutionarily isolated species that contribute highly to the assemblage of features or options. Such indices can help making choices about what to protect where it is not possible or too costly to protect everything. They represent a bridge between measurement of diversity and value. A few important examples have been developed within the economics literature. Weitzman's index (1992, 1993, 1998) can be informally described as arising from an evolutionary tree. The diversity of a set of species can be summarized as the sum of the lengths of the branches in its evolutionary tree. The resulting measure of diversity can be related to the probability that a set of organisms contains an instance of a genetic attribute not found in the common ancestor. A similar approach is developed in Solow and Polasky (1994) whose diversity measure is based on the probability that a set of species contains a given characteristics which may turn out to be valuable.

3. THE FUNCTIONS AND VALUE OF BIODIVERSITY

The interaction between organisms, populations and communities in the natural environment supports ecological services that are essential to the functioning of all living systems and to the existence of human societies. Soil retention and formation, flood control, climate regulation, waste assimilation, biological control of pests and diseases, supply of raw materials for industrial use, water regulation and supply are a few among the services that biodiversity contributes to maintain[3] (see Box 1).

For any ecological function or process, the expression 'functional biodiversity' refers to the combination of species needed for that function or process to be preserved (Schindler, 1990; Kaufman *et al.*, 1998). Whereas the measures of diversity mentioned above capture differences in the genetic characteristics of species or the structural characteristics of ecosystems, the concept of functional diversity captures differences in the functions performed by species (the 'division of labour' in ecological systems). Rather than focussing on each species in isolation, it concentrates on the interrelation of species living together. Within a community some species may have a functional role that is critical to the stability and survival of the community, and their loss is potentially far more disruptive than a mere reduction in the corresponding number of species present (Bisby, 1995).

3.1. Biodiversity, ecosystem stability and resilience

Biodiversity loss may not only reduce the equilibrium stock of resources, but also affect the resilience of ecosystems – their capacity to respond to stresses or shocks imposed by habitat conversion, predation, harvesting and pollution. Ecosystem resilience can be measured either as the time taken by a perturbed system to return to its original state (Pimm, 1984), or by the size of the perturbation that can be absorbed before the system loses its ability to recover towards its original condition and converges instead on another equilibrium state (Holling, 1973).

For many ecosystems there may be a range of conditions, for instance in terms of population stocks, over which the system remains stable. However, if any population falls below a critical threshold, the self-organization of the ecosystem as a whole may be fundamentally altered (Pielou, 1993). It has been frequently observed in productive grassland, for instance, that it retains its basic structure and functioning while the composition of plant species changes as a consequence of livestock grazing; at some point it switches to a shrub-dominated semidesert when cumulative disturbance reaches a critical intensity. The disruption of the ecosystem and of its functions may be irreversible or only slowly reversible, even if the source of disturbance at some point is removed.

The impact of a biodiversity decline on resilience may take the form of a reduction of a resource's stability domain, sometimes associated with the appearance of critical thresholds in terms of a minimum viable population size which overstepped may cause a resource collapse and even extinction. This means that the structural change induced by biodiversity loss has the potential to make the system less predictable and more difficult to manage (Dalmazzone, 1998).

The vulnerability of ecological processes to this sort of structural changes is a function of the number of alternative species that can "take over" a particular function when an ecosystem is perturbed. Functional diversity, in other words, contributes to determine the resilience of natural systems (Schindler, 1990). The combination of species in an ecosystem in fact enables the system both to provide a flow of services under given environmental conditions, and to maintain that flow if environmental conditions change (Heywood, 1995). In areas with markedly seasonal climate, or high inter-annual variation, stability and resilience, and hence economic productivity, depend on the diversity of species available to make the most of spatial and temporal variation in environmental conditions (Perrings, 2000).

An important implication is that in agriculture, forest, fish, and grassland management, policies aimed at reducing natural variability of an ecosystem's critical structuring variables (plants, insects, forest fires, fish populations) may induce a gradual loss of functional diversity and hence an increased sensitivity to disturbance.

Box 1. Ecosystem services

Ecosystem services are now widely studied, both in terms of scientific assessment and classification of ecosystem functions, and of economic analyses of their value. One possible classification could group them in the following general types:

REGULATION FUNCTIONS
Gas regulation (CO_2/O_2 balance, ozone layer...), climate regulation, prevention of disturbance, water regulation and supply, soil retention and formation, flood control, nutrient regulation and waste treatment, pollination, biological control.

HABITAT FUNCTIONS
Refugium, nursery for species (suitable living space and reproductive habitat).

PRODUCTION FUNCTIONS
Food, drinking water, water for industrial uses, supply of raw materials, genetic resources, medicinal resources, energy and fuel.

INFORMATION FUNCTIONS
Scientific, aesthetic and cultural information, education, recreation.

Source: adapted from de Groot et al. (2002), Costanza et al. (1997).

3.2. Biodiversity and economic productivity

Diversity in several cases fosters not only the stability and resilience of ecosystems in the face of disturbance, but also their productivity. Experimental grasslands research, to remain with the example used above, has shown that productivity increases with plant biodiversity. The main limiting nutrient, oil mineral nitrogen, is used more effectively over a range of environmental conditions the greater the diversity of species (Tilman and Downing, 1994; Tilman et al., 2005). In addition, evidence has been generated by a number of scientific studies that the ecological services associated with biodiversity generate economic benefits often in excess of those obtained from habitat conversion to agriculture, forest plantation, fish farming and so on. Balmford *et al.* (2002), for instance, present documented evidence referred to five case studies (use of tropical forests in Malaysia and Cameroon, draining of freshwater marshes in Canadian agricultural areas, conversion of mangrove systems for aquaculture in Thailand, and fishing exploitation of reefs in the Philippines). In all cases, despite a very conservative analysis, the short term economic benefits of habitat conversion are substantially outweighed by the foregone value of the services provided by the original natural system (storm protection, carbon sequestration, water filtration, tourism, sustainable recreational hunting and fishing, long term productivity). In the current stage of development, further habitat destruction and the consequent biodiversity loss often arise as a result of myopic decisions aimed at maximising, at best, short term private benefits.

In agro-ecosystems, biotechnology has in many cases accomplished genetic improvements of crop varieties that, mainly thanks to herbicide tolerance and resistance to insects, have yielded remarkable increases in productivity. The diffusion of the genetically modified, high yield varieties has however resulted in a reduction in the incentive to conserve wild relatives, in the abandonment of traditional varieties bred to meet local conditions, and a dramatic diffusion of monoculture. In addition, it has lead to increased application of herbicides that (besides being linked to cancer and other human health issues) harm beneficial insects including predators, bees and soil decomposers.

The adoption of crops with a narrow genetic base, in short, appeared as a powerful way of increasing economic productivity but has also seriously contributed to biodiversity loss in agro-ecosystems. If average yields have increased, the lower diversity has led to an increase in the variance of yields – and hence in the risk carried by farmers. Biodiversity conservation, in this sense, also has an insurance value. The insurance value of displaced breeds is likely to be highest for low income farmers who have less access to markets for insurance against crop failure. Furthermore, the widespread monoculture of high yield varieties causes the risks to farm incomes to become correlated across large areas. In low income countries, governments often act as insurers of last resort. That source of protection is however substantially undermined if the risks to farm incomes become highly correlated due to the adoption of the same crop varieties everywhere (Perrings and Stern, 2000; Di Falco and Perrings, 2003).

3.3. Components of economic value

The value of biodiversity depends on many things besides its ecological significance, and different societies place different values on biological resources. Culture, social preferences, technology and distribution of income influence perceptions of value. Understanding the sources of value helps understanding what is driving the current rates of biodiversity loss and what may be done to conserve it.

The (private) value of a resource is defined as the value of goods and services an individual is prepared to forego for being able to use that resource now and in the future, and it corresponds to the value of the stream of services provided over time. This is referred to as 'use value'. The use value of biodiversity to human beings has several components:

- The 'direct value' is represented by the demand for particular species for consumption or as inputs in production. Examples are forest wood as fuel for household or raw material for construction; fish for nutrition, as well as much less essential but valued goods such as ivory, leopard furs, Chinese musk.
- The 'indirect value' is given by the ecosystem services from biophysical cycles supported by combinations of organisms that may have no immediate direct use themselves. It was estimated that the economic value of the water storage role of China's forests was (in 1998) 7.5 million Yuan, which was three times the value of the wood in those forests (Dudley and Stolton, 2003).
- The 'option value' is the value of ensuring the availability of a natural resource for possible future use – the value of keeping the option open.
- The 'quasi-option value' is the value of the potential increase in knowledge that can be derived from a species, for example when it may turn out that it contains information necessary to develop a new drug or vaccine.

There are also sources of value independent from direct or indirect current use:

- the 'bequest value' we may give to conserving species and habitats for future generations, and
- the existence value, which is the intrinsic value of a species, separate from its instrumental value to humans – for some, a moral right to existence to the non-human biota (Randall, 1991; Sagoff, 1992).

3.4. Value, prices, social welfare

In some cases the price of a resource is a good approximation of its (use) value. Markets for ecological services, however, generally do not exist (all the more so for future services): it is generally unfeasible to exact a compensation from those who benefit from such services, and therefore they are not part of the private valuation of the resource. If nobody can appropriate the benefits of supplying such services by charging a market price, as far as the market mechanism is concerned they will be underprovided or not provided at all.

A further obstacle for the price mechanism to automatically regulate the provision of this kind of services and to protect them from depletion is the incomplete knowledge and, in many cases, the fundamental uncertainty that still affects our understanding of ecological dynamics. This bears heavily particularly on indirect values. In addition, much work is needed in linking ecosystem processes to human welfare. For many ecosystem services (say, protection from floods and land disruption in case of extreme weather events, water filtration, biological control of human and animal pests and pathogens, conservation of soil productivity, and many others) it is feasible, in principle, to identify the connection between specific natural resources under threat and services that also are of extreme economic importance, and to come to meaningful estimates of their use value[4]. Although some interesting examples exist, the economic literature capable of quantifying the contribution of ecological services to human welfare is still at an embryonic stage.

For all of the above reasons, the market value of biodiversity (which informs the countless independent decisions causing biodiversity loss) does not reflect the change in social welfare associated with that loss (Perrings et al., 1992).

4. SOCIAL AND ECONOMIC FACTORS DRIVING BIODIVERSITY LOSS

Humans impact on biodiversity in a number of ways. Proximate causes include land conversion and loss, destruction or fragmentation of habitats associated with the expansion of agriculture, mining, forestry, and the construction of infrastructures. Ecologists estimate that less than one tenth of 1% of naturally occurring species is directly exploited by humans. The major threat to the loss of species is not the intentional human exploitation, through harvesting or hunting, of wild living resources (although in some cases of economically very valuable species it does play a critical role), but our impact on habitats. This also includes agricultural practices involving the use of chemicals as fertilizers and pesticides, the removal of hedgerows, ponds, and wetlands to maximise adjacent arable land, and the diffusion of surface irrigation schemes which tend to convert micro-habitats into a uniform agricultural landscape thus favouring fewer crop species and varieties. Other causes of biodiversity loss are air, water and soil pollution from industry, transport, and waste generating consumption activities such as tourism.

Different forms of environmental change related to human activities can also drive species to extinction: changes in atmospheric composition, hydrology and nutrient inputs, desertification, as well as changes in biological interaction through the introduction of alien species that outcompete the native ones and drive them to extinction. The latter is the second major source of biodiversity loss after habitat destruction. Introductions can be unintentional, carried for instance with the ballast waters of transoceanic ships or piggybacking on merchandise packaging, but have often been intentional imports meant to support agriculture and fisheries. Climate change too has a potentially serious role as a cause of biodiversity loss, whose extent however is still being researched.

All these are proximate causes. What are, however, the underlying causes that have led the expansion of the human niche to happen at the expense of the diversity of other co-existing species?

Population growth, and hence population density, is a critical factor behind the increase in the level of stress imposed on natural systems, and the consequent loss of resilience by such systems. The carrying capacity of ecosystems in terms of any species, including humans, is not unlimited. Humans, in addition, have developed over time, within complex socio-economic structures, a capacity to transform their environment immensely beyond that of any other species.

Poverty and underdevelopment may also play an important role. Half of the world's poor live in ecologically fragile rural areas – tropical forests, upland areas, arid and semi-arid regions (World Bank, 1992). Such lands face a more than average risk of soil erosion, salinization and permanent degradation. Poverty often becomes at the same time both a cause and an effect of environmental degradation, thus generating a vicious cycle.

The lines of causation from poverty to environmental degradation are complex: poverty is actually a mechanism through which other factors (for instance, population growth, inappropriate property rights) lead to degradation. Poor people tend to rely directly on natural resources (soil for crop or livestock, woodlands for fuel and building material, rivers for water, and so on) for their survival. Fragile, marginal natural capital generally requires investment to be maintained productive (fertilizing, irrigation, tree planting, management of livestock numbers). However, poor people tend to face a higher than average uncertainty about their future, as a consequence of poor health, diet and living conditions, and hence apply high discount rates to their intertemporal choices. As a result, investments that tend to yield low returns although on a long term basis, such as those required by the maintenance of natural capital, tend not to be undertaken – or undertaken less than it would be optimal. The lack of credit available to poor people is an additional factor lowering the rate of investment in natural capital. Finally, migrations, war and refugee flows often break down social and cultural rules of resource management, leading to further deforestation and conversion of marginal lands. The concurrence of factors among the above often leads to "mining the future" as the only perceived option for survival.

Policy failures frequently contribute to worsen the picture. This happens whenever policies usually aimed at objectives different from environmental protection (for instance, assigning responsibility to farmers over the land) introduce perverse incentives to overuse or mismanage natural resources. In many countries, for example, formal legal rights are given only when the land has been cleared. In others, monetary subsidies for land clearance exacerbate the incentives to deforestation. Temporary new land tenure rights may give incentive to over-exploitation. In several instances, upon receiving contracted forest land, farmers, fearful that the land use policy might change again, have immediately responded by felling all the trees. Government migration policies and even external aid may, while pursuing their prime objective, act as sources of bias on the conservation incentives.

The development and implementation of conservation policies, on the other hand, is often hindered by institutional failures such as a bad coordination or conflicts between different agencies or hierarchical levels, assignments of powers over the environment that do not reflect the comparative advantages of different levels of government, and corruption.

5. POLICIES FOR BIODIVERSITY CONSERVATION

What can be done? A range of policy instruments can be used to slow down or reverse the trend of biodiversity loss: land use policies, the establishment of property rights, the elimination or reduction of price distortions, trade policies attentive to the pathways of introduction of alien species, the establishment of protected areas and safe minimum standards of conservation, pollution control policies, the management of harvesting policies, international agreements, and in general a careful consideration of all the incentives governing the choices that may affect biodiversity.

5.1. Property rights

Although biodiversity in general is a public good, local populations of endangered species and local ecosystems are often both excludable and rival in consumption. If it is possible to generate markets over such resources without adversely affecting important social objectives, this may then prevent over-exploitation of those resources. Assignment of property rights should be accompanied by measures aimed at providing people with the means to make long-term investments in conservation (security of land tenure, access to credit, technology). Since poverty is often one of the drivers behind the over-exploitation of natural resources, conservation policies may need to include forms of income transfer.

5.2. Price reforms

For those environmental resources for which markets exist or can be created, policies that act on resource prices may be an effective tool for protection. Price liberalization is often advocated as a means to eliminate price distortions due to government intervention and to narrow the gap between private and social value. Particularly in agricultural and factor markets government intervention has often driven prices below those ruling in the world market. Typically, for example, for decades ministries of forestry have been organized and run mainly as suppliers of raw materials for industry (more or less like ministries of mining), and procurement prices set at very low, fixed rates – leading to constant overcutting, shortages and waste at all levels in the system. Where administered prices reduce producers' income, they may discourage investment in land conservation.

There is a considerable scope for policies (both in domestic price intervention and in international trade agreements) to reduce discrepancies between private and social costs. Price liberalization is however neither a necessary nor a sufficient condition for aligning private and social cost: export/import parity prices fix only the lower bound of a resource's social opportunity cost.

5.3. Incentives

People's decentralized behaviour, including choices affecting the environment, tends to respond to incentive mechanisms. The benefits of local biodiversity conservation can be enhanced, for example, by removing subsidies that artificially inflate the benefits from depletion and forest conversion (e.g. granting of titles for land clearing, under-pricing of timber concessions etc.). Agricultural subsidies in many instances play an important role in escalating overproduction and unsustainable practices. According to Pearce (2003), subsidies (mainly agricultural) to biodiversity-threatening activities worldwide could be estimated at $648-808 billion. Damage may occur also as an indirect result of policy failures, as discussed in section 4.

Establishing mechanisms for the involvement of local communities in the conservation of environmental public goods is proving an effective way of amending the structure of incentives that influence conservation choices. Zimbabwe's wildlife community programmes (Barbier, 1992) and India's Joint Forest Management Committees are cases in point (Somanathan et al., 2006).

Decentralization can bring about potential benefits such as a more direct access to local information, recognition of heterogeneous local preferences and socio-economic conditions, as well as initiative and creativity on the part of the citizenry and sub-national governments. At the same time it needs to be thoughtfully designed, making sure that the transfer of competence to the local level is accompanied by an appropriate transfer of resources for management, research and monitoring, or by appropriate fund raising mechanisms. In the attempt to make local conservation financially autonomous there is a risk to create pressures to exploit ecological resources for profit or revenue raising (e.g. money-making activities within natural reserves), so that the incentives to protect biodiversity are overweighed. It is also crucial that the staff involved in nature conservation be well motivated. If the remote location of a natural reserve, for example, makes the job unattractive, this needs to be compensated for.

5.4. Protected areas

Habitat protection through the establishment of natural reserves is one of the main available tools used by governments in their attempt to do something about biodiversity loss.

Biodiversity hotspots are not automatically the best choice in designating the location of natural reserves: taking complementarity into account and protecting areas with lower diversity but more endemic species may allow identifying a combination of sites that maximises the number of covered species.

The selection of sites for establishing reserves and of species to protect should be the result of cost-effective conservation strategies that combine ecological and economic information. Biological richness alone is not necessarily a good criterion: a site choice that considers also the economic value of the land in alternative uses can sometimes achieve the same level of species coverage for a fraction of the cost (or a larger coverage given the available budget for conservation) (Ando et al., 1998). Working lands can also often be consistent with some habitat needs and conservation objectives (Polasky et al., 2005).

The governance issue related to decentralization mentioned in the previous section applies straightforwardly to protected areas: in many countries worldwide, the expansion of local conservation activities has been accompanied by fiscal decentralization. This has often left local governments with a growing budget, non sustainable in the long run if local conservation efforts are not accompanied by a commensurate increase in financial support by central governments, or by an increased capacity to generate revenues (tax power).

Institutions should ensure that the local providers of global public goods are compensated for the benefits they offer to the national and international community – also because often biodiversity hotspots are in the poorer areas, where the same isolation that has slowed down environmental damage has also acted as a barrier to economic development. If this is not taken into account, conservation will encounter resistance by local communities, who see environmental policies depriving them of resources that they consider theirs. At the national level, for example, green taxes can be used to require downstream beneficiaries of ecological services to help cover the costs of ecosystem protection (richer downstream provinces transferring funds to upland poorer areas, for example). Proper compensation is needed however also at a global level, between industrialised countries and developing countries that host most of the remaining biological diversity.

5.5. Trade and biological invasions

Biological invasions are the second most important cause of biodiversity loss after habitat destruction. The movement of people and goods is usually taken to be the main driver of the process. Even though, from the very early stages of agriculture, crops and animals have been intentionally transported from one region of the world to another, the scale of species introduction by humans has vastly increased in recent years. It is likely that invasions, other things being equal, be an increasing function of the extroversion of an economy – its openness to the movement of goods and services (trade) and of people.

Foreign plants introduced for agricultural or forestry purposes, being 'pest-free' in their new environment, can be especially productive and profitable, but can become a threat to biodiversity when they naturalize and spread in the wild or penetrate conservation areas, replacing diverse local plant communities. Proposed new crops should require an import risk analysis by technical experts, and a balancing of ecological risks and economic advantages. Aid programmes as well need to consult conservation authorities to prevent the introduction of alien organisms with long term costs exceeding the benefits. The same holds for animals introduced as livestock or for hunting purposes and fishery releases both for aquaculture and in the wild.

Trade prohibition based on international regulations, inspections, quarantine and public education can be used as means to reduce also the unintentional introductions that can take place through merchandise containers, ballast water of ships, tourists luggage and so on, working in a way similar to the sanitary measures used to prevent the introduction of diseases in animals traded for agricultural and other purposes (Wittenberg and Cock, 2001).

6. INTERNATIONAL COOPERATION

At the supra-national level, mechanisms are needed that will allow host countries to capture the global benefits of biodiversity protection. The Joint Implementation and Clean Development Mechanisms, best known for the role they have been assigned in the implementation of the Kyoto Protocol, can be effective instruments also for international cooperation aimed at protecting biodiversity. Current mechanisms to promote international investment in local conservation include environmental funds, debt-for-nature swaps, and the Global Environment Facility (GEF), run by the World Bank, UNDP and UNEP, that grants support projects to protect the global environment in developing countries.

Besides flexibility and funding mechanisms, international agreements are essential for providing a general framework aimed at coordinating conservation efforts and compensating the unequal burdens. The Convention on Biological Diversity (CBD) opened for signature at the June 1992 UN Conference on Environment and Development and entered into force on 29 December 1993, is the most important international agreement on biodiversity. The stated objective is promoting the conservation of biodiversity, a sustainable use of its components, and a fair and equitable sharing of benefits from the utilization of genetic resources. Approved at the same time as the Kyoto Protocol it has not however gained comparable attention by governments, the media and the public. It includes 188 parties (all of the world nations except Andorra, Brunei, Iraq, Somalia, Timor-Leste, USA, Vatican).

The CBD is very different from other international agreements on conservation. It sets goals rather than obligations (like the Convention on International Trade in Endangered Species of Wild Flora and Fauna – CITES) or targets (like the EU Habitat directive). It tends to outline the policies that governments should follow, and requires them to spell them out explicitly, where they can be scrutinized by the public, environmental groups and other countries. The emphasis is on action at the national level: it contains no agreed list of protected species or habitats, which are left up to national governments.

Key obligation for the Parties to the Convention is the preparation of a national biodiversity strategy (art. 6), including planning, steps to implement the strategy, who should do what, how it would be funded[5]. A common element in some well developed national strategies is the establishment of a "Clearing-House Mechanism" to ensure that all governments have ready access to the information and technologies needed for their work on biodiversity. The responsibility for conserving biodiversity is shared among provincial, territorial and federal governments (art. 6 allows countries to undertake separate regional strategies or plans).

The Cartagena Protocol on Biosafety to the Convention on Biological Diversity parallels, in the CBD, the Kyoto Protocol to the United Nations Convention on Climate Change. It seeks to protect biological diversity from the potential risks posed by living modified organisms resulting from modern biotechnology. It establishes a procedure for ensuring that countries are provided with the information necessary to make informed decisions before agreeing to the import of such organisms into their territory. The Protocol entered into force on 11 September 2003, ninety days after receipt of the 50th instrument of ratification (art. 37)[6].

Finally, the Millennium Ecosystem Assessment (2005) produces an account of the current state of biodiversity and of the conservation polices worldwide, playing a role similar to the reports of the Intergovernmental Panel on Climate Change: informing public and policy makers about the current state and trends, the scenarios, and the available policy responses to one of the most crucial forms of global environmental change, involving potentially enormous costs for future generations and a serious threat to development opportunities.

Silvana Dalmazzone
Department of Economics,
University of Turin, Italy

NOTES

[1] The text of the Convention is available on the CBD website, www.biodiv.org. See UNEP (1992).
[2] The distinction between α, β and γ-*diversity* was first introduced by Whittaker (1960, 1972).
[3] China's forested uplands, for example, "provide vitally important watershed protection to the country's lowland river valleys by absorbing rainfall and slowing the rate of runoff, preventing soil erosion and reducing the severity of both flood and drought – which contributes to make possible China's intensive irrigated agricultural system" (WWF, 2005).
[4] One possible way of proceeding is estimating replacement costs. For example, water filtration plants for both industrial and residential use, able to substitute for the natural water filtration lost due to pollution, land conversion and the disappearance of wetlands, imply investment costs in the order of magnitude of tens of millions to billions of Euros, depending on the dimension of the basin to be served. Replacement costs do not correspond to a calculation of the value of the resource needed in order not to lose the service: one can think of cases when a service of little value is extremely costly to substitute for, in which case replacement costs would overestimate the true economic value. They must be applied with caution. Still, they can represent a feasible instrument to provide decision makers with an evaluation of the benefits from investments in conservation that can sometime be of critical importance.
[5] China's Action Plan (by the National Environmental Protection Agency, with the participation of the Chinese Academy of Sciences, Ministry of Agriculture, Ministry of Construction, Ministry of Finance, Ministry of Forestry, Ministry of Public Security, State Oceanic Administration, State Planning Commission, State Science and Technology Commission), which includes estimates on endangered species, is available at www.bpsp-neca.brim.ac.cn/books/actpln_cn/index.html.
[6] An updated list of countries that have ratified the Cartagena Protocol is available at www.biodiv.org/biosafety. China signed the Protocol in August 2000, ratified it in June 2005, and allowed it to enter into force on September 6th, 2005.

REFERENCES

Ando A., Camm J., Polasky S. and Solow A., 1998. Species Distributions, Land Values, and Efficient Conservation, *Science*, New Series, Vol. 279, No. 5359, pp. 2126-2128.
Balmford A., Bruner A., Cooper P., Costanza R., Farber S., Green R. E., Jenkins M., Jefferiss P., Jessamy V., Madden J., Munro K., Myers N., Naeem S., Paavola J., Rayment M., Rosendo S., Roughgarden J., Trumper K., and Turner R. K., 2002. "Economic reasons for saving wild nature", *Science* 297, pp. 950-953.
Barbier E., 1992. "Community based development in Africa", in Swanson, T. and Barbier, E. (eds.), *Economics for the Wilds: Wildlife, Diversity, and Development*. Island Press, Washington DC, pp. 107-118.
Bisby F.A., 1995. "Characterization of biodiversity", in Heywood, V. (ed.) 1995. *The Global Biodiversity Assessment*, Cambridge, Cambridge University Press, pp. 21-106.
Brock W.A. and Xepapadeas A., 2003. "Valuing Biodiversity from an Economic Perspective: A Unified Economic, Ecological, and Genetic Approach," *American Economic Review*, Vol. 93(5), pp. 1597-1614.
Costanza R., d'Arge R., de Groot R., Farber S., Grasso M., Hannon B., Limburg K., Naeem S., O'Neill S.V., Paruelo J., Raskin R.G., Sutton P., and van den Belt M., 1997. "The Value of the World's Ecosystem Services and Natural Capital", *Nature*, Vol. 387, No. 6230.

Dalmazzone S., 1998. "Ecological Resilience and Economic Sustainability", in Acutt M. and P. Mason (eds.), *Environmental Valuation, Economic Policy and Sustainability: Recent Advances in Environmental Economics*, Edward Elgar, Cheltenham, UK, 1998, pp. 171-188.
De Groot R.S., Wilson M., Boumans A. and Roelof M.J., 2002. "Typology for the classification, description and valuation of ecosystem functions, goods and services, *Ecological Economics* 41(3), pp. 393-408.
Di Falco S. and Perrings C., 2003. "Crop Genetic Diversity, Productivity and Stability of Agroecosystems. A Theoretical and Empirical Investigation." *Scottish Journal of Political Economy*. 50, pp. 207-216.
Dudley N. and Stolton S. (eds.), 2003. *Running Pure: The importance of forest protected areas to drinking water*. Research report for the World Bank and WWF Alliance for Forest Conservation and Sustainable Use, http://assets.panda.org/downloads/runningpurereport.pdf
Heywood V. (ed.), 1995. *The Global Biodiversity Assessment*, Cambridge, Cambridge University Press.
Holling C.S., 1973. "Resilience and stability of ecological systems", *Annual Review of Ecological Systems* 4, pp. 1-24.
Holling C.S., Schindler D.W., Walker B.W. and Roughgarden J., 1995. "Biodiversity in the functioning of ecosystems: An ecological synthesis", in Perrings C., Mäler K.-G. Folke C. Holling C.S. Jansson B.-O. (eds.), 1995, pp. 44-83.
Kaufman J. H., Brodbeck D. and Melroy O. R., 1998. "Critical biodiversity", *Conservation Biology*, 12, 251.
Land Institute, 2005. *Annual Report 2005*, Salina, Kansas, U.S.
Loreau M., Naeem S, Inchausti P., Bengtsson J., Grime J.P., Hector A., Hooper D.U., Huston M.A., Raffaelli D., Schmid B., Tilman D. and Wardle D.A., 2001. "Biodiversity and ecosystem functioning-current knowledge and future challenges", *Science* 294(5543), pp. 804-843.
Metrick A. and Weitzman M.L., 1998. Conflicts and Choices in Biodiversity Preservation, *Journal of Economic Perspectives*, Vol. 12, No. 3 (Summer, 1998), pp. 21-34.
Millennium Ecosystem Assessment, 2005. *Ecosystems and Human Well-Being: Multiscale Assessments. Findings of the Sub-global Assessments Working Group*, Millennium Ecosystem Assessment Series. Washington, D.C., Island Press.
Mitsch William J., 1991. "Ecological engineering: Approaches to sustainability and biodiversity in the U.S. and China", in R. Costanza, 1991, *Ecological Economics, The Science and Management of Sustainability*, NY, Columbia University Press, pp. 428-448.
Pearce D.W., 2003. "Environmentally harmful subsidies: barriers to sustainable development", in P. Prinsen-Geerligs, M. Patterson, A. Cox M. Tingay (eds.), *Environmentally Harmful Subsidies: Policy Issues and Challenges*, Paris, OECD.
Perrings C., Folke C. and Mäler K.G., 1992. "The ecology and economics of biodiversity loss: The research agenda". *Ambio* XXI (3).
Perrings C., Mäler K.-G. Folke C. Holling C.S. and Jansson B.-O. (eds.), 1995. *Biodiversity Loss: Economic and Ecological Issues*, Cambridge, Cambridge University Press.
Perrings C. (ed.), 2000. *The Economics of Biodiversity Conservation in Sub-Saharan Africa*, Cheltenham, UK, E. Elgar.
Perrings C. and Stern D., 2000. "Modelling Loss of Resilience in Agroecosystems: Rangelands in Botswana", *Environmental and Resource Economics* 16(2), pp. 185-210.
Pielou E., 1993. *Ecological Diversity*, New York: John Wiley.
Pimm S. L., 1984. "The complexity and stability of ecosystems", *Nature* 307, pp. 321-326.
Polasky, S. N., Nelson E., Lonsdorf E., Fackler P., and Starfield A., 2005. "Conserving species in a working landscape: land use with biological and economic objectives.", *Ecological Applications* 15(4), pp. 1387-1401.
Polasky S., 2006. "Biodiversity Conservation and Ecosystem Services", Plenary Lecture, *Third World Congress of Environmental and Resource Economists*, Kyoto, July 2006.
Randall A., 1991. "The value of biodiversity", *AMBIO* 20(2), pp. 64-68.
Rohde R.A. and Muller R.A., 2005. "Cycles in fossil diversity", *Nature* 434, pp. 208-210.
Sagoff M., 1992. "Has Nature a Good of Its Own?", in Costanza R., Norton B.G., Haskell B.D. (eds.), *Ecosystem health: New goals for environmental management*. Washington, D.C.: Island Press, pp. 57-71.
Schindler D.W., 1990. "Experimental perturbations of whole lakes as tests of hypotheses concerning ecosystem structure and function, *Oikos* 57, pp. 25-41.

Simpson R.D., 2002. "Definitions of biodiversity and measures of its value", Discussion Paper 02-62, *Resources for the Future*, Washington D.C. (available at www.rff.org).
Solow A. and Polasky S., 1994. "Measuring biological diversity", *Environmental and Ecological Statistics* 1(2), pp. 95-107.
Somanathan E., Prabhakar R., and Bhupendra Singh Mehta, 2006. "Does decentralization work? Forest conservation in the Himalayas", paper presented at the 3rd *World Congress of Environmental and Resource Economists*, Kyoto, 3-7 July 2006. http://www.webmeets.com/ERE/WC3/Prog/
Tilman D., Polasky S. and Lehman C., 2005. Diversity, productivity and temporal stability in the economies of human and nature, *Journal of Environmental Economics and Management*, 49(3), pp. 405-426.
Tilman D. and Downing J.A., 1994. "Biodiversity and stability in grasslands". *Nature* 367, pp. 363-365.
UNEP (1992), *Convention on Biological Diversity*, Secretariat of the CBD, www.biodiv.org/convention/convention.shtml
UNEP, 2000. *Sustaining Life on Earth*, Secretariat of the CBD, www.biodiv.org
Weitzman M., 1992. "On diversity", *Quarterly Journal of Economics* 107(2), pp. 363-405.
Weitzman M., 1993. "What to Preserve? An Application of Diversity Theory to Crane Conservation", *Quarterly Journal of Economics* 108(1), pp. 157-183.
Weitzman M., 1998. "Recombinant Growth", *Quarterly Journal of Economics* 113(2), pp. 331-360.
Whittaker R.H., 1960. "Vegetation of the Siskiyou Mountains, Oregon and California", *Ecological Monographs* 30, pp. 279-338.
Whittaker R.H., 1972. "Evolution and measurement of species diversity", *Taxon* 21, pp. 213-251.
Wittenberg R. and Cock M.J.W. (eds.), 2001. *Invasive Alien Species: A toolkit of best prevention and management practices*, Global Invasive Species Programme, Oxon, UK: CAB International.
World Bank, 1992. *World Development Report 1992*, Oxford.
World Wide Fund for Nature, 2005. "China's Biological Treasures", WWF China Newsletter, www.wwfchina.org

G. SCARASCIA-MUGNOZZA AND M. E. MALVOLTI

FORESTRY AND RURAL DEVELOPMENT: GLOBAL TRENDS AND APPLICATIONS TO THE SINO-ITALIAN CONTEXT

Abstract. International Agreements, as those on biodiversity, sustainable development and climate change, have recognized the role of forests at the global scale identifying also the risk of the impact of environmental changes on forests' stability and functions. A sound management of natural and man-made forests and their extension can help counteracting the risks of global change. Humankind is increasingly challenging environmental degradation: deforestation, land degradation and erosion of biological diversity. Foresters are urgently required to improve productivity and sustainability of planted forests in an effort to preserve natural forests. In Europe as well as in other Countries, a large development of tree plantations is expected to occur mainly in the agricultural areas since planting trees will ensure a sustained, high productivity of woody biomass and interesting revenues to farmers. The role of forestry and agro-forestry is then discussed also in relation to the Sino-Italian scientific cooperation in this field.

1. TRENDS IN WORLD FORESTRY: A GLOBAL ISSUE WITH WIDE LOCAL DIVERSITIES

The world land surface is covered by about 30% with forest ecosystems, corresponding to a total surface of 3.5 Gha (i.e. 10^9 ha). This figure is approximately half of the forest cover existing in the world when the human population started the agricultural revolution 10,000 years ago; at that time, mainly in the countries of oldest civilization as China and Europe, a large portion of the existing forests disappeared or were extensively degraded. Presently, European forest land (former USSR excluded) covers a surface area of about 200 million ha corresponding to 35% of the total surface area of Europe, a percentage similar also to the relative forest cover in Italy. In Europe, as well as in Italy, the forest cover is presently undergoing an historical change, a true inversion of the shrinking trend that was observable for the last centuries: the European forest surface is finally expanding steadily, since 1950, at an annual rate of circa 0.5%, because of the abandonment of marginal agricultural farmland, particularly in the mountain and hilly regions. However, in most of the world, and particularly in the tropical regions, forest cover is still decreasing at a very fast rate with millions of hectares that every year are degraded to open forests or even bare lands.

Most of the forest land in Europe is presently made up of productive forests, which are managed to produce a variety of goods and services. Among those, production of timber, paper and pulp still represent important objectives of forest management contributing also to sustain European related industries and to supply the European demand for these products. As such, forests are a fundamental part of the European social and economical life.

The relative role of the different functions varies across Europe: in fact, while wood production and recreation is prominent in Northern and Central Europe, soil protection against erosion and landslides, and landscape improvement have a relevant role in Southern Europe. International Agreements, as those on biodiversity, sustainable development and climate change, have clearly recognized the role of forests at the global scale but have also identified the risk of the impact of environmental changes on forests stability and functions. However, a sound management of natural and man-made forests and the extension of their surface can help counteracting the risks of global change. After the UN Conference on Environment and Development in Rio de Janeiro (1992), the European forestry ministers embraced the principles contained in Agenda 21, i.e., the need and utility of internationally agreed '*criteria and indicators for the conservation, management and sustainable development of all types of forests*' (MCPFE-Ministerial Conference on the Protection of Forests in Europe, Helsinki 1993); these objectives form the base of the pan-European criteria for sustainable forest management (SFM). These criteria and indicators of SFM were adopted at the Third Ministerial Conference in Lisbon (1998) and now represent the foundation for the definition and implementation of SFM protocols for all European countries. Also, the '*maintenance and appropriate enhancement of forest resources and their contribution to global C cycles*' was identified as one and important criterion of the SFM, quite relevant also for the compliance to the Climate Change Convention and to the Kyoto Protocol. This Protocol deserves much consideration, beside for its specific role in reducing greenhouse gas (GHG) emissions, also because it is one of the first examples of assessing and quantifying the economical value of environmental services provided by forest ecosystems.

2. FORESTS, FOREST PLANTATIONS AND AGRO-FORESTRY

Humankind is increasingly facing and challenging environmental degradation: deforestation, forest and land degradation and erosion of biological diversity are major threats for natural forests. Foresters are urgently required to strengthen forest management practices and improve productivity and sustainability of planted forests in an effort to preserve natural forests. In fact, while natural forests suffer from deforestation, the importance of forest plantations is increasing, occupying 190 Mha worldwide with annual establishment rates of 8.5-10.5 Mha, and an annual net gain of 1.96 Mha year^{-1} between 1965 and 1990 (FAO, 2001).

In Europe, a large development of tree plantations is expected to occur mainly in the agricultural areas because of the highly developed and structured landscape of this continent; furthermore, planting trees in farms will ensure a sustained, high productivity of woody biomass and interesting revenues to European farmers.

Current interest in agro-forestry relates mostly to the wider general focus on sustainable agriculture for environmental protection. This focus is often in great contrast to the urgent economic needs of rural people. A partial solution to this problem has come through funding from the European Union for sustainable agriculture and afforestation of arable lands. Trees, mostly valuable broadleaved (hardwood) species like walnut, cherry and ash, were planted back into farmland using public funds and with the aim of producing high quality timber, which is in very short supply on the European market. Afforestation of agricultural land has covered a substantial surface in recent years, in many countries. As an example, more than 100,000 ha of agricultural lands were afforested in Italy during the '90s. However, the appropriate tree species and cultivation techniques should be carefully selected to match the quite different site conditions existing in Europe; the danger is that a significant proportion of those plantations could be unsuccessful. Alternative and new cultural models, such as agro-forestry and mixed models were studied for replacing hardwood plantation forestry. These studies are in connection with the Mediterranean tradition of mixed cultural systems, which are now marginal. Researches show that both cultural models have numerous advantages in comparison with traditional forestry plantations. Tree growth and timber quality are often improved due to enhanced tree care, better site quality and synergisms among plant/system components. Technical advantages are augmented by ecological ones, such as improved biodiversity, soil erosion control and reduced fire risk; however, farmers' reactions to innovative agro-forestry systems should also be carefully examined. Eventually, agro-forestry can be more effective than pure cultivation for the restoration of degraded agro-ecosystems and for the preservation of rural landscape, creating biological corridors useful for the reproduction of flora and fauna natural species.

3. NEW ISSUES AND ENVIRONMENTAL ASPECTS

The international community, by acknowledging forests, both natural and planted, as an increasing source of environmental and socio-economical benefits (Del Lungo, 2003), has taken action with multiple initiatives:

- At the 1992 Earth Summit in Rio de Janeiro, world leaders agreed on a comprehensive strategy for "sustainable development" to preserve natural forests and to expand planted.
- The Kyoto Protocol acknowledges afforestation and forest management as a means to reduce Greenhouse Gas (GHGs) through the Clean Development Mechanism (CDM), International Emission Trading (IET) and Joint Implementation (JI).
- The United Nations Convention for Biological Diversity (UNCBD) acknowledges planted forests, if soundly planned and managed, can be positive contributors to preserve natural ecosystems and funding is available through the Global Environment Facility (GEF).
- The United Nations Convention for Combating Desertification (UNCCD) recommends national policies to support the rehabilitation and restoration of degraded ecosystems by planting through the Global Mechanism (GM).

According to the Kyoto Protocol, carbon trading allows industries in developed countries to offset their emissions of carbon dioxide by investing in reforestation and clean energy projects in developing countries with several objectives: erosion control, reclaiming barren land, industrial roundwood production or household wood supply. Planting forests projects can offer investors the same carbon benefits as industrial tree plantations and at lower risk. Many industries will prefer to buy "socially responsible" carbon credits, as long as the cost is competitive and countries are strengthening opportunities arising under the CDM while promoting increased bio-fuel utilisation and conversion efficiency to substitute for fossil fuels. The potential environmental benefits of planted forests can be enormous. According to some researches (see Del Lungo, 2003), converting low-yielding crop and pasture lands to higher-yielding agro-forestry would sequester 5 to 50 tons of carbon per hectare per year. Furthermore, it has been estimated that the afforestation of 20% of present arable land of the European Union would increase soil carbon stocks by about 5% over a century and would offset European carbon emissions by about 4% (Smith et al., 1997), figures comparable with the Kyoto obligations.

Another option to reduce anthropogenic carbon emissions is to replace fossil fuels by the use of bio-fuels. In fact, in 1994 the Declaration of Madrid set the objective of substituting 15% of real primary energy demand in the EU with renewable energy sources; presently, biomass supplies about 3.5% of the total energy consumption within Europe, with contributions as high as 13 to 17% in such countries as Austria, Finland and Sweden. However, in 1998 the European Biomass Conference stated, as a feasible objective, to raise within the year 2010 the energy production from biomass up to 8.5%, with a twofold increase of the present day's relative contribution. Approximately 50% of this biomass production can be obtained from forest tree plantations with an overall requirement of 7 Mha of land withdrawn from surplus agricultural crops. These objectives are also in line with the new directions of the European agricultural policy, contained in the Agenda 2000, which give a more prominent role to agro-environmental measures.

Early research with species that regenerate after coppicing has shown that poplar and willow are the most reliable species for use as energy crops (Tabbush and Parfitt, 1996). While producing bio-fuels with no net carbon emissions from their combustion, short rotation plantations of poplars and willows are able to sequester atmospheric carbon into the soil. However, the carbon budget is even more favourable when yield, maximised by the sound selection of most suitable clones and the adoption of the best crop management techniques.

The management practices currently employed in Europe were developed with the aim of optimising cost-benefit ratios in relation to the objectives defined above. However, in the context of the current debate over the implementation of the Kyoto Protocol, it has become apparent that they may have a fundamental role to play in the calculation of the national and continental C budgets, in addition to reductions in greenhouse gas (GHG) emissions.

The Intergovernmental Panel on Climate Change (IPCC) special report on Land Use, Land Use Change and Forestry has provided a reference benchmark in identifying the most important forestry practices in relation to carbon sequestration and carbon fluxes (IPCC, 2000). The IPCC report defines forest management as "the application of biological, physical, quantitative, managerial, social and policy principles to the regeneration, tending, utilisation, and conservation of forests to meet specified goals and objectives, while maintaining forest productivity.

Management intensity spans the range from wilderness set-asides to short-rotation woody cropping systems. Forest management encompasses the full cycle of regeneration, tending, protection, harvest, utilization and access".

It can therefore be argued that the implementation of this criterion for SFM, based on the principles of Agenda 21, requires a thorough knowledge of the effects that forest management practices have on ecosystem C cycles. Research on this topic is therefore urgently required.

The ability of managed forest and agro-forestry systems to sequester carbon at the regional and global scale will be strongly influenced by the responses of trees and tree communities to global change, particularly to the predicted increase of atmospheric CO_2 concentration (Jarvis, 1989), when tree stands influence climate at regional and global scale, therefore affecting the process of environmental change at different scales (IGBP-International Geosphere Biosphere Program 1998). Despite the key role played by trees and forests within the terrestrial biosphere, we still have limited information on the responses of whole forest systems and agro-forestry system to enhanced CO_2 and other aspects of global change (nutrient levels, land-use change) because of the complex web of interactions (Norby et al., 1999).

The studies conducted at the whole-tree and community scale under elevated CO_2 indicate that there will be a marked increase of primary production, mainly allocated into below-ground biomass (IPCC, 2000). However, the proportionality of this response may well depend on soil nutrients and water availability, tree species, genotypes and management. Another critical point concerns the implications of below-ground carbon allocation for long-term carbon storage (Schlesinger and Lichter, 2001). The enhanced carbon transfer to the root system may result mainly in an enhanced root respiration or, otherwise, in an increase of root dry matter, mycorrhizal activity and subsequent transfer of carbon to soil C pools.

Soil carbon pools may be enlarged and restored by managing agricultural and forest soil and by the use of surplus agricultural land for reforestation or high yield woody crops for biomass and bio-energy. Recently, bio-energy woody crops have been reported to have the greatest potential for carbon mitigation, for their capability to sequester carbon and, contemporarily, to substitute fossil fuel carbon (Smith et al., 2000). Thornley and Cannell (2000) have also used a mechanistic forest model to propose that annual removal of forest biomass (similar to a coppiced system) may provide the best compromise between stemwood productivity and belowground carbon sequestration.

Planting mixed broadleaved tree species may further stimulate C sequestration in soil after conversion; broadleaved species seem to be the most effective in storing soil C (Guo and Gifford, 2002), although when the litter layer is also accounted for, spruce accumulates C in soil at a faster rate than oak (Vesterdal et al., 2002). In Northern Italy, Del Galdo et al. (2003) showed that the negative impact of agriculture on soil organic C-storage could be reversed when crop soils are converted into woodland. After 20 years from tree planting, the afforested soil had gained 5.2 tC ha^{-1} in the top 10 cm, corresponding to 24% of the soil organic C present in the same layer of a soil kept under crop. Furthermore, afforestation resulted in significant sequestration of new C and stabilization of old C in physically protected soil organic matter (SOM) fractions, mainly associated with micro-aggregates.

An interesting contribution of Short Rotation Forest (SRF) plantations to environmental amelioration is the ability of the trees to remediate soil and water pollution by extracting, immobilizing or metabolizing various pollutants. No experimental data exist on this subject warranting the importance of research in this field; however, it can be reasonable to hypothesize that the large increase of C-assimilation under elevated CO_2 will improve phytoremediation properties of poplar plantations because of their faster growth, higher water and nutrient uptake, increased root productivity and greater root exudates, that will augment mass and metabolic activities of microbial and fungal soil communities. Finally, soil erosion control and slope stabilization could be beneficially influenced by elevated CO_2 because of the large increase of root mass and their penetration into deeper soil layers.

4. COOPERATION BETWEEN CHINA AND ITALY IN FORESTRY AND AGRO-FORESTRY

4.1. The context

China is a large country covering a surface of 9,602,700 Km^2; about 100 Mha are cultivated for agricultural crops, while forests cover about 115 Mha (Zhaohua et al., 1997). Woody resources are important for China since they provide about 40% of rural energy, material for construction, pulp and paper, as well as fodder for animals. In the last decade the demand for timber is rapidly expanding following the same trend as industrialization, but at the same time is growing also the environmental pressure and the resources exploitation. As it is well known, this is one of the principal reasons of soil erosion and desertification. Therefore, there is a large demand for developing sound management strategies of the rich and diversified natural resources of China.

Considering this need, in recent years, scientific cooperation between European Countries and China on agricultural issues was strongly strengthened in order to promote sustainable development. In this sense, the recent ratification of the Kyoto Protocol on the reduction of CO_2 and pollutant emissions and the scientific programs focused on sustainable development recently expressed by the Chinese Minister for Science and Technology, highlight the interest of the Chinese Government regarding the environment.

The National Research Council (CNR) of Italy, with the Italian Ministry of Foreign Affairs, has been particularly active in establishing scientific agreements with complementary Chinese partners (Chinese Academy of Sciences, CAS, Chinese Academy for Agricultural Science, CAAS and Chinese Academy of Forestry, CAF) for promoting the introduction and the adoption of innovative agricultural models aimed at improving agricultural production while preserving or restoring the environment, and reducing the risks for human health.

The possible benefits allowed from the cooperation were focused and remarked in the framework of various meeting and congresses organized in collaboration with the Italian Embassy in Beijing. Those conferences identified the necessity of carrying out scientific research on natural resources (biodiversity conservation, desertification reduction and sustainable utilisation of water resources), sustainable management (development of new strategies for the defence of agro-forestry systems from abiotic and biotic stresses, forest fire prevention), forests and agro-forestry ecology (natural forest monitoring, evaluation of the growing cycle of artificial forest) and technology, ecology and forestry strategies development (improvement of technological characteristics of artificial forest wood, wood quality, genetic drainage, economic evaluation of agro-forestry activities). The conferences stressed the need of considering the above mentioned subjects the core of large research programmes to be developed at international level.

Also several ASEM (the Asia-Europe Meeting) summits encouraged the establishment of active cooperation between Europe and Asia on forest conservation and sustainable development by integrating agricultural and forestry activities with the wide participation of policy-makers, universities, research institutes and enterprises.

The Chinese economical system has rapidly converted from a structure mainly devoted to an internal market and based on agricultural systems focused on home consumption, to an industrially developed system that requires a new type of agriculture able to create innovative chains for high value products. This aspect has received great attention since Chinese participation in the WTO.

During the last decade agricultural land use system has changed as highlighted in the document presenting FAO data during the period 1990-2001. Wheat, barley and rice have reduced by 15-20% the total cultivated area and among cereals, only maize slightly increased the cultivated area. The different trends between cultivated area and total production is probably due to an intensification of agricultural practices as demonstrated by the increased use of fertilisers (average increase of 27% for nitrogen, phosphorus and potassium) and the concentration of cultivated land in the most favourable areas (fertile soils, plain lands). This process is determining an initial abandonment of wide unfavourable agricultural areas where environmental problems such as desertification and soil erosion are becoming manifest.

Population is also concentrating more in urban areas where an average increment of 48% has been recorded during the period 1990-2001 followed by a small reduction of rural population (average decrease of 3%). Consequently, in the near future labour is likely to be less available in agricultural activities while it should be more easily accessible for industrial activities. This aspect appears substantiated by comparing the active agricultural population (increase of about 2%) with the population which has no agricultural activity (increment of about 34%).

Intensive agriculture (mono-culture, specialised orchards, etc.) even if profitable in the short term is not sustainable in the long term because it requires high level of external inputs and induces environmental constraints (biodiversity reduction, microclimate alterations, soil erosion, desertification, etc.) as already visible in several agricultural areas of industrialised countries. All these environmental and human constrains represent the main reasons for requiring the identification and development of innovative agricultural models able to be efficient also within modern industrial set. Among the cultural practices able to meet all these needs, agro-forestry systems or mixed trees plantation appear the most suitable because of their capacity for producing several kinds of benefits, both economic and ecological.

The potentials and benefits of integrating forestry into farms is, as above exposed, becoming recognized also in Europe, where EU identified the high potential of woody species and offered financial incentives for adopting innovative cultural models in which crops and trees are integrated at different levels within farms.

Considering the Chinese context, new agricultural practices should be introduced in rural areas in order to avoid economic and ecological problems such as the uncontrolled diffusion of agricultural pollutants (i.e. weed-killers and pesticides), the increasing desertification and the indiscriminate use of water resources. Practices should be designed considering different spatial integration of trees and crops in order to generate high value products for food consumption and industrial processing, taking into account also the environmental services. The generation of new productions should be advantageous also for industrial countries that will improve both their commercial exchanges with China and the global environmental quality.

In this way environmental sustainability and economic competitiveness will be the main features of the practices as requested by the development strategies.

4.2. Objectives and applications

The cooperation has been developed in order to deal with themes of common interest at practical and scientific level. The overall aim of the project is to valorize and apply the scientific knowledge implemented through the cooperation between Europe and China as well as to increase the economic relationships between the two continents.

More in detail, the following themes have been approached:

- Conservation and revaluation of local germplasm.
- Test the efficiency of agro-forestry systems to protect the soil from the degradation processes (wind and hydrological erosion).
- Development of cultural agro-forestry systems in order to obtain diversified products (quality wood/food) for industrial and local uses.
- Improvement of the welfare in rural areas in order to contrast the country abandonment with particular attention to marginal areas.

Walnut (*Juglans regia* L.) has been considered as principal agro-forestry species to be used in the field trials, as pure and intercropped plantations. This choice was made because the species is well known both in Europe and China with an old traditional cultivation both for fruit and wood production. China is the major producer/consumer of nuts in the world (about 285,000 ton yr^{-1} on 650,000 ha of cultivation); China is then the strongest competitor in terms of export of nuts in comparison with USA (255,000 ton yr^{-1} on 81,000 ha) and Italy (15,000 ton on 4,000 ha). In addition, the establishment of plantations for wood quality production only is foreseen in the new Chinese agricultural plan. In this situation the relationships, agreements and coordination between Eastern and Western Countries are then very important.

4.3. Some examples of cooperative research

The above mentioned scientific agreements between CNR and CAF were extremely useful for exchanging information and knowledge between scientists and local policy makers as well as for establishing solid relationship between Italian and Chinese Organisms. According to these considerations practical and basic researches were carried out in four Chinese regions characterized by different economic and social conditions: Xinjiang, Shandong, Henan and Hebey.

The *Autonomous Province of Xinjiang* (North-West China) is China's largest region, with a total area of about 1,646,800 km^2, approximately one-sixth of China's total land area. This Province is included in the Western Development Strategy Program of the Chinese Government. The region is rich in natural resources (some remaining natural forests of the world are here) even if it has wide, particularly arid zones (less than 80 mm of total rainfall per year with varying temperature from -30°C to +30°C) with poor and sandy soil. More then ¾ of Xinjiang territory are occupied by two deserts: the Taklamakan Desert at West with flat zones around the Tarim basin (this river rises from the Karakorum Mountains and dies in the desert); the Guarbantinggut Desert, hilly and rocky, around the Junggar basin. Xinjiang's climate is dry and continental, with warm summers and long, cold winters. North of the Tien Shan in the Junggar slope, the city of Urumqi has an average annual precipitation of only 178 mm: in January the mean temperature is -11°C, while in July it is 25°C. The area south of the Tien Shan, in the Taklamakan, has received no precipitation in some years. Agriculture is the traditional economic base of Xinjiang, and the principal origin of income and employment. Farming occurs where rivers flow onto the basins and are harnessed for irrigation (about 3,4000,000 ha). Agriculture is tending towards modernization.

The Chinese and Italian research projects were well inserted in the context of soil conservation and in the political support for rural development of Xinjiang region: the experimental fields have been established in the Southern part of the Province, in the Aksu and Keshen Counties, close to the desert of Taklamakan (the second in size in the world) within the Tarim basin. This choice was made because the local administration, through tax reductions and free distribution of tree plants such as walnut, is promoting the conversion of irrigated zones from cotton cultivation to agro-forestry. The established agro-forestry systems intercropped tree species at medium/long term (walnut and almond), with fruit-culture at short term (peach and plum), and annual crops (soy, maize, horticulture). The intercropped species were decided in agreement with local researchers belonging to the Xinjiang Forest Academy and farmers' representatives. There are industries to process agricultural products.

Water management was a focal point in order to avoid wastes due to obsolete irrigation techniques (underground and flowing) on culture at low conversion index. An important aspect is the struggle against desertification in order to acquire new zones for future forest and agricultural activities. Autochthonous trees and shrub species have been employed to combat desertification and improve soil fertility. They are *Populus euphratica, Tamerix* spp., *Ziziphus jujuba,* and other species resistant to salinity, drought and temperature variation.

At the present, some aspects of agro-forestry system optimisation are still in progress: the effects of species intercropping on growth, the proper choice of varieties, the different breeding of plants considering the final use, plants architecture, irrigation management, the eco-physiological and microclimate evaluation and the valorisation of products (fruit and wood), following international standards.

The *Henan Province* (Central China) is characterized by two different climatic zones from North to South; the temperate and the sub-tropical ones. The Yellow River runs through the whole province from West (hilly and mountain areas) to East (flat zones). There is an area of 167,000 km^2 with a monthly mean temperature of 22-30°C in July and 0-6°C in January. The annual precipitation varies from 500 to 1200mm. There are large areas of hilly land integrated with branch rivers and flat lands in the west and central parts. Many pieces of flat and large farmland recline along the Yellow River in which wheat and corn double cropping system is practiced year after year and there are wide spanning poplar shelterbelts too. Agro-forestry systems (Fig. 1) of winter wheat intercropped with various tree species are well developed and the dominating trees include candlenut, poplar, Chinese jujube and persimmon.

Figure 1. Agro-forestry in China, Henan Province: plantation including walnut, wheat and peanut. In the background Poplars (P. euphratica) used as shelterbelts.

Even if Henan's economy is based on agriculture, and in spite of the richness of vegetation particularly in temperate and sub-tropical areas, the high population density makes this province very poor. The Government is trying to transform the old economy based on monoculture system into forestry and agro-forestry.

The research activities have been developed in two different areas:

Guangshan County is located at the northern feet of Dabie Mountain and the whole county has a total area of 1831 km^2. Agriculture is the main stay of the economy and the main crops are rice, wheat, beans, rapeseed, etc. The forest coverage rate is 18.3% and the main economic forests are tea, Chinese chestnut, and ginkgo. Frequent natural disasters coupled with weak fighting capacity and severe soil erosion and water loss render the local economy poor. Large areas are characterized by low-yields, slopping and flooded fields, poor soil fertility, low and unstable grain yields. Some advanced agricultural techniques, even when available, are difficult to spread over wide areas because of the difficulty in popularizing agricultural techniques, insufficient new force, low speed in updating knowledge, and unstable technical capacity. Besides, poor circulation structures, difficulty in transportation and high freight charges have increased production costs and reduced the competitive power of the products.

Luoning County, the second area for experimentation, is located in the ecotone between the Yellow river valley and the Taihang Mountain, whose climate is semiarid monsoon. There are many different kinds of agro-forestry systems and several plant species, including tree, crop and Chinese medicine plant species. The main tree species are walnut, plum, apricot, apple, poplar and the tree coverage was 20.5% in 2003. The main crop species are wheat, cotton, corn, and peanut. The main Chinese medicine plant species are *Isatis indigotica* Fort, *Dioscorea zingiberensis* c. H. Wright, *Macrocarpium nakai*. Agro-forestry systems were established according to the natural and economic status in the regions and located on terracing fields, strongly eroded by three factors: wind, rain and human impact. Association of walnut plants for wood and fruit, intercropped with wheat, cotton, corn, and peanut was considered. For that purpose different pruning techniques were applied both on Chinese germplasm and on some of Italian provenance tested here to study their behaviour, in particular environmental conditions and intercropping with new crops. The first results showed that the Italian product had a better architecture as plant for wood, but the growth percentage (stem diameter and annual apical shoot) is lower than the one observed for Chinese individuals.

In the Henan locations the above experimental plantations (Fig. 2) were established in 2002. Farmers observed that the model permitted different harvest times, from spring to past autumn, and copied the system spreading the models in other counties. The measurement of growing aptitude, as well as the effect on soil fertilization and protection is still in progress. From a social point of view, since the experimental field trials of Luoning County became an example of rehabilitation of eroded land, the local government improved the infrastructures in order for cars to have access.

Figure 2. Afforestation in a hilly area, Henan Province. Cooperation between Chinese and Italian researchers.

In the *Hebei Province,* the *Raoyang District* is located in the northern part of the North China Plain extending into the Inner Mongolian Plateau. Beijing, the capital of China, and Tianjin, the important trading port in the north of the country, are situated in the centre of the province, although they are not a part of it. It adjoins Liaoning and Inner Mongolia in the north, Shanxi Province in the west and Henan and Shandong Provinces in the south. The Bohai Sea lies to the east of the province. Its coastline extends 487 km, and the total area of the province is 190,000 km^2. The topography of the province declines from northwest to southeast. Because of its position, at South with respect to Beijing, the forestry and agro-forestry plantations are established not only for production but also to protect the capital from sand storms. In the province, there are more than 3,000 species of plants, of which more than 140 are fibre plants, more than 1,000 are medicinal herbs and plants, and more than 100 are timbers. There are 300 or more kinds of forage grass, 140 kinds of oily plants, and more than 450 varieties of cultivated plants. Among these, the output of cotton makes up one-seventh of the nation's total and that of maize and fruits makes one-tenth of that total. There are 215 varieties of farming products in 12 major categories that can be listed as brand names and best products, or native or rare products. There are more than 20 varieties of medicinal herbs and materials for export.

The objective of the tree plantations is to create shelterbelts, mitigation of microclimate and provide farmers with increasing income by fruit crops (peaches, pears, apples). Generally in this region poplars are widely used, including transgenic trees.

The common agro-forestry project established experimental plantation, intercropping poplars with walnut and cherry in order to obtain quality wood and fruits at short term as well as wood at medium term (poplars).

The *Shandong Province* with capital Jinan is a significant coastal province in East China, having a population of 90.79 millions (March 2001). Located in the lower reaches of the Yellow River, it borders on the Bohai Sea and the Yellow Sea. Its area is about 156,700 km^2. The Province has a warm-temperate monsoonal climate, with hot, rainy summers and dry, sunny winters. Shandong has a mean annual temperature of 12°C-14°C, increasing from the north-eastern seaboard to the west and the south, and a mean annual precipitation of 500-900 mm or more, increasing from northwest to southeast. The province has large deposits of petroleum, natural gas, coal, iron, diamonds and bauxite. Its gold output ranks first in China. The Shengli Oilfield, the country's second biggest crude-oil producer, is located in north-western Shandong near where the Yellow River discharges into the Bohai Gulf. Xindian in Zibo is a rising chemical industrial centre. Shandong has a long history of tobacco and tussah growth. It is also a famous temperate fruit grower (apples of Yantai, pears of Laiyang, small dates of Leling and watermelons of Dezhou). The offshore waters are rich in fish, prawns and kelp. It is also a major producer of wheat, corn, cotton and other crops. Its peanut production accounts for a quarter of the nation's total. The sectors encouraged for investments are, among others: household appliance, metallurgical, light and textile industries, food processing, agricultural planting, breeding, and product processing, tree-felling and timber-processing, traditional and chemical medicine and health care. Agriculture is considered as one of the most important tools to protect environment (soil and water), human health and contrast industrial pollutions.

Several agricultural zones, mainly located in hilly areas (Fig. 3), have been already neglected and agriculture is mainly concentrated in plain fertile areas with artificial irrigation and intensive cultural practices. In the neglected areas, environmental problems such as soil erosion and desertification are becoming clear. About 20,000 ha of new tree plantations have been planned and about 2,000 ha have already been established in most favourable sites. Walnut represents the most commonly used species for fruit production. The item of the common research has been the improvement of the economic and environmental efficiency of new plantations, considering also the potential timber production and other woody products as an interesting integration of local economy. Moreover an important aspect was the possibility of developing industries, for fruits and wood processes, establishing also joint ventures with Italian and European enterprises.

In *Zoucheng Municipality* the experimental plantation was realized in 2003 on four hectares. The principal species was walnut for wood production consociated with poplar species and intercropped with herbaceous species. The growth of the apical shout, the stem diameter and the bud burst time were registered. Plants showing valuable apical dominance were labelled in order to select genotypes with suitable characters for wood production. A very active cooperation with Zoucheng Municipality was established so that several Chinese students spent 6 months in the

Italian CNR Institutes for scientific training in the field of agro-forestry, forest genetics and biochemistry.

5. CONCLUSIONS

Forests and forest plantations, particularly those established on farmlands can greatly contribute to rural development in European countries as well as in China. Benefits from trees and forests span from woody biomass for industry as well as for energy to environmental protection, erosion control and landscape improvement (Fig. 4); recently, new ecological services have been internationally attributed to forest systems namely carbon sequestration and climate change mitigation: this is a newly expanding field of technical innovation and application that deserves much attention because it will open new markets for trading eco-certificates and eco-labels. Also, a common effort in the evaluation, conservation and sustainable utilization of natural tree resources has been started with particular emphasis on the species that can be utilized in agro-forestry plantations. It is worth mentioning that some of the tree species utilized in these plantations can provide, at the same time, a valuable food production that can significantly integrate farmers' revenues.

Figure 3. Natural genetic resources in China, Fruit Gullis Protected area, Xinjiang Province.

It is also interesting to note that even though socio-economic conditions are still quite different between Europe and China, the importance and role of such man-made forests are quite similar and have many points in common. This observation will warrant the mutual significance and usefulness of the joint research activities undertaken by China and Italy in the field of forestry and agro-forestry systems.

Giuseppe Scarascia Mugnozza
University of Tuscia, Viterbo, Italy
IBAF-CNR, Porano (TR), Italy

Maria Emilia Malvolti
IBAF-CNR, Porano (TR), Italy

REFERENCES

Del Galdo I., Six J., Peressotti A. and Cotrufo M.F., 2003. Assessing the impact of land-use change on soil C sequestration in agricultural soils by means of organic matter fractionation and stable C isotopes. *Global Change Biology* 9, pp. 1204-1213.

Del Lungo A., 2003. *Planted forests in the world: an analysis of available data for silvicultural parameters and regional trend.* Doctorate dissertation, University of Tuscia, Viterbo.

FAO, 2001. *Global Forest Resources Assessment 2000.* Main Report. FAO, Forestry Paper 140.

Guo L.B. and Gifford R.M., 2002. Soil carbon stocks and land use change: a meta analysis. *Global Change Biology* 8, pp. 345-360.

IGBP Terrestrial carbon working group, 1998. The terrestrial carbon cycle: implications for the Kyoto protocol. *Science* 280, pp. 1393-1394.

IPCC 2000, Watson R.T., Noble L.R., Bolin B., Ravindranath N.H., Verardo D.J. & Dokken D.J. (eds.), 2000. *Land Use, Land Use Change and Forestry. Report for the Intergovernmental Panel on Climate Change*, Cambridge University Press, Cambridge.

Jarvis P.G., 1989. Atmospheric carbon dioxide and forest. *Phil. Trans. R. Soc. Lond.* B324, pp. 369-392.

Norby R.J., Wullschleger S.D., Gunderson C.A., Johnson D.W. and Ceulemans R., 1999. Tree responses to rising CO_2 in field experiments: implications for the future forest. *Plant Cell Environ* 22, pp. 683-714.

Post W.M. and Kwon K.C., 2000. Soil carbon sequestration and land-use change: processes and potential. *Global Change Biology* 6, pp. 317-327.

Schlesinger W.H. and Lichter J., 2001. Limited carbon storage in soil and litter of experimental forest plots under increased atmospheric CO_2. *Nature* 411, pp. 466-469.

Smith P., Smith J.U., Powlson D., McGill W.B., Arah J.R.M., Chertov O.G., Coleman K., Franko R.G., Kilei-Gunnewiek H., Komarov A.S., Li C., Molina J.A.E., Mueller T., Parton W.J., Thornley J.H.M. and Whitmore A.P.P., 1997. A comparison of the performance of nine soil organic matter models using datasets from seven long-term experiments. *Geoderma* 81, pp. 153-225.

Smith P., Powlson D.S., Smith J.U., Falloon P. and Coleman K., 2000. Meeting Europe's climate change commitments: quantitative estimates of the potential for carbon mitigation by agriculture. *Global Change Biology* 6, pp. 525-539.

Tabbush P.M. and Parfitt R., 1996. Poplar and willow clones for short rotation coppice. *Research Information Note 278*, Forestry Commission, Edinburgh.

Thornley J.H.M. and Cannell M.G.R., 2000. Managing forests for wood yield and carbon storage: a theoretical study. *Tree Physiology* 20, pp. 474-484.

Vesterdal L., Ritter E. and Gundersen P., 2002. Change in soil organic carbon following afforestation of former arable land. *For Ecol Manag* 169, pp. 137-147.

Zhaohua Z., Maoyi F. and Sastry C.B., 1997. Agroforestry in China – An overview. In: *Overall study in agroforestry systems in China*, International Development Research Centre, Ottawa, Canada.

INDEX

3R philosophy 367, 371, 373, 378, 379

ADEME 296
Agriculture 3, 4, 7, 9, 38, 49, 53, 59, 117, 118, 141, 145, 203, 205, 223, 226, 406, 410, 415, 421-428, 431-439, 441-446, 452, 455, 456, 458, 462, 464, 468, 471, 475, 477, 479
Agroecology 421, 422, 424, 427, 428
Agroecosystem performance indicators (APIs) 425, 426
Agro-forestry 467-480
Air Mass 135, 197
Air Quality Monitoring System 131, 132, 154
Alternating Current (AC) 201, 252, 255, 260
Alternate Cycles Process (ACP) 99, 100, 106-110
Anabolism 366
Catabolism 366
Anaerobic fermentation (AF) 99-104, 110, 226
Aquifer 51, 54-59, 61
ATO, *see* optimal territorial basin
Authorization of the Regulator (AEEG) 294
Automatic Vehicle Localization 154, 155

Benzene 94, 126-128, 131, 135, 142, 146
Bhophal 34
Biodiversity 26, 43, 355, 358, 421, 427, 431-433, 451-464
Biodiversity loss 451-464
Biofuel 221, 226, 227
Biogas 100, 102, 110, 166, 167, 221-224, 226, 228, 373, 443
Biologically Early Warning Systems (BEWS) 91
Biological nutrient removal (BNR) 99-104, 110
Biological productivity 33
Biomass 54, 91, 110, 221-235, 242, 327, 385, 425, 453, 467, 468, 470, 471, 480
Biopellet 230
Biotechnology 434, 456, 463
Birds Directive 86
Brownfield 388, 393, 397-410
Brownfields redevelopment 382, 389, 397, 400, 401, 403

Brownfields remediation 397, 398, 400-406, 409
Brundtland Report 21, 23, 24
Benzene, Toluene, Xylene (BTX) 126-128, 131, 134, 137
Bus Rapid Transit 145, 149

Carbon dioxide, *see* CO_2
Carnot cycle 271
CER 325, 329
Circular economy 335, 336, 347, 348, 367, 406
Clean Development Mechanism (CDM) 295, 311, 312, 317-331, 462, 469, 470
Cleaner Production Promotion Law 347
Closed Circuit TeleVision (CCTV) 148, 149, 151
Cluster development 388
CO_2 36, 43, 164-167, 196, 208, 209, 212, 221, 223, 228, 230, 233, 250, 265, 274, 279, 280, 282-284, 290 301, 311, 312, 314, 324, 325, 327, 328, 348, 455, 469, 471, 472
Coefficient of Performance (COP) 271, 272
Combined Heat and Power System (CHP) 247, 251-253, 282, 283
Command and control (CAC) 3, 7, 8, 10, 14, 287, 288, 289, 299, 312
Committee for European Standardization (CEN) 225
Common Implementation Strategy (CIS) 87, 88, 94
Compact city 369, 370, 393
Composting 159, 161, 165-167, 272, 442, 443
Comprehensive Environmental Response, Compensation and Liability Act (CERCLA) 399, 400
Computational fluid dynamic (CFD) 126, 127, 130, 135, 137
Conference on Environment and Development 21, 26, 451, 463, 468
Conference on the Human Environment 20
Contamination 34, 169, 188, 365, 369, 373, 374, 397-400, 402, 403, 405-408, 437, 442
Convention on Biological Diversity 26, 27, 437, 451, 463

483

Convention on International Trade in Endangered Species of Wild Flora and Fauna 463
Convention on the Law of the Sea 26
Convention to Combat Desertification 437
Conventional Activated Sludge Processes (CASPs) 100, 107-110
Convergence 302, 303, 368
Cultural diversity 33, 367, 368, 375

Deforestation 43, 459, 467, 468
Depletion 50, 55-57, 116, 167, 168, 217, 341, 371, 435, 439, 457, 460
Design Builder 266
Design for Environment (DFE) 325, 338, 340, 341, 346
Direct combustion 221, 228, 229, 232, 235
Direct Current (DC) 201, 252, 260, 457
Dissolved Oxygen 90, 99, 100, 106
Distributed Power (DP) 249, 250, 252-254, 257-259
DPSIR 376
Drinking water 49, 50, 59, 60, 65-68, 71, 83, 84, 86, 114, 118, 119, 455

Earth summit 319, 469
Eco-building 263
Eco-efficiency 3, 6, 338, 341, 346
Ecological government 368
Ecological Quality Ratios (EQR) 87, 88
Ecological service 371, 440, 451, 454, 457, 458, 462, 480
Eco-Management and Audit Scheme (EMAS) 13, 351-361, 372
Ecosystem 24, 38, 54, 83-86, 216, 336, 343, 366-368, 374, 375, 378-380, 398, 399, 404, 407, 415, 419-421, 424, 426-428, 433, 436, 451, 452, 454, 455, 457, 462-471
ECOTECT 266
Electric energy 195, 197, 224, 228, 232, 233, 281
Electricity generation 207, 211, 217, 247, 301
Electricity Market Operator 294, 296
Emission 3, 8-12, 36, 43, 83, 85, 86, 94, 96, 125-132, 134-138, 141, 142, 145-149, 152, 153, 155, 156, 159, 161, 164, 166, 167, 169, 176, 178, 184, 196, 197, 200, 210, 221, 225, 227, 229-232, 235, 250, 265, 271, 279, 280, 283, 284, 287, 288, 290, 292, 295, 301, 306, 311-314, 317-331, 342, 358, 371, 372, 374, 375, 378, 379, 389, 393, 466-470, 472
Emission Trading 10
Energy bill 287
Energy consumption 9, 100, 107-110, 147, 155, 170, 204, 215, 223, 265, 276, 279, 281, 282, 289, 290, 295, 296, 306, 327, 370, 371, 375, 470
Energy conversion 224, 235, 265, 270, 271

Energy efficiency 287-293, 295-299, 302, 319, 327, 328, 372, 379
Energy Efficiency Commitment Scheme (EEC) 289
Energy Plus 266
Energy recovery 101, 159-161, 163, 164, 187, 227
Environmental certification 345
Environmental economics 403
Environmental Impact Assessment (EIA) 44, 372, 395, 405
Environmental legislation 3, 5, 12, 14, 142, 176, 180
Environmental management 3, 7, 13, 17, 29, 83, 143, 369, 415
Environmental Management System (EMS) 143, 243, 351-355, 358, 359, 372
Environmental policy 3-14, 21, 24, 142, 174, 329, 354, 355, 359, 378, 379, 462
Environmental protection 6, 11, 22, 24, 25, 27, 28, 128, 131, 133, 141, 152-154, 171, 203, 311, 330
Environmental Compatibility Assessment 392
Erosion 43, 58, 371, 425, 435, 439, 441, 459, 464, 467-469, 472-474, 477, 479, 480
ESCO 292, 294
European Commission DGXVI structural funds 37
European Hydrogen Network 251
European Spatial Development Perspective (ESDP) 385
European Union (EU) 3, 4, 6, 8, 9, 11-13, 44, 45, 80, 83, 85, 117, 131, 188, 231, 232, 257, 272, 287, 288, 299, 301, 319, 345, 352, 384, 385, 391, 409, 426, 468, 470
European Waste Catalogue (CER) 163, 325, 329
Evaporation 49-51, 55, 59, 126, 271, 272
Evapotranspiration 50, 55, 59, 61
Extended Producer Responsibility (EPR) 171, 175, 178, 180, 183, 344, 345
Exxon Valdez 34

Field boxes 199, 200
Forestry 223, 224, 226, 227, 235, 319, 423, 436, 458, 460, 472, 467-481
Fossil fuel 54, 159, 167, 168, 209, 216, 221, 223, 224, 234, 265, 309, 310, 348, 371, 425, 470, 471
Framework Convention on Climate Change 26, 319, 437
Fuel 54, 125, 126, 136, 147, 148, 151, 159, 167, 168, 196, 209, 216, 221-223, 233, 234, 251-253, 265, 290-293, 298, 303, 309, 310, 327, 328, 343, 370, 371, 375, 425, 426, 455, 457, 459, 470, 471
Fuel cell 250-253
Fur Seals Case 23, 24

INDEX 485

Gabcikovo-Nagymaros Case 27
Gas Chromatography/Mass Spectrometer 91
Gasification 164, 221, 226, 229, 233, 234
Gate Bipolar Transistor 255
GDP 36, 144, 335, 336, 339, 340, 347, 434, 438
General Agreement on Tariffs and Trade (GATT) 28
Genetically modified organism (GMO) 416
Gentini Case 18, 19
Geographic Information System (GIS) 95, 128, 155, 359, 392
Geostatistical Analysis 128
Geothermal electricity 216, 217
Geothermal Energy 207, 212, 214-216, 218, 265, 301
Global Environmental Facility (GEF) 463, 469
Global mechanism (GM) 469
Globalization 25, 302, 303, 335
Green space 371, 384, 392
Greenhouse Gas (GHG) 11, 12, 54, 131, 142, 167, 223, 279, 301, 306, 312, 319, 320, 322, 324, 327, 331, 468-470
Grid 128, 130, 195, 196, 198, 201, 202, 216, 217, 238, 242, 244, 246, 247, 282, 304, 306, 307, 309
Groundwater 38, 49-62, 84-86, 121, 399, 403, 435
Group of Evaluation of Medicinal Plants (GEPM) 43, 44

Habitat function 455
Haloxylon ammodendron 441
Hazardous hospital waste 188, 191
Heating Ventilation Air Conditioning (HVAC) 282, 283
Human metabolism 338
Hybrid 146, 148, 149, 198, 203, 205, 211, 242, 249, 253, 259
Hydrocarbon 126, 146, 164, 229
Hydroelectric 216
Hydrolysis 59

Incentive 3, 4, 7-10, 12, 61, 73, 80, 113, 147, 214, 224, 246, 287, 288, 291, 292, 301-304, 306-308, 310-312, 314, 315, 317, 318, 320, 322, 323, 328, 330, 345, 368, 371, 393, 399, 405, 410, 423, 451, 456, 459-461, 474
Incineration 159-161, 163, 164, 167, 173, 174, 178, 180, 185, 232, 373
Industrial Ecology 335-348
Infilling development 388
Initial Environmental Analysis (IEA) 221, 354, 355, 358, 359
Insulated Gate Bipolar Transistor (IGBTs) 255
Integrated Product Policy Approach (IPP) 13, 15
Intelligent transport system (ITS) 27, 28, 136, 141, 148-156, 370

Intergovernmental Panel on Climate Change (IPCC) 317, 320, 329, 463, 470, 471
International Court of Justice (ICJ) 17, 19
International Emission Trading 331
International federation of Organic Agriculture Movements (IFOAM) 427
Inverse cycle machine 271
IPAT equation 339
Isms 33
ISO 94, 198, 201, 272, 372
Italian Ministry of the Environment, Land and Sea (IMELS) 71-73, 131, 152, 431, 438, 440, 443, 444
Italian Supervising Committee on the Use of Water Resources 66

Johannesburg Conference 22, 23, 25, 26
Johannesburg Declaration 25, 26, 30
Joint Implementation (JI) 295, 311, 312, 323, 324, 462, 469

Kyoto Protocol 39, 301, 311, 312, 318-322, 324, 328, 331, 451, 463, 468-470, 472
k-ε turbulence model 130, 135

Land and Environmental Sustainability Indicator 392
Landfill 100, 108, 161, 162, 164, 166, 167, 169, 173-181, 183, 226, 228, 326, 327, 373
Leakage 59, 60, 126, 256
Liability 4, 12, 13, 397, 399, 400, 402, 405, 407, 408, 410
Liberalization 116, 301, 302, 460
Life cycle analysis 340, 341
Life Cycle Assessment (LCA) 161, 167-169, 342
LIOR 223
Lisbon Strategy 144
LNG 147
Low Emission Zone 130, 136

Material Flow Analysis 335, 340, 342, 347
Market 7-11, 14, 15, 34-36, 65, 72, 73, 76, 116, 117, 171-175, 177, 178, 180-185, 196, 221, 225-227, 231, 232, 234, 237, 243, 245, 246, 252, 265, 271, 287-289, 291-299, 301
Master equation 339
Material Flow Analysis (MFA) 335, 340, 342, 347
MB 288-290, 439
MBR 99, 100, 108-110
Measurement and Verification of energy savings (M&V) 294-296
Mediterranean Geothermal Belt 215
Metabolic model 365, 375-379
Methyl Tert-Butyl Ether 93
Millennium Ecosystem Assessment 452, 463

Mobility 12, 141, 142, 146-150, 370, 382, 388, 389, 394, 395, 399, 402, 443
Models of Territorial Development (MTD) 369, 370
Montreal Protocol 327, 437, 439
MOS Field Effect Transistor (MOSFET) 255
MOSAV project 118, 119
MRENet 423
Multi-criteria decision-aid (MCDA) 39
Multicriteria evaluation 34-37, 39, 43, 44
Municipal Solid Waste (MSW) 99-101, 110, 164, 166, 168, 169, 222, 223

Natural inversion 59
Negligence 13
Net Present cost Value (NPV) 259
Never In My Back-Yard (NIMBY) 39
New Urban Metabolic Model (NUMM) 365, 377-379

OECD 331, 402, 434-436, 441
OFGEM 293, 296
Olympic Games 153, 217, 218
Optimal Territorial Basin (ATO) 71-78, 115-117
Organic agriculture 426, 427
Organic food 437, 445
Organic Fraction of Municipal Solid Waste (OFMSW) 99-103, 105, 110, 166
Over-the-counter (OTC) 132, 294
Oxidation Reduction Potential (ORP) 99, 106, 107

Parallel resonance 256
Partecipative Mutli-criteria Evaluation (PMCE) 39
Participatory approach 3, 7
Phase-Locked Loop (PLL) 255
Photovoltaic (PV) 195-206, 242, 249, 253-256, 258, 281, 291, 299
Piano Assetto Territorio (PAT) 351, 453-460
PIF project 118, 119, 122
Point of Common Coupling (PCC) 255, 256
Pollution 4, 8-14, 19, 20, 37, 55, 56, 59-61, 83, 85, 87, 89, 113, 114, 118, 119, 125-132, 134, 136-138, 141, 142, 144, 146, 147, 149, 150, 152-156, 161, 167, 178, 216-218, 256, 287, 289, 301, 306, 310-315, 335, 336, 338, 340, 345-348, 355, 357-359, 365, 370, 371, 373-375, 381, 384, 385, 393, 402, 404, 410, 432, 439, 440, 452, 454, 458, 460, 464, 472
Populus euphratica 476
Power plant 167, 200, 202, 204, 207, 209-211, 216, 217, 221, 222, 225, 227, 230, 232, 233, 243, 257, 305, 306, 310, 347
Power system 198, 204, 229, 251, 252, 254, 258, 259
Prestige 34

Price reforms 51, 460
Products to Services 345
Programmed inflation 72
Property rights 8, 176, 311, 383, 399, 408, 451, 459, 460
Protected areas 86, 451, 459, 461, 480
Public Transport Management System (PTMS) 155, 156
PURPA 307
Pyrolysis 164, 221, 226, 229, 235

Quality Elements (QE) 83, 87-93

R&D 9, 14, 36, 229, 241, 307, 438
Radio Data Systems (RDS) 151
Radio Frequency Identification Devices (RFID) 148, 149 151
Recycling 5, 9, 10, 65, 114, 117, 159-163, 167, 168, 172-175, 177-181, 183-185, 187, 355, 359, 360, 365, 367, 372-274, 398, 403, 406, 407, 442, 444
Reference Conditions (RC) 87, 88, 230
Renewable energy sources (RES) 40, 42, 247, 314
Renewable energy technologies (RETs) 301, 302, 306-310, 313, 314
Renewable energy resource (RES-E) 265, 301-303, 306-310, 313, 314, 365
Resilience 33, 367, 405, 454, 455, 458
Rio Conference 21-23, 25, 26
Rio Declaration 21, 22, 24-26
River Basin Management Plan (RBMP) 89, 96
Rural development 423, 425, 446, 467, 475, 480

Salinization 56-59, 440, 459
Series resonance 256, 259
Seveso 34
Short Rotation Forest (SRF) 222, 226, 472
Shrimps-Turtles Case 27, 28
Sino-Italy Environment and Management Building (SIEEB) 279
Sino-Italian Cooperation Program for Environmental Protection 131, 152, 431, 438, 439, 441-445, 467
Sistema Integrato Fusina Ambiente 119-122
SME 179
Social welfare 317, 457, 458
Social Multi-Criteria Evaluation (SMCE) 39, 41-45
Soil Organic Matter (SOM) 472
Solar 195-201, 203-206, 223, 237, 254, 265-279, 280, 281, 301, 310, 425, 426
Solid bio-fuels (SBF) 221, 222, 224-231
Solid recovered fuel (SRF) 225
Spread city 369
Stakeholder Multi-criteria decision-aid (SMCDA) 39

INDEX

Static Compensator (STATCOM) 258, 260
Stockholm Convention 437
Stockholm Declaration 20-22
Strategic Environmental Assessment 351-359, 361, 372, 395
Struvite crystallization process (SCP) 99, 100, 103
Subsidence 56, 217
Subsidiarity 3, 11, 390
Substance Flow Analysis (SFA) 342
Superfund 399, 400
Sustainable development 3, 5-9, 13, 17, 21-29, 33, 51, 57, 144, 160, 169, 317, 318, 320, 322, 324, 327-331, 335, 341, 351, 352, 357, 369, 375, 377, 379, 382, 385, 387, 389, 391, 398, 404, 406, 409, 415, 431, 433, 435, 438, 442, 444, 445, 467-469, 472, 473
Sustainable Mobility Program 147, 148
Sustainable planning 385, 386, 388, 391
Sustainable Urban Development (SUD) 142, 365, 370
System marginal price 305, 309, 310

Tamerix 476
Tax 3, 8-10, 177-179, 183, 246, 288, 290, 306, 307, 310, 311, 329, 400, 403, 405, 410, 461, 462, 475
Thematic Strategy on Urban Environment 142, 385
Toluene 126-128, 131
Total economic value (TEV) 34, 404
Towards a thematic strategy for urban environment 385
Tradable Green Certificates (TGCs) 301, 308-310, 313, 315
Tradable pollution permits (TPPs) 10, 11, 301, 310-313, 315
Traffic Limited Zones 148
Traffic Message Channel (TMC) 151
Trail Smelter Case 19
Turbulence 130, 135, 240

Ultrafiltration membrane (MBR) 99, 100, 108-110
UN 20-23, 25-28, 62, 319, 331, 340, 342, 345, 415, 423, 451, 463, 468, 469
UNESCO water portal 53, 56, 60-62
United Nations Framework Convention on Climate Change (UNFCC) 26, 319, 331, 437

Urban ecology 365, 366, 368
Urban metabolism 366, 367
Urban planning 381, 383, 387
Urban sprawl 142, 380, 405
US Federal Public Utilities Regulatory Act (PURPA) 307

Volatile Organic Compounds (VOC) 126-128, 137
Voluntary environmental agreements 14
Voluntary participation 7, 13, 321

Waste 5, 9, 49, 61, 62, 70-72, 95, 99-102, 104, 107, 108, 110, 113, 116, 121, 159-169, 171-185, 187-192, 210, 216, 217, 221-223, 228, 232, 265, 271, 282, 301, 327, 335, 342-346, 348, 355, 356, 360, 365-367, 371-374, 376, 378-380, 385, 389, 392, 393, 395, 407, 408, 442, 443, 454, 455, 458, 460
Waste electrical and electronic equipment (WEEE) 162, 345
Waste Water Treatment Plant (WWTP) 101, 104, 107, 121
Waste-to-energy 161
Water Framework Directive (WFD) 68, 82, 84, 86-89, 91, 96
Water policy 62, 65, 71, 75, 79, 115
Water quality 54, 56, 57, 60, 61, 68, 70, 83-87, 93, 99, 358, 385
Water scarcity 51, 56, 60
Water treatment 62, 117
Wind atlas 238, 239
Wind energy 39, 41, 237, 240-242, 244-246, 257, 259, 441
Wind turbine 39, 237, 238, 240-246, 258
Woody-fuel 227, 228, 231
World Council on Environment and Development (WCED) 6, 317
World Trade Organization (WTO) 27, 28, 346, 437, 473

XX(g) 28
Xylene 126-128, 131

Ziziphus jujuba 476
ZTL, see Traffic Limites Zones